Nonlinear Equations: Numerical Methods for Solving

Nonlinear Equations: Numerical Methods for Solving

Edited by Eli Hill

STATES
ACADEMIC PRESS
www.statesacademicpress.com

States Academic Press,
109 South 5th Street,
Brooklyn, NY 11249, USA

Visit us on the World Wide Web at:
www.statesacademicpress.com

ISBN: 978-1-63989-723-0

Cataloging-in-Publication Data

Nonlinear equations : numerical methods for solving / edited by Eli Hill.
p. cm.
Includes bibliographical references and index.
ISBN 978-1-63989-723-0
1. Differential equations, Nonlinear. 2. Numerical analysis. 3. Nonlinear theories.
4. Differential equations. I. Hill, Eli.
QA377 .N663 2023
515.35--dc23

Table of Contents

Preface

Nonlinear equations are those that do not form a straight line. They may form a bell curve, S-curve, or any other nonlinear shape on a graph. These equations have a variable slope and are more complicated as compared to linear equations. A nonlinear equation has a degree of two or greater than two, but not less than two. As the value of the degree increases, so does the curvature of the graph. In general, nonlinear equations cannot be solved analytically. Two criteria can be taken into account when choosing a method for solving nonlinear equations, namely, method convergence and the cost of calculating of the method. Nonlinear equations are utilized for managing projects and making predictions. Some examples of nonlinear equations include S-curve graphs, regression analysis, and logarithmic scales. Important numerical methods for solving nonlinear equations include first order methods, Newton-Raphson method, and bisection method. This book explores all the important aspects of the numerical methods used for solving nonlinear equations. It is a resource guide for experts as well as students.

This book is a comprehensive compilation of works of different researchers from varied parts of the world. It includes valuable experiences of the researchers with the sole objective of providing the readers (learners) with a proper knowledge of the concerned field. This book will be beneficial in evoking inspiration and enhancing the knowledge of the interested readers.

In the end, I would like to extend my heartiest thanks to the authors who worked with great determination on their chapters. I also appreciate the publisher's support in the course of the book. I would also like to deeply acknowledge my family who stood by me as a source of inspiration during the project.

Editor

1

Study of a High Order Family: Local Convergence and Dynamics

Cristina Amorós [1], Ioannis K. Argyros [2], Ruben González [1], Á. Alberto Magreñán [3], Lara Orcos [4] and Íñigo Sarría [1,*]

1 Escuela Superior de Ingeniería y Tecnología, Universidad Internacional de La Rioja, 26006 Logroño, Spain; cristina.amoros@unir.net (C.A.); ruben.gonzalez@unir.net (R.G.)
2 Department of Mathematics Sciences, Cameron University, Lawton, OK 73505, USA; iargyros@cameron.edu
3 Departamento de Matemáticas y Computación, Universidad de La Rioja, 26004 Logroño, Spain; angel-alberto.magrenan@unirioja.es
4 Facultad de Educación, Universidad Internacional de La Rioja, 26006 Logroño, Spain; lara.orcos@unir.net
* Correspondence: inigo.sarria@unir.net

Abstract: The study of the dynamics and the analysis of local convergence of an iterative method, when approximating a locally unique solution of a nonlinear equation, is presented in this article. We obtain convergence using a center-Lipschitz condition where the ball radii are greater than previous studies. We investigate the dynamics of the method. To validate the theoretical results obtained, a real-world application related to chemistry is provided.

Keywords: high order; sixteenth order convergence method; local convergence; dynamics

---

1. Introduction

A well known problem is that of approximating a locally unique solution x^* of equation

$$F(x) = 0, \tag{1}$$

where F is a differentiable function defined on a nonempty convex subset D of S with values in Ω, where Ω can be $\mathbb{R}$ or $\mathbb{C}$. In this article, we are going to deal with it.

Mathematics is always changing and the way we teach it also changes as it is presented in [1,2]. In the literature [3–8], we can find many problems in engineering and applied sciences that can be solved by finding solutions of equations in a way such as (1). Finding exact solutions for this type of equation is not easy. Only in a few special cases can we find the solutions of these equations in closed form. We must look for other ways to find solutions to these equations. Normally we resort to iterative methods to be able to find solutions. Once we propose to find the solution iteratively, it is mandatory to study the convergence of the method. This convergence is usually seen in two different ways, which gives rise to two different categories, the semilocal convergence analysis and the local convergence analysis. The first of these, the semilocal convergence analysis, is based on information around an initial point, which will provide us with criteria that will ensure the convergence of an iteration procedure. On the other hand, the local convergence analysis is generally based on information about a solution to find values of the calculated radii of the convergence balls. The local results obtained are fundamental since they provide the degree of difficulty to choose the initial points.

We must also deal with the domain of convergence in the study of iterative methods. Normally, the convergence domain is very small and it is necessary to be able to extend this convergence domain without adding any additional hypothesis. Another important problem is finding more accurate estimates of error in distances. $\|x_{n+1} - x_n\|$, $\|x_n - x^*\|$. Therefore, to extend the domain without the need for additional hypotheses and to find more precise estimates of the error committed, in addition to the study of dynamic behavior, will be our objectives in this work.

The iterative methods can be applied to polynomials, and the dynamic properties related to this method will give us important information about its stability and reliability. Recently in some studies, authors such as Amat et al. [9–11], Chun et al. [12], Gutiérrez et al. [13], Magreñán [14–16], and many others [8,13,17–30] have studied interesting dynamic planes, including periodic behavior and other anomalies detected. For all the above, in this article, we are going to study the parameter spaces associated with a family of iterative methods, which will allow us to distinguish between bad and good methods, always speaking in terms of their numerical properties.

We present the dynamics and the local convergence of the four step method defined for each $n = 0, 1, 2, \ldots$ by

$$\begin{aligned} y_n &= x_n - \alpha F'(x_n)^{-1} F(x_n) \\ z_n &= y_n - C_1(x_n) F'(x_n)^{-1} F(y_n) \\ v_n &= z_n - C_2(x_n) F'(x_n)^{-1} F(z_n) \\ x_{n+1} &= z_n - C_3(x_n) F'(x_n)^{-1} F(v_n), \end{aligned} \tag{2}$$

where $\alpha \in \mathbb{R}$ is a parameter, x_0 is an initial point and $C_i : \mathbb{R} \to \mathbb{R}$, $i = 1, 2, 3$ are continuous functions given. Numerous methods of more than one step are particular cases of the previous method (2). For example, for certain values of the parameters this family can be reduced to:

- Artidiello et al. method [31]
- Petković et al. method [32]
- Kung-Traub method [29]
- Fourth order King family
- Fourth order method given by Zhao et al. in [33]
- Eighth order method studied by Dzunic et al. [34].

It should be noted that to demonstrate the convergence of all methods after the method (2), in all cases Taylor expansions have been used as well as hypotheses involving derivatives of order greater than one, usually the third derivative or greater. However, in these methods only the first derivative appears. In this article we will perform the analysis of local convergence of the method (2) using hypotheses that involve only the first derivative of the function F. In this way we save the tedious calculation of the successive derivatives (in this case the second and third derivatives) in each step. The order of convergence (COC) is found using and an approximation of the COC (ACOC) using that do not require the usage of derivatives of order higher than one (see Remark 1). Our objective will also be able to provide a computable radius of convergence and error estimates based on the Lipschitz constants.

We must also realize that there are a lot of iterative methods to approximate solutions of nonlinear equations defined in $\mathbb{R}$ or $\mathbb{C}$ [32,35–38]. These studies show that if the initial point x_0 is close enough to the solution x^*, the sequence $\{x_n\}$ converges to x^*. However, from the initial estimate, how close to the solution x^* should it be? In these cases, the local results do not provide us with information about the radius of the convergence ball for the corresponding method. We will approach this question for the method (2) in Section 2. Similarly, we can use the same technique with other different methods.

2. Method's Local Convergence

Let us define, respectively, $U(v, \rho)$ and $\bar{U}(v, \rho)$ as open and closed balls in S, of radius $\rho > 0$ and with center $v \in \Omega$.

To study the analysis of local convergence of the method (2), we are going to define a series of conditions that we will name (C):

(C_1) $F : D \subset \Omega \to \Omega$ is a differentiable function.

We know that exist a constant $x^* \in D$, $L_0 > 0$, such that for each $x \in D$ is fulfilled

(C_2) $F(x^*) = 0$, $F'(x^*) \neq 0$.

(C_3) $\|F'(x^*)^{-1}(F'(x) - F'(x^*))\| \leq L_0\|x - x^*\|$

Let $D_0 := D \cap U(x^*, \frac{1}{L_0})$. There exist constants $L > 0$, $M \geq 1$ such that for each $, y \in D_0$

(C_4) $\|F'(x^*)^{-1}(F'(x) - F'(y))\| \leq L\|x - y\|$

(C_5) $\|F'(x^*)^{-1}F'(x)\| \leq M$.

There exist parameters γ_i and continuous nondecreasing functions $\psi_i : [0, \gamma_i) \to \mathbb{R}$ such that $i = 0, 1, 2, 3$:

(C_6) $\gamma_{i+1} \leq \gamma_i \leq \dfrac{1}{L_0}$

and

(C_7) $\psi_i(t) \to a +\infty$ or a number greater than 0 as $t \to \gamma_i^{-1}$. For $\alpha \in \mathbb{R}$, consider the functions

$$q_j : [0, \gamma_j) \to \mathbb{R} \quad j = 0, 1, 2, 3 \text{ by}$$

$$q_j(t) = \begin{cases} M|1 - \alpha|, & j = 0 \\ \\ M^{i+j}|1 - \alpha|\prod_{i=0}^{j}\psi_1(t) \cdots \psi_j(t), & j = 1, 2, 3 \end{cases}$$

(C_8) $p_j := q_j(0) < 1, \quad j = 0, 1, 2, 3,$

(C_9) $C_i : \Omega \to \Omega$ are continuous functions such that for each $x \in D_0$, $\|C_i(x)\| \leq \psi_i(\|x - x^*\|)$ and

(C_{10}) $\bar{U}(x^*, r) \subset D$ for some $r > 0$ to be appointed subsequently.

We are going to introduce some parameters and some functions for the local convergence analysis of the method (2). We define the function g_0 on the interval $[0, \frac{1}{L_0})$ by

$$g_0(t) = \frac{1}{2(1 - L_0 t)}(Lt + 2M|1 - \alpha|)$$

and parameters r_0, ϱ_A by

$$r_0 = \frac{2(1 - M|1 - \alpha|)}{2L_0 + L}, \quad \varrho_A = \frac{2}{2L_0 + L}.$$

Then, since $p_0 = M|1 - \alpha| < 1$ by (C_8), we have that $0 < r_0 < \varrho_A$, $g_0(r_1) = 1$ and for each $t \in [0, r_1)$ $0 \leq g_0(t) < 1$. Define functions g_i, h_i on the interval $[0, \gamma_i)$ by

$$g_i(t) = (1 + \frac{M\psi_i(t)}{1 - L_0 t})g_{i-1}(t)$$

and

$$h_i(t) = g_i(t) - 1$$

for $i = 1, 2, 3$. We have by (C_8) that $h_i(0) = p_j - 1 < 0$ and by (C_6) and (C_7) $h_i(t) \to a$ positive number or $+\infty$. Applying the intermediate value theorem, we know that functions h_i have zeros in the interval $[0, \gamma_i)$. Denote by r_i the smallest such zero. Set

$$r = \min\{r_j\}, \quad j = 0, 1, 2, 3. \tag{3}$$

Therefore, we can write that

$$0 \leq r < r_A \tag{4}$$

moreover for each $j = 0, 1, 2, 3$, $t \in [0, r)$

$$0 \leq g_j(t) < 1. \tag{5}$$

Now, making use of the conditions (C) and the previous notation, we will show the results of local convergence for the method (2).

Theorem 1. *Let us assume that* (C) *conditions hold, if we take the radius* r *in* (C_{10}) *that has been defined previously. Then, the sequence* $\{x_n\}$ *generated by our method* (2) *and considering* $x_0 \in U(x^*, r) \setminus \{x^*\}$ *is well defined, remains in the ball* $U(x^*, r)$ *for each* $n \geq 0$ *and converges to the solution*x^**. On the other hand, we see that the estimates are true:*

$$\|y_n - x^*\| \leq g_0(\|x_n - x^*\|)\|x_n - x^*\| < \|x_n - x^*\| < r, \tag{6}$$

$$\|z_n - x^*\| \leq g_1(\|x_n - x^*\|)\|x_n - x^*\| < \|x_n - x^*\|, \tag{7}$$

$$\|v_n - x^*\| \leq g_2(\|x_n - x^*\|)\|x_n - x^*\| < \|x_n - x^*\| \tag{8}$$

and

$$\|x_{n+1} - x^*\| \leq g_3(\|x_n - x^*\|)\|x_n - x^*\| < \|x_n - x^*\|, \tag{9}$$

where the "g" functions are defined previously. Furthermore, for

$$T \in [r, \frac{2}{L_0}) \tag{10}$$

the unique solution of equation $F(x) = 0$ *in* $\bar{U}(x^*, T) \cap D$ *is the bound point* x^**.*

Proof. Using mathematical induction we shall prove estimates (6) and (10). By hypothesis $x_0 \in U(x, r) \setminus \{x^*\}$, the conditions (C_1), (C_3) and (3), we have that

$$\|F'(x^*)^{-1}(F'(x_0) - F'(x^*))\| \leq L_0\|x_0 - x^*\| < L_0 r < 1. \tag{11}$$

Taking into account the Banach lemma on invertible functions [5,7,39] we can write that $F'(x_0)^{-1} \in L(S, S)$ and

$$\|F'(x_0)^{-1}F'(x^*)\| \leq \frac{1}{1 - L_0\|x_0 - x^*\|}. \tag{12}$$

consequently, y_0 is well defined by the first substep of the method (2) for $n = 0$. We can set using the conditions (C_1) and (C_2) that

$$F(x_0) = F(x_0) - F(x^*) = \int_0^1 F'(x^* + \theta(x_0 - x^*))(x_0 - x^*)d\theta. \tag{13}$$

Remark that $\|x^* + \theta(x_0 - x^*) - x^*\| = \theta\|x_0 - x^*\| < r$, so $x^* + \theta(x_0 - x^*) \in U(x^*, r)$. Then, using (13) and condition (C_5), we have that

$$\|F'(x^*)^{-1}F(x_0)\| \leq \|\int_0^1 F'(x^*)^{-1}F'(x^* + \theta(x_0 - x^*))(x_0 - x^*)d\theta\| \leq M\|x_0 - x^*\|. \tag{14}$$

In view of conditions (C_2), (C_4), (3) and (5) (for $j = 0$) and (12) and (14), we obtain that

$$
\begin{aligned}
\|y_0 - x^*\| &= \|x_0 - x^* - F'(x_0)^{-1}F(x_0) + (1-\alpha)F'(x_0)^{-1}F(x_0) \\
&\leq \|x_0 - x^* - F'(x_0)^{-1}F(x_0)\| + |1-\alpha|\|F(x_0)^{-1}F'(x^*)\|\|F'(x^*)^{-1}F(x_0)\| \\
&\leq \|F(x_0)^{-1}F'(x^*)\|\|\int_0^1 F'(x^*)^{-1}(F'(x^* + \theta(x_0 - x^*)) - F'(x_0))(x_0 - x^*)d\theta \\
&\quad + \frac{|1-\alpha|M\|x_0 - x^*\|}{1 - L_0\|x_0 - x^*\|} \\
&\leq \frac{L\|x_0 - x^*\|^2}{2(1 - L_0\|x_0 - x^*\|)} + \frac{|1-\alpha|M\|x_0 - x^*\|}{1 - L_0\|x_0 - x^*\|} \\
&= g_0(\|x_0 - x^*\|)\|x_0 - x^*\| < \|x_0 - x^*\| < r,
\end{aligned}
\tag{15}
$$

which evidences (6) for $n = 0$ and $y_0 \in U(x^*, r)$. Then, applying (C_9) condition, (3) and (5) (for $j = 1$), (12) and (14) (for $y_0 = x_0$) and (15), we achieve that

$$
\|z_0 - x^*\| \leq g_1(\|x_0 - x^*\|)\|x_0 - x^*\| \leq \|x_0 - x^*\|, \tag{16}
$$

which displays (7) for $n = 0$ and $z_0 \in U(x^*, r)$. In the same way, we show estimates (8) and (9) for $n = 0$ and $v_0, x_1 \in U(x^*, r)$. Just substituting x_0, y_0, z_0, v_0, x_1 by $x_k, y_k, z_k, v_k, x_{k+1}$ in the preceding estimates, we deduct that (6)–(9). Using the estimates $\|x_{k+1} - x^*\| \leq c\|x_k - x^*\| < r$, $c = g_3(\|x_0 - x^*\|) \in [0, 1)$, we arrive at $\lim_{k\to\infty} x_k = x^*$ and $x_{k+1} \in U(x^*, r)$. We have yet to see the uniqueness, let $y^* \in \bar{U}(x^*, T)$ be such that $F(y^*) = 0$. Define $B = \int_0^1 F'(y^* + \theta(x^* - y^*))d\theta$. Taking into account the condition (C_2), we obtain that

$$
\|F'(x^*)^{-1}(B - F'(x^*))\| \leq \frac{L_0}{2}\|y^* - x^*\| \leq \frac{L_0}{2}T < 1. \tag{17}
$$

Hence, $B \neq 0$. Using the identity $0 = F(y^*) - F(x^*) = B(y^* - x^*)$, we can deduct that $x^* = y^*$. □

Remark 1.

1. *Considering* (10) *and the next value*

$$
\begin{aligned}
\|F'(x^*)^{-1}F'(x)\| &= \|F'(x^*)^{-1}(I + F'(x) - F'(x^*))\| \\
&\leq \|F'(x^*)^{-1}(F'(x) - F'(x^*))\| + 1 \\
&\leq L_0\|x_0 - x^*\| + 1
\end{aligned}
$$

we can clearly eliminate the condition (10) *and* M *can be turned into*

$$
M(t) = 1 + L_0 t \text{ or what is the same } M(t) = M = 2, \text{ because } t \in [0, \frac{1}{L_0}).
$$

2. *The results that we have seen, can also be applied for* F *operators that satisfy the autonomous differential equation [5,7] of the form*

$$
F'(x) = P(F(x)),
$$

where P *is a known continuous operator. As* $F'(x^*) = P(F(x^*)) = P(0)$, *we are able to use the previous results without needing to know the solution* x^*. *Take for example* $F(x) = e^x - 1$. *Now, we can take* $P(x) = x + 1$. *However, we do not know the solution.*

3. *In the articles [5,7] was shown that the radius ϱ_A has to be the convergence radius for Newton's method using* (10) *and* (11) *conditions. If we apply the definition of r_1 and the estimates* (8)*, the convergence radius r of the method* (2) *it can no be bigger than the convergence radius ϱ_A of the second order Newton's method. The convergence ball given by Rheinboldt [8] is*

$$\varrho_R = \frac{2}{3L_1}. \tag{18}$$

In particular, for $L_0 < L_1$ or $L < L_1$ we have that

$$\varrho_R < \varrho_A$$

and

$$\frac{\varrho_R}{\varrho_A} \to \frac{1}{3} \quad as \quad \frac{L_0}{L_1} \to 0.$$

That is our convergence ball r_1 which is maximum three times bigger than Rheinboldt's. The precise amount given by Traub in [28] for ϱ_R.

4. *We should note that family* (3) *stays the same if we use the conditions of Theorem 1 instead of the stronger conditions given in [15,36]. Concerning, for the error bounds in practice we can use the approximate computational order of convergence (ACOC) [36]*

$$\xi = \frac{In\frac{\|x_{n+2}-x_{n+1}\|}{\|x_{n+1}-x_n\|}}{In\frac{\|x_{n+1}-x_n\|}{\|x_n-x_{n-1}\|}}, \qquad \text{for each } n = 1,2,\ldots$$

or the computational order of convergence (COC) [40]

$$\xi^* = \frac{In\frac{\|x_{n+2}-x^*\|}{\|x_{n+1}-x^*\|}}{In\frac{\|x_{n+1}-x^*\|}{\|x_n-x^*\|}}, \qquad \text{for each } n = 0,1,2,\ldots$$

And these order of convergence do not require higher estimates than the first Fréchet derivative used in [19,23,32,33,41].

Remark 2. *Let's see how we can choose the functions in the case of the method* (2)*. In this case we have that*

$$\overline{C_1}(\frac{F(y_n)}{F(x_n)}) = C_1(x_n), \quad \overline{C_2}(\frac{F(y_n)}{F(x_n)}, \frac{F(z_n)}{F(y_n)}) = C_2(x_n), \quad \overline{C_3}(\frac{F(y_n)}{F(x_n)}, \frac{F(z_n)}{F(y_n)}, \frac{F(v_n)}{F(z_n)}) = C_3(x_n)$$

To begin, the condition (C_3) *can be eliminated because in this case we have $\alpha = 1$. Then, if $x_n \neq x^*$, the following inequality holds*

$$\|(F'(x^*)(x_n - x^*))^{-1}\left[F(x_n) - F(x^*) - F'(x^*)(x_n - x^*)\right]\|$$

$$\leq \|x_n - x^*\|^{-1}\frac{L_0}{2}\|x_n - x^*\| = \frac{L_0}{2}\|x_n - x^*\| < \frac{L_0}{2}r < 1.$$

Hence, we have that

$$\|F'(x_n)^{-1}F(x^*)\| \leq \frac{1}{\|x_n - x^*\|(1 - \frac{L_0}{2}\|x_n - x^*\|)}.$$

Consequently, we get that

$$\begin{aligned} \|\frac{F(y_n)}{F(x_n)}\| &= \|F'(x_n)^{-1}F'(x^*)\|\|F'(x^*)^{-1}F(y_n)\| \\ &\leq \frac{M\|y_n - x^*\|}{\|x_n - x^*\|(1 - \frac{L_0}{2}\|x_0 - x^*\|)} \\ &\leq \frac{Mg_0(\|x_n - x^*\|)}{1 - L_0\|x_n - x^*\|}. \end{aligned} \tag{19}$$

Similarly, we obtain

$$\|F(y_n)^{-1}F'(x^*)\| \leq \frac{1}{\|y_n - x^*\|(1 - \frac{L_0}{2}\|y_n - x^*\|)},$$

$$\|\frac{F(z_n)}{F(y_n)}\| \leq \frac{M(1 + \frac{M\psi_1(\|x_n - x^*\|)}{1 - L_0\|x_n - x^*\|})}{1 - \frac{L_0}{2}g_0(\|x_n - x^*\|)\|x_n - x^*\|}, \tag{20}$$

$$\|F(z_n)^{-1}F'(x^*)\| \leq \frac{1}{\|z_n - x^*\|(1 - \frac{L_0}{2}\|y_n - x^*\|)},$$

and

$$\|\frac{F(z_n)}{F(y_n)}\| \leq \frac{M(1 + \frac{M\psi_2(\|x_n - x^*\|)}{1 - L_0\|x_n - x^*\|})}{1 - \frac{L_0}{2}g_0(\|x_n - x^*\|)\|x_n - x^*\|}, \tag{21}$$

Let us choose C_i, $i = 1, 2, 3, 4$ *as in [31]:*

$$C_1(a) = 1 + 2a + 4a^3 - 3a^4 \tag{22}$$

$$C_2(a, b) = 1 + 2a + b + a^2 + 4ab + 3a^2b + 4ab^2 + 4a^3b - 4a^2b^2 \tag{23}$$

and

$$C_3(a, b, c) = 1 + 2a + b + c + a^2 + 4ab + 2ac + 4a^2b + a^2c + 6ab^2 + 8abc - b^3 + 2bc. \tag{24}$$

As these functions, they fulfill the terms imposed in Theorem 1 in [31], So, we have that the order of convergence of the method (2) *has to reach at least order 16.*

Set

$$a = a(t) = \frac{Mg_0(t)}{1 - L_0t}, \tag{25}$$

$$b = b(t) = \frac{M(1 + \frac{M\psi_1(t)}{1 - L_0t})}{1 - \frac{L_0}{2}t}, \tag{26}$$

$$c = c(t) = \frac{M(1 + \frac{M\psi_2(t)}{1 - L_0t})}{1 - \frac{L_0}{2}t}, \tag{27}$$

and

$$\gamma_i = \frac{1}{L_0}, \quad i = 0, 1, 2, 3.$$

Then it follows from (19)–(24) *that functions* ψ_i *can be defined by*

$$\psi_1(t) = 1 + 2a + 4a^3 + 3a^4 \tag{28}$$

$$\psi_2(t) = 1 + 2a + b + a^2 + 4ab + 3a^2b + 4ab^2 + 4a^3b + 4a^2b^2 \tag{29}$$

and

$$\psi_3(t) = 1 + 2a + b + c + a^2 + 4ab + 2ac + 4a^2b + a^2c + 6ab^2 + 8abc + b^3 + 2bc. \tag{30}$$

3. Dynamical Study of a Special Case of the Family (2)

In this article, the concepts of critical point, fixed point, strange fixed point, attraction basins, parameter planes and convergence planes are going to be assumed. We refer the reader to see [5,7,16,38] to recall the basic dynamical concepts.

In this third section we will study the complex dynamics of a particular case of the method (2), which consists in select:

$$C_1(x_n) = F'(y_n)^{-1}F'(x_n),$$

$$C_2(x_n) = F'(z_n)^{-1}F'(x_n)$$

and

$$C_3(x_n) = F'(y_n)^{-1}F'(x_n).$$

Let be a polynomial of degree two with two roots, that they are not the same. If we apply this operator on the previous polynomial and using the Möebius map $h(z) = \frac{z-A}{z-B}$, we obtain

$$G(z,\alpha) = \frac{z^8(1-\alpha+z)^8}{(-1-z+\alpha z)^8}. \tag{31}$$

The fixed points of this operator are:

- 0
- ∞
- And 15 more, which are:
  - 1 (related to original ∞).
  - The roots of a 14 degree polynomial.

In Figure 1 the bifurcation diagram of all fixed points, extraneous or not, is presented.

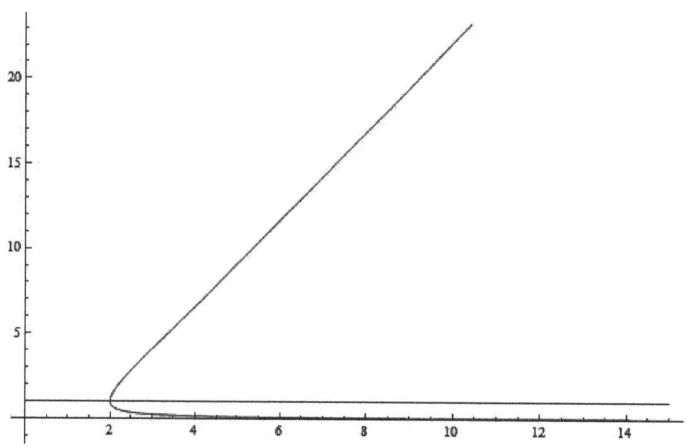

Figure 1. Fixed points's bifurcation diagram.

Now, we are going to compute the critical points, i.e., the roots of

$$G'(z,\alpha) = -\frac{8(-1+\alpha-z)^7 z^7 (-1+\alpha-2z-z^2+\alpha z^2)}{(-1-z+\alpha z)^9}$$

The free critical points are: $cp_1(\alpha) = -1+\alpha$, $cp_2(\alpha) = \frac{1-\sqrt{-(-2+\alpha)\alpha}}{-1+\alpha}$ and $cp_3(\alpha) = \frac{1+\sqrt{-(-2+\alpha)\alpha}}{-1+\alpha}$. We also have the following results.

Lemma 1.

(a) *If* $\alpha = 0$

(i) $cp_1(\alpha) = cp_2(\alpha) = cp_3(\alpha) = -1.$

(b) *If* $\alpha = 2$

(i) $cp_1(\alpha) = cp_2(\alpha) = cp_3(\alpha) = 1.$

You can easily verify that for every value of α *we have to* $cp_2(\alpha) = \frac{1}{cp_3(\alpha)}$

It is easy to see that there is only one independent critical point. So, we assume that $cp_2(\alpha)$ is the only free critical point without loss of generality. Taking $cp_2(\alpha)$, we perform the study of the parameter space associated with the free critical point. This will allow us to find the some members of the family, and we want to stay with the best members.

We are going to show different planes of parameters. In Figure 2 we show the parameter spaces associated to critical point $cp_2(\alpha)$. Now let us paint a point of cyan if the iteration of the method starting in $z_0 = cp_1(\alpha)$ converges to the fixed point 0 (related to root A) or if it converges to ∞ (allied to root B). That is, the points relative to the roots of the quadratic polynomial will be painted cyan and a point is painted in yellow if the iteration converges to 1 (related to ∞). Therefore, all convergence will be painted cyan. On the other hand, convergence to strange fixed points or cycles appears in other colors. As an immediate consequence, all points of the plane that are not cyan are not a good choice of α in terms of numerical behavior.

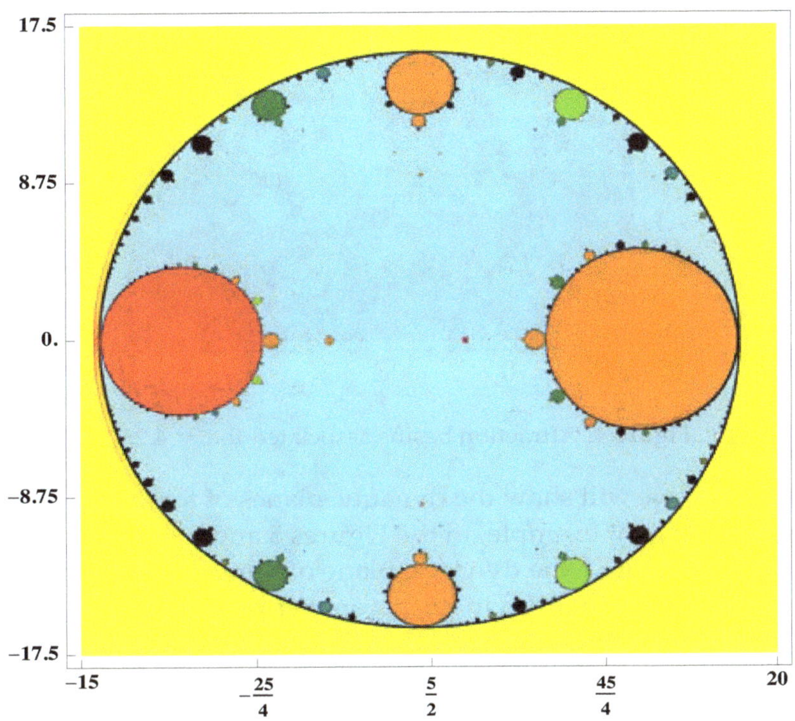

Figure 2. Parameter space of the free critical point $cp_2(\alpha)$.

Once we have detected the anomalies, we can go on to describe the dynamic planes. To understand the colors we have used in these dynamic planes, we have to indicate that if after a maximum of 1000 iterations and with a tolerance of 10^{-6} convergence has not been achieved to the roots, we have painted in black. Conversely, we colored in magenta the convergence to 0 and colored in cyan the convergence to ∞. Then, the cyan or magenta regions identify the convergence.

If we focus our attention on the region shown in Figure 2, it is clear that there are family members with complicated behaviors. We will also show dynamic planes in Figures 3 and 4, of a family member with convergence regions to any of the strange fixed points.

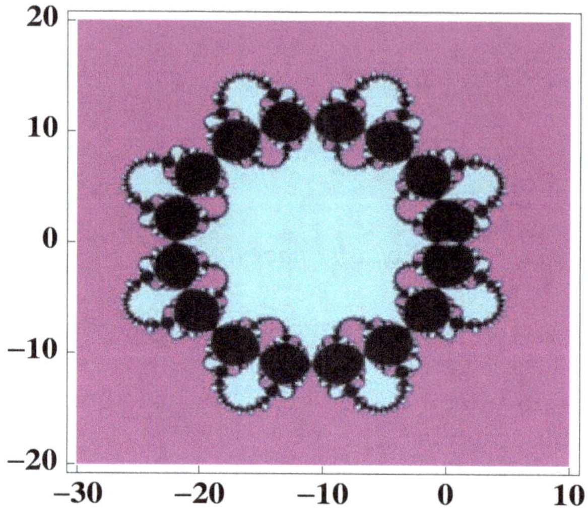

Figure 3. Attraction basins associated to $\alpha = -10$.

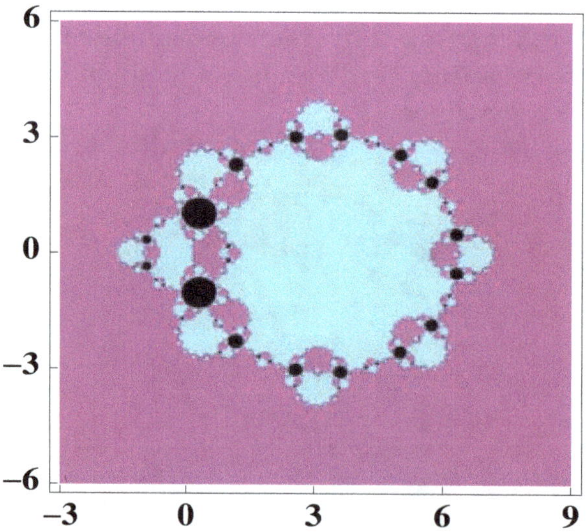

Figure 4. Attraction basins associated to$\alpha = 4.25$.

In the following figures, we will show the dynamic planes of family members with convergence to different attracting n-cycles. For example, in the Figures 5 and 6, we see the dynamic planes to an attracting 2-cycle and in the Figure 7 the dynamic plane of family members with convergence to an attracting 3-cycle that was painted in green in the parameter planes.

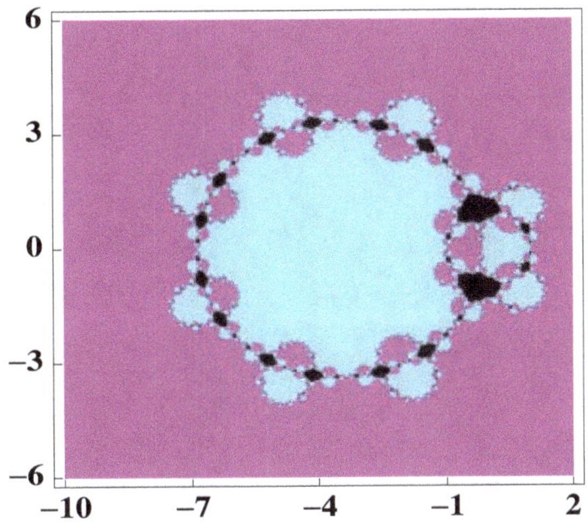

Figure 5. Attraction basins associated to $\alpha = -2.5$.

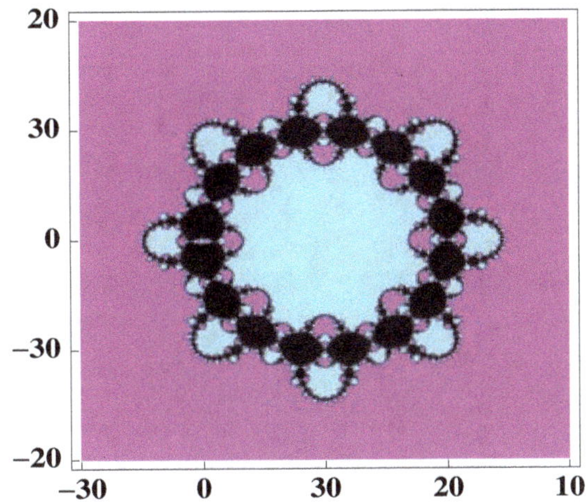

Figure 6. Attraction basins associated to $\alpha = 11$.

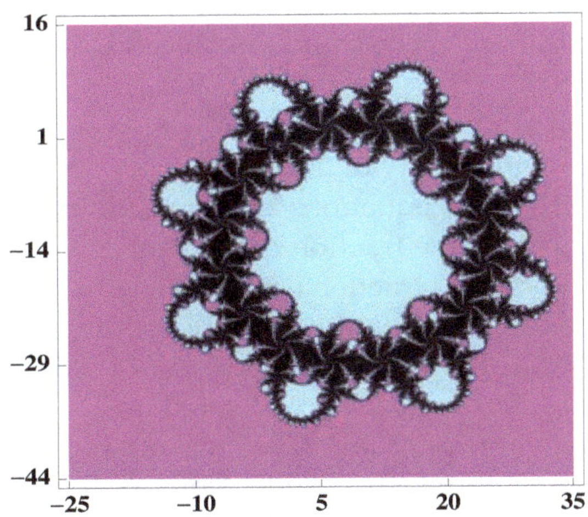

Figure 7. Attraction basins associated to $\alpha = 10 - 13i$.

Other particular cases are shown in Figures 8 and 9. The basins of attraction for different α values in which we see the convergence to the roots of the method can be seen.

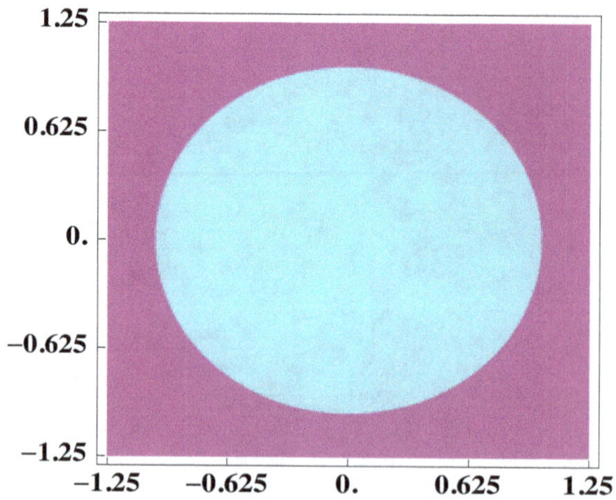

Figure 8. Attraction basins associated to $\alpha = 0.5$.

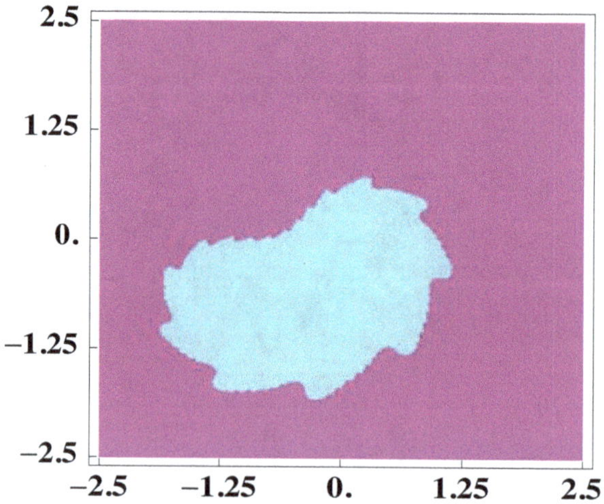

Figure 9. Attraction basins associated to $\alpha = -0.5i$.

4. Example Applied

Next, we want to show the applicability of the theoretical part previously seen in a real problem. Chemistry is a discipline in which many equations are handled. In this concrete case, let us consider the quartic equation that can describe the fraction or amount of the nitrogen-hydrogen feed that is turned into ammonia, which is known as fractional conversion and is shown in [42,43].

If the pressure is 250 atm. and the temperature reaches a value of 500 °C, the previous equation reduces to: $g(x) = x^4 - 7.79075x^3 + 14.7445x^2 + 2.511x - 1.674$. We define S as all real line, D as the interval $[0,1]$ and $\xi = 0$. We consider the function F defined on D. If we now take the functions $\psi_i(t)$ with $i = 1,2,3$ and choosing the value of as $\alpha = 1.025$, we obtain: $L_0 = 2.594\ldots$, $L = 3.282\ldots$. It is clear that in this case $L_0 < L$, so we improve the results. Now, we compute $M = 1.441\ldots$. Additionally, computing the zeros of the functions previously defined, we get: $r_0 = 0.227\ldots$, $\varrho_A = 0.236\ldots$, $r_1 = 0.082\ldots$, $r_2 = 0.155\ldots$, $r_3 = 0.245\ldots$, and as a result of it we get $r = r_1 = 0.082\ldots$. Then we can guarantee that the method (2) converges for $\alpha = 1.025$ due to Theorem 1. The applicability of our family of methods is thus proven.

Author Contributions: All authors have contributed equally in writing this article. All authors read and approved the final manuscript

References

1. Tello, J.I.C.; Orcos, L.; Granados, J.J.R. Virtual forums as a learning method in Industrial Engineering Organization. *IEEE Latin Am. Trans.* **2016**, *14*, 3023–3028. [CrossRef]
2. LeTendre, G.; McGinnis, E.; Mitra, D.; Montgomery, R.; Pendola, A. The American Journal of Education: Challenges and opportunities in translational science and the grey area of academic. *Rev. Esp. Pedag.* **2018**, *76*, 413–435. [CrossRef]
3. Argyros, I.K.; González, D. Local convergence for an improved Jarratt–type method in Banach space. *Int. J. Interact. Multimed. Artif. Intell.* **2015**, *3*, 20–25. [CrossRef]
4. Argyros, I.K.; George, S. Ball convergence for Steffensen–type fourth-order methods. *Int. J. Interact. Multimed. Artif. Intell.* **2015**, *3*, 27–42. [CrossRef]
5. Argyros, I.K.; Magreñán, Á.A. *Iterative Methods and Their Dynamics with Applications: A Contemporary Study*; CRC-Press: Boca Raton, FL, USA, 2017.
6. Behl, R.; Sarría, Í.; González-Crespo, R.; Magreñán, Á.A. Highly efficient family of iterative methods for solving nonlinear models. *J. Comput. Appl. Math.* **2019**, *346*, 110–132. [CrossRef]
7. Magreñán, Á.A.; Argyros, I.K. *A Contemporary Study of Iterative Methods: Convergence, Dynamics and Applications*; Elsevier: Amsterdam, The Netherlands, 2018.
8. Rheinboldt, W.C. An adaptive continuation process for solving systems of nonlinear equations. *Pol. Acad. Sci.* **1978**, *3*, 129–142. [CrossRef]
9. Amat, S.; Busquier, S.; Plaza, S. Dynamics of the King and Jarratt iterations. *Aequ. Math.* **2005**, *69*, 212–223. [CrossRef]
10. Amat, S.; Busquier, S.; Plaza, S. Chaotic dynamics of a third-order Newton-type method. *J. Math. Anal. Appl.* **2010**, *366*, 24–32. [CrossRef]
11. Amat, S.; Hernández, M.A.; Romero, N. A modified Chebyshev's iterative method with at least sixth order of convergence. *Appl. Math. Comput.* **2008**, *206*, 164–174. [CrossRef]
12. Chicharro, F.; Cordero, A.; Torregrosa, J.R. Drawing dynamical and parameters planes of iterative families and methods. *Sci. World J.* **2013**, *2013*, 780153. [CrossRef] [PubMed]
13. Gutiérrez, J.M.; Hernández, M.A. Recurrence relations for the super-Halley method. *Comput. Math. Appl.* **1998**, *36*, 1–8. [CrossRef]
14. Kou, J.; Li, Y. An improvement of the Jarratt method. *Appl. Math. Comput.* **2007**, *189*, 1816–1821. [CrossRef]
15. Li, D.; Liu, P.; Kou, J. An improvement of the Chebyshev-Halley methods free from second derivative. *Appl. Math. Comput.* **2014**, *235*, 221–225. [CrossRef]
16. Magreñán, Á.A. Different anomalies in a Jarratt family of iterative root-finding methods. *Appl. Math. Comput.* **2014**, *233*, 29–38.
17. Budzko, D.; Cordero, A.; Torregrosa, J.R. A new family of iterative methods widening areas of convergence. *Appl. Math. Comput.* **2015**, *252*, 405–417. [CrossRef]
18. Bruns, D.D.; Bailey, J.E. Nonlinear feedback control for operating a nonisothermal CSTR near an unstable steady state. *Chem. Eng. Sci.* **1977**, *32*, 257–264. [CrossRef]
19. Candela, V.; Marquina, A. Recurrence relations for rational cubic methods I: The Halley method. *Computing* **1990**, *44*, 169–184. [CrossRef]
20. Candela, V.; Marquina, A. Recurrence relations for rational cubic methods II: The Chebyshev method. *Computing* **1990**, *45*, 355–367. [CrossRef]
21. Ezquerro, J.A.; Hernández, M.A. New iterations of R-order four with reduced computational cost. *BIT Numer. Math.* **2009**, *49*, 325–342. [CrossRef]
22. Ezquerro, J.A.; Hernández, M.A. On the R-order of the Halley method. *J. Math. Anal. Appl.* **2005**, *303*, 591–601. [CrossRef]
23. Ganesh, M.; Joshi, M.C. Numerical solvability of Hammerstein integral equations of mixed type. *IMA J. Numer. Anal.* **1991**, *11*, 21–31. [CrossRef]
24. Hernández, M.A. Chebyshev's approximation algorithms and applications. *Comput. Math. Appl.* **2001**, *41*, 433–455. [CrossRef]
25. Hernández, M.A.; Salanova, M.A. Sufficient conditions for semilocal convergence of a fourth order multipoint iterative method for solving equations in Banach spaces. *Southwest J. Pure Appl. Math.* **1999**, *1*, 29–40.

26. Jarratt, P. Some fourth order multipoint methods for solving equations. *Math. Comput.* **1966**, *20*, 434–437. [CrossRef]
27. Ren, H.; Wu, Q.; Bi, W. New variants of Jarratt method with sixth-order convergence. *Numer. Algorithms* **2009**, *52*, 585–603. [CrossRef]
28. Traub, J.F. *Iterative Methods for the Solution of Equations*; Prentice-Hall Series in Automatic Computation: Englewood Cliffs, NJ, USA, 1964.
29. Wang, X.; Kou, J.; Gu, C. Semilocal convergence of a sixth-order Jarratt method in Banach spaces. *Numer. Algorithms* **2011**, *57*, 441–456. [CrossRef]
30. Cordero, A.; Torregrosa, J.R.; Vindel, P. Dynamics of a family of Chebyshev-Halley type methods. *Appl. Math. Comput.* **2013**, *219*, 8568–8583. [CrossRef]
31. Artidiello, S.; Cordero, A.; Torregrosa, J.R.; Vassileva, M.P. Optimal high order methods for solving nonlinear equations. *J. Appl. Math.* **2014**, *2014*, 591638. [CrossRef]
32. Petković, M.; Neta, B.; Petković, L.; Džunić, J. *Multipoint Methods for Solving Nonlinear Equations*; Elsevier: Amsterdam, The Netherlands, 2013.
33. Zhao, L.; Wang, X.; Guo, W. New families of eighth-order methods with high efficiency index for solving nonlinear equations. *Wseas Trans. Math.* **2012**, *11*, 283–293.
34. Dźunic, J.; Petković, M. A family of Three-Point methods of Ostrowski's Type for Solving Nonlinear Equations. *J. Appl. Math.* **2012**, *2012*, 425867. [CrossRef]
35. Chun, C. Some improvements of Jarratt's method with sixth-order convergence. *Appl. Math. Comput.* **1990**, *190*, 1432–1437. [CrossRef]
36. Cordero, A.; Torregrosa, J.R. Variants of Newton's method using fifth-order quadrature formulas. *Appl. Math. Comput.* **2007**, *190*, 686–698. [CrossRef]
37. Ezquerro, J.A.; Hernández, M.A. Recurrence relations for Chebyshev-type methods. *Appl. Math. Optim.* **2000**, *41*, 227–236. [CrossRef]
38. Magreñán, Á.A. A new tool to study real dynamics: The convergence plane. *Appl. Math. Comput.* **2014**, *248*, 215–224. [CrossRef]
39. Rall, L.B. *Computational Solution of Nonlinear Operator Equations*; Robert E. Krieger: New York, NY, USA, 1979.
40. Weerakon, S.; Fernando, T.G.I. A variant of Newton's method with accelerated third-order convergence. *Appl. Math. Lett.* **2000**, *13*, 87–93. [CrossRef]
41. Cordero, A.; García-Maimó, J.; Torregrosa, J.R.; Vassileva, M.P.; Vindel, P. Chaos in King's iterative family. *Appl. Math. Lett.* **2013**, *26*, 842–848. [CrossRef]
42. Gopalan, V.B.; Seader, J.D. Application of interval Newton's method to chemical engineering problems. *Reliab. Comput.* **1995**, *1*, 215–223.
43. Shacham, M. An improved memory method for the solution of a nonlinear equation. *Chem. Eng. Sci.* **1989**, *44*, 1495–1501. [CrossRef]

2

A Seventh-Order Scheme for Computing the Generalized Drazin Inverse

Dilan Ahmed [1], Mudhafar Hama [2], Karwan Hama Faraj Jwamer [2] and Stanford Shateyi [3,*]

1 Department of Mathematics, College of Education, University of Sulaimani, Kurdistan Region, Sulaimani 46001, Iraq
2 Department of Mathematics, College of Science, University of Sulaimani, Kurdistan Region, Sulaimani 46001, Iraq
3 Department of Mathematics, University of Venda, Private Bag X5050, Thohoyandou 0950, South Africa
* Correspondence: stanford.shateyi@univen.ac.za

Abstract: One of the most important generalized inverses is the Drazin inverse, which is defined for square matrices having an index. The objective of this work is to investigate and present a computational tool in the form of an iterative method for computing this task. This scheme reaches the seventh rate of convergence as long as a suitable initial matrix is chosen and by employing only five matrix products per cycle. After some analytical discussions, several tests are provided to show the efficiency of the presented formulation.

Keywords: drazin inverse; generalized inverse; iterative methods; higher order; efficiency index

JEL Classification: 15A09; 65F30.

1. Introduction

Drazin, in the pioneering work in [1], proposed and generalized a different type of outer inverse in associative rings and semigroups that does not possess the reflexivity feature but commutes with the element. The significance of this type of inverse and its calculation was then discussed wholly in [2]. Accordingly, several authors attempted to propose procedures for the calculation of generalized inverses. See, e.g., [3–5].

It is recalled that the smallest non-negative integer k such that [6]

$$\text{rank}(A^k) = \text{rank}(A^{k+1}), \tag{1}$$

is named as the matrix A's index and is shown by $\text{ind}(A)$. Furthermore, assume that A is an $N \times N$ matrix with complex entries. The Drazin inverse of A, shown by A^D, is the unique matrix X reading the following identities [7]:

1. $A^k X A = A^k$,
2. $XAX = X$,
3. $AX = XA$.

Throughout the paper, with A^*, $\mathcal{R}(A)$, $\mathcal{N}(A)$, and $\text{rank}(A)$, we show the conjugate transpose, the range, the null space, and the rank of $A \in \mathbb{C}^{N\times N}$, respectively [8]. It is remarked that if $\text{ind}(A) = 1$, X is named as the g-inverse or group inverse of A. In addition, if A is nonsingular, then it is easily seen that

$$indA = 0, \text{ and } A^D = A^{-1}. \tag{2}$$

For the square system $Ax = b$, the general solution can be represented using the concept of the Drazin inverse as follows [7]:

$$x = A^D b + (I - AA^D)z, \tag{3}$$

wherein $z \in \mathcal{R}(A^{k-1}) + \mathcal{N}(A)$.

Methods in the category of iteration schemes for computing generalized inverses like the Drazin inverse are quite sensitive to the choice of the initial approximation. In fact, the convergence of the Schulz-type method can only be observed if the initial value is chosen correctly [9,10]. This selection can be done under special care on some criteria, as already discussed, and given in the literature for different types of outer inverses. For some in-depth discussions about this, one may refer to [11].

Authors in [12] showed that iterative Schulz-type schemes can be used for finding the Drazin inverse of square matrices both possessing real or complex spectra. Actually, the authors investigated the initial matrix below:

$$X_0 = \alpha A^l, \;\; l \geq \mathrm{ind}(A) = k, \tag{4}$$

wherein the parameter α should be selected such that the criterion $\|I - AX_0\| < 1$ is read. Employing the starting value (4) yields an iterative scheme for computing the famous Drazin inverse with second-order convergence.

It is in fact necessary to apply an appropriate initial matrix when calculating the Drazin inverse. One way is as follows [12,13]:

$$X_0 = \frac{2}{\mathrm{Tr}(A^{k+1})} A^k, \tag{5}$$

wherein $\mathrm{Tr}(\cdot)$ stands for the trace of an arbitrary square matrix. Another fruitful initial matrix which could lead in converging sequence of matrices for computing the generalized Drazin inverse can be written as

$$X_0 = \frac{1}{2\|A\|_2^{k+1}} A^k. \tag{6}$$

The Schulz method of the quadratic convergence rate for doing this task can be defined by [14]

$$X_{n+1} = X_n(2I - AX_n), \;\; n = 0, 1, 2, \cdots, \tag{7}$$

where I is the identity matrix and requires only two matrix products to achieve this rate per cycle.

Authors in [15] re-studied Chebyshev's method for calculating $A^{\dagger}_{MN}$ using a suitable initial value as follows:

$$X_{n+1} = X_n(3I - AX_n(3I - AX_n)), \;\; n = 0, 1, 2, \cdots, \tag{8}$$

with a third-order of convergence having three matrix products per cycle. Another scheme, having a cubic rate of convergence and a greater number of matrix products, was given by the same authors as follows:

$$X_{n+1} = X_n \left[I + \frac{1}{2}(I - AX_n)(I + (2I - AX_n)^2) \right], \;\; n = 0, 1, 2, \cdots. \tag{9}$$

A general procedure for having p-order methods with a p number of matrix–matrix products was given in [16] (Chapter 5). For instance, the authors presented the following fourth-order scheme:

$$X_{n+1} = X_n(I + B_n(I + B_n(I + B_n))), \;\; n = 0, 1, 2, \cdots, \tag{10}$$

in which $B_n = I - AX_n$.

The main goal and motivation for investigating novel or useful matrix schemes for computing the Drazin inverse is not only because of the applications of such solvers in different kinds of mathematical problems [17,18] but also to improve the computational efficiency index. In fact, the hyperpower structure as given for a sample case in (10) requires a p number of matrix–matrix products to achieve a

p rate of convergence. This leads to more and more inefficient methods from this class of methods, particularly when the order increases.

As such, motivated by extending efficient methods of higher orders for calculating generalized inverses, here we focus on a seventh-order scheme and discuss how we can reach this higher rate by employing only five matrix by matrix products. This will reveal a higher efficiency index for the discussed scheme in computing the Drazin inverse. For more studies and investigations in this field and related issues of generalized matrix inverses, interested readers are guided to [19–22].

The organization of this paper is as follows. In Section 1, we quickly review the definition, literature, and the need for development of higher order schemes. Section 2 is devoted to extending and proposing an efficient iterative expression for the Drazin inverse. It is derived that the scheme requires only five matrix–matrix products to achieve this rate.

Theoretical discussions with some concrete proofs are provided in Section 3, while Section 4 is oriented on the application of this scheme for computing the Drazin inverses. Results will reveal the effectiveness of this scheme for calculating the Drazin inverse. Lastly, some concluding comments are given in Section 5.

2. Derivation of an Efficient Formulation

Here, the aim is to present a competitive formulation for a member of the hyperpower family of iterations, so as to not only gain a high rate of convergence but also improve the computational efficiency index. In fact, we must factorize the formulation so as to gain the same high convergence rate but a lower number of matrix products.

Toward this objective, let us take into account the following seventh-order method from the family of hyperpower iteration schemes [23]:

$$X_{n+1} = X_n(I + B_n + B_n^2 + B_n^3 + B_n^4 + B_n^5 + B_n^6). \tag{11}$$

It is necessary to emphasize that we are looking for a seventh-order scheme and not a higher-order one, since we wish to hit several targets at the same time. First, the derived scheme for the Drazin inverse must be efficient, viz., it must improve the computational efficiency index of the existing solvers, as will be shown at the end of this section. Second, very high-order schemes might occasionally become hard for coding purposes, and this limits their application, so we aim to have this order be high but not so high. Besides, higher-order schemes mean that fewer stopping criteria (the computation of matrix norms) should be calculated per cycle, which is useful in terms of the elapsed time.

Now, to improve the performance of (11), we factorize (11) in what follows:

$$X_{n+1} = X_n \left(I + B_n(I + B_n)(I - B_n + B_n^2)(I + B_n + B_n^2)\right). \tag{12}$$

However, the formulation (12) needs six matrix–matrix multiplications, which is still not that useful for improving the computational index of efficiency theoretically. As such, doing more factorization would yield the following scheme:

$$X_{n+1} = X_n \left(I + (B_n + B_n^2)(I - B_n + B_n^2)(I + B_n + B_n^2)\right). \tag{13}$$

The scheme (13) requires five matrix products per cycle to achieve the seventh order of convergence. Noting that one reason for the need to propose and have an efficient higher scheme in the category of matrix Schulz-type methods is also in the fact that Schulz-type schemes of lower orders are quite slow at the initial stage of iterates, and this could yield a greater computational burden for finding the Drazin inverse [24]. In fact, it sometimes takes several iterates for the scheme to arrive at its convergence phase, and due to imposing some stopping termination based on matrix norms, this might add some more elapsed time for the application of lower-order schemes.

It is recalled that the definition of index of efficiency is given by [25]:

$$EI = \rho^{\frac{1}{\kappa}}, \tag{14}$$

wherein ρ and κ are the convergence rate and the total cost per cycle, respectively.

Hence, the efficiency indexes of different methods are reported by

$$EI(7) = 2^{\frac{1}{2}} \simeq 1.41421, \; EI(8) = 3^{\frac{1}{3}} \simeq 1.44225, \tag{15}$$

$$EI(9) = 4^{\frac{1}{4}} \simeq 1.41421, \; EI(12) = 7^{\frac{1}{6}} \simeq 1.38309, \; EI(13) = 7^{\frac{1}{5}} \simeq 1.47577. \tag{16}$$

This shows that we have achieved our motivation by improving the efficiency index for calculating the Drazin inverse via a competitive formulation.

3. Seventh Rate of Convergence

Let us now recall some of the well-known lemmas we need in the rest of this section.

Proposition 1 ([26]). *Assume that $M \in \mathbb{C}^{n\times n}$ and $\varepsilon > 0$ are given. There is at least one matrix norm $\|\cdot\|$ such that*

$$\rho(M) \leq \|M\| \leq \rho(M) + \epsilon, \tag{17}$$

wherein $\rho(M)$ shows the collection of all of M's eigenvalues (in the maximum of absolute value sense).

Proposition 2 ([26]). *If $P_{L,M}$ shows the projector on a space L on space M, then*
(i) $P_{L,M}Q = Q$ if and only if $\mathcal{R}(Q) \subseteq L$,
(ii) $QP_{L,M} = Q$ if and only if $\mathcal{N}(Q) \supseteq M$.

The proof of the main theorem concerning the convergence as well as its rate of (13) for calculating the generalized Drazin inverse is now addressed as follows.

Theorem 1. *Consider that $A \in \mathbb{C}^{N\times N}$ is a square singular matrix. In addition, let the initial value X_0 be selected via (4) or (5). Thence, the matrices $\{X_n\}_{n=0}^{n=\infty}$ generated via (13) satisfy the following error estimate for calculating the Drazin inverse:*

$$\|A^D - X_n\| \leq \|A^D\| \|I - AX_0\|^{7^n}. \tag{18}$$

In addition, the convergence order is seven.

Proof. To prove that the sequence is converging, we first take into consideration that

$$\begin{aligned} R_{n+1} &= I - AX_{n+1} \\ &= I - A(X_n \left(I + (B_n + B_n^2)(I - B_n + B_n^2)(I + B_n + B_n^2)\right)) \\ &= I - A(X_n \left(I + B_n(I + B_n)(I - B_n + B_n^2)(I + B_n + B_n^2)\right)) \\ &= I - A(X_n \left(I + B_n + B_n^2 + B_n^3 + B_n^4 + B_n^5 + B_n^6\right)) \\ &= (I - AX_n)^7 \\ &= R_n^7, \end{aligned} \tag{19}$$

wherein $R_n = I - AX_n$. Employing a matrix norm on (19), we obtain that

$$\|R_{n+1}\| \leq \|R_n\|^7. \tag{20}$$

Since X_0 is selected as in (4) or (5), we have

$$\mathcal{R}(X_0) \subseteq \mathcal{R}(A^k). \tag{21}$$

This could now be stated as

$$\mathcal{R}(X_n) \subseteq \mathcal{R}(X_{n-1}). \tag{22}$$

Thus, we can conclude that

$$\mathcal{R}(X_n) \subseteq \mathcal{R}(A^k),\ n \geq 0. \tag{23}$$

In a similar way, by defining the scheme by the left-multiplying of X_n, we can state that

$$\mathcal{N}(X_n) \supseteq \mathcal{N}(A^k),\ n \geq 0. \tag{24}$$

Now, an application of the definition of the Drazin inverse yields

$$AA^D = A^D A = P_{\mathcal{R}(A^k),\mathcal{N}(A^k)}. \tag{25}$$

Proposition 2, along with (23), (24), and (25) could lead to

$$X_n AA^D = X_n = A^D AX_n,\ n \geq 0. \tag{26}$$

To complete the proof, we proceed in what follows. The error matrix $\delta_n = A^D - X_n$ satisfies

$$\begin{aligned} \delta_n &= A^D - X_n \\ &= A^D - A^D AX_n \\ &= A^D (I - AX_n) \\ &= A^D R_n. \end{aligned} \tag{27}$$

Using (20), we obtain the following inequality

$$\|\delta_n\| = \|A^D\| \|R_n\| \leq \|A^D\| \|R_0\|^{7^n}, \tag{28}$$

which is an affirmation of (18). Employing (28) and Proposition 2, one gets that

$$\begin{aligned} A\delta_{n+1} &= AA^D - AX_{n+1} \\ &= AA^D - I + I - AX_{n+1} \\ &= AA^D - I + R_{n+1}. \end{aligned} \tag{29}$$

Note that the idempotent matrix AA^D is the projector on $\mathcal{R}(A^k)$ along $\mathcal{N}(A^k)$, where $\mathcal{R}(A^k)$ denotes the range of A^k, and $\mathcal{N}(A^k)$ is the null space of A^k. Considering (19) and applying several simplifications, one obtains

$$A\delta_{n+1} = AA^D - I + R_n^7. \tag{30}$$

Now, by taking into account the following feature,

$$(I - AA^D)^t = (I - AA^D),\ t \geq 1, \tag{31}$$

we can get that

$$\begin{aligned} (I - AA^D)A\delta_n &= (I - AA^D)A(A^D - X_n) \\ &= X_n - AA^D V_n \\ &= 0. \end{aligned} \tag{32}$$

We obtain, for each $t \geq 1$, that (here we use (32) in simplifications)

$$\begin{aligned}(R_n)^t + AA^D - I &= (I - AV_n)^t + AA^D - I \\ &= (I - AA^D + AA^D - AV_n)^t + AA^D - I \\ &= \left((I - AA^D) + A\delta_n\right)^t + AA^D - I \\ &= I - AA^D + (A\delta_n)^t + AA^D - I \\ &= (A\delta_n)^t.\end{aligned} \tag{33}$$

From (33) and (30), we have

$$A\delta_{n+1} = (A\delta_n)^7. \tag{34}$$

Taking matrix norms from both sides yields

$$\|A\delta_{n+1}\| \leq \|A\delta_n\|^7. \tag{35}$$

Considering (35) and the second criterion of (1), we obtain that

$$\begin{aligned}\|\delta_{n+1}\| &= \|X_{n+1} - A^D\| \\ &= \|A^D A V_{n+1} - A^D A A^D\| \\ &= \|A^D (A V_{n+1} - A A^D)\| \\ &\leq \|A^D\| \|A\delta_{n+1}\| \\ &\leq \|A^D\| \|\delta_n\|^7.\end{aligned} \tag{36}$$

The relations in (36) yield the point that $X_n \to A^D$ as $n \to +\infty$ with the seventh order of convergence. The proof is complete. □

4. Computational Tests

The purpose of this section is to investigate the efficiency of our competitive formulation for computing the Drazin inverse, both theoretically and numerically. For such a task, we employ the challenging schemes (7), (8), and (13).

Here, we have simulated the tests in Mathematica 11.0, [27,28] and the time shown is in seconds. Noting that the compared methods are programmed in the same environment using the hardware CPU Intel Core i5 2430–M, 16 GB RAM in Windows 7 Ultimate with an SSD hard disk.

Test Problem 1. *The aim of this test is to testify the computation of the Drazin inverse for the following input [12]:*

$$A = \begin{bmatrix} 2 & 0.4 & 0 & 0 & 0 & 0 & 0 & 0 & 0 & 0 & 0 & 0 \\ -2 & 0.4 & 0 & 0 & 0 & 0 & 0 & 0 & 0 & 0 & 0 & 0 \\ -1 & -1 & 1 & -1 & 0 & 0 & 0 & 0 & -1 & 0 & 0 & 0 \\ -1 & -1 & -1 & 1 & 0 & 0 & 0 & 0 & 0 & 0 & 0 & 0 \\ 0 & 0 & 0 & 0 & 1 & 1 & -1 & -1 & 0 & 0 & -1 & 0 \\ 0 & 0 & 0 & 0 & 1 & 1 & -1 & -1 & 0 & 0 & 0 & 0 \\ 0 & 0 & 0 & -1 & -2 & 0.4 & 0 & 0 & 0 & 0 & 0 & 0 \\ 0 & 0 & 0 & 0 & 2 & 0.4 & 0 & 0 & 0 & 0 & 0 & 0 \\ 0 & -1 & 0 & 0 & 0 & 0 & 0 & 0 & 1 & -1 & -1 & -1 \\ 0 & 0 & 0 & 0 & 0 & 0 & 0 & 0 & -1 & 1 & -1 & -1 \\ 0 & 0 & 0 & 0 & 0 & 0 & 0 & 0 & 0 & 0 & 0.4 & -2 \\ 0 & 0 & 0 & 0 & 0 & 0 & 0 & 0 & 0 & 0 & 0.4 & 2 \end{bmatrix},$$

at which $k = \text{ind}(A) = 3$. *Here, the Drazin inverse is expressed by*

$$A^D = \begin{bmatrix} 0.25 & -0.25 & 0. & 0. & 0. & 0. & 0. & 0. & 0. & 0. & 0. & 0. \\ 1.25 & 1.25 & 0. & 0. & 0. & 0. & 0. & 0. & 0. & 0. & 0. & 0. \\ -1.66406 & -0.992187 & 0.25 & -0.25 & 0. & 0. & 0. & 0. & -0.0625 & -0.0625 & 0. & 0.15625 \\ -1.19531 & -0.679687 & -0.25 & 0.25 & 0. & 0. & 0. & 0. & -0.0625 & 0.1875 & 0.6875 & 1.34375 \\ -2.76367 & -1.04492 & -1.875 & -1.25 & -1.25 & 1.25 & 1.25 & 1.25 & 1.48438 & 2.57813 & 3.32031 & 6.64063 \\ -2.76367 & -1.04492 & -1.875 & -1.25 & -1.25 & 1.25 & 1.25 & 1.25 & 1.48438 & 2.57813 & 4.57031 & 8.51563 \\ 14.1094 & 6.30078 & 6.625 & 3.375 & 5. & -3. & -5. & -5. & -4.1875 & -8.5 & -10.5078 & -22.4609 \\ -19.3242 & -8.50781 & -9.75 & -5.25 & -7.5 & 4.5 & 7.5 & 7.5 & 6.375 & 12.5625 & 15.9766 & 33.7891 \\ -0.625 & -0.3125 & 0. & 0. & 0. & 0. & 0. & 0. & 0.25 & -0.25 & -0.875 & -1.625 \\ -1.25 & -0.9375 & 0. & 0. & 0. & 0. & 0. & 0. & -0.25 & 0.25 & -0.875 & -1.625 \\ 0. & 0. & 0. & 0. & 0. & 0. & 0. & 0. & 0. & 0. & 1.25 & 1.25 \\ 0. & 0. & 0. & 0. & 0. & 0. & 0. & 0. & 0. & 0. & -0.25 & 0.25 \end{bmatrix}.$$

The results are obtained by applying the stop termination

$$||X_{n+1} - X_n||_1 \leq 10^{-6}, \tag{37}$$

and now by employing the definition in Section 1, we have for (13),

$$\begin{aligned} ||A^{k+1}X_{n+1} - A^k||_\infty &\simeq 3.69638 \times 10^{-12}, \\ ||X_{n+1}AX_{n+1} - X_{n+1}||_\infty &\simeq 8.43992 \times 10^{-10}, \\ ||AX_{n+1} - X_{n+1}A||_\infty &\simeq 3.75205 \times 10^{-10}. \end{aligned} \tag{38}$$

It is also necessary to mention that the domain of validity for the proposed formulation (13) is not only limited to the Drazin inverse, and if a suitable initial approximation is used, under some assumptions we can construct a converging sequence of matrix iterates for other types of generalized inverses.

Test Problem 2. *In this test, we compare the results of various schemes for computing the regular inverse using the initial matrix*

$$X_0 = \frac{1}{||A||_F} A^*, \tag{39}$$

the stopping condition (37), and the following complex matrices constructed in Mathematica:

```
N = 5000; no = 25;
ParallelTable[
  A[j] = SparseArray[
    {Band[{-100, 1100}] -> RandomReal[20], Band[{1, 1}] -> 2.,
     Band[{1000, -50}, {N - 20, N - 25}] -> {2.8, RandomReal[] + I},
     Band[{600, 150}, {N - 100, N - 400}] -> {-RandomReal[3], 3. + 3 I}
     },
    {N, N}, 0.],
  {j, no}
  ];
```

The plot of the large sparse matrices in Test Problem 2 is plotted in Figure 1, showing the sparsity pattern of these matrices, while the pattern of sparsity for the inverse matrix is provided in Figure 2. Figure 3 shows the clear superiority of the proposed formulation in computing the inverse of large sparse matrices.

Here, we give a simple Mathematica implementation of (13) in solving Test Problem 2:

```
For[j = 1, j <= number, j++,
  {
   X = A[j]/(Norm[A[j], "Frobenius"]^2);
   k = 1;
   X1 = 20 X;
   Time[j] = Part[
     While[k <= 75 && N[Norm[X - X1, 1]] >= 10^(-6),
        X1 = SparseArray[X];
        XX = Id - A[j].X1;
        X2 = XX.XX;
        X =
         Chop@
          SparseArray[
           X1.(Id + (XX + X2).(Id - XX + X2).(Id + XX + X2))];
        k++]; // AbsoluteTiming,
     1];
   }];
```

To apply our scheme in modern applications of numerical linear algebra getting involved with sparse large matrices, one may use some commands such as `SparseArray[]` for tackling matrices in sparse forms and subsequently reduce the computational effort and time for preserving the sparsity pattern and finding an approximate inverse. Such applications may occur in various types of problems like the ones in [29,30].

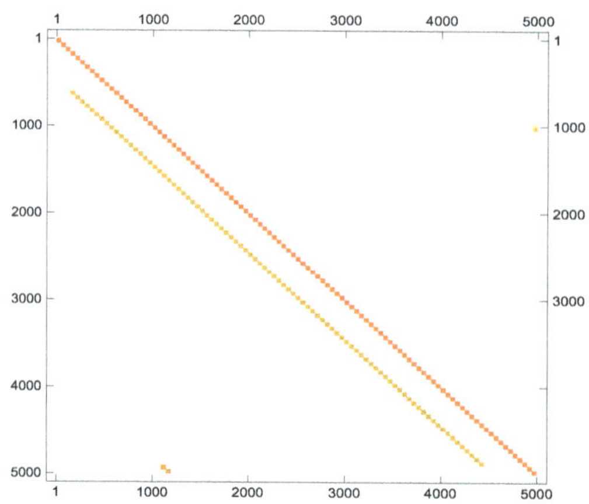

Figure 1. The sparsity pattern of matrices in Test Problem 2.

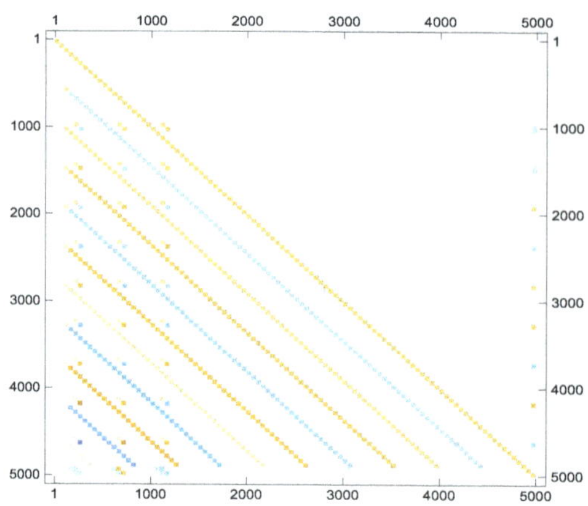

Figure 2. The sparsity pattern of the inverse matrix $X = A_{25}^{-1}$ in Test Problem 2.

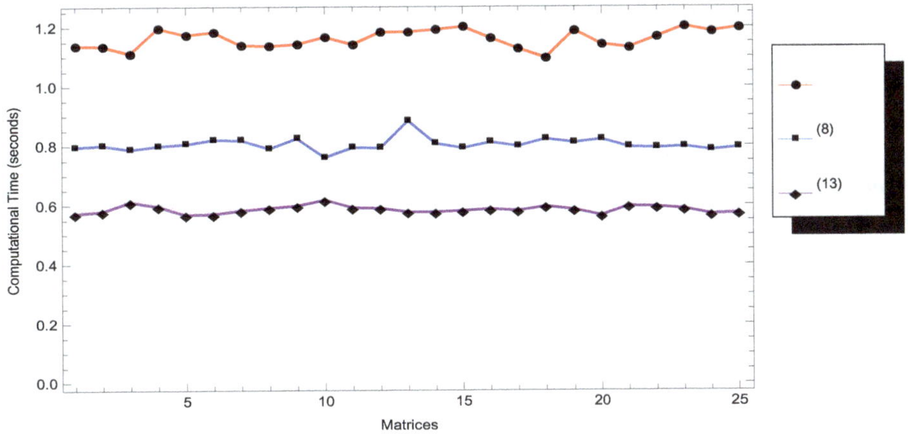

Figure 3. The CPU time required for different matrices in Test Problem 2.

5. Conclusions

Following the motivation of proposing and extending efficient iteration schemes for computing generalized inverses, and particularly for Drazin inverses, in this work we have extended and discussed theoretically how we could achieve a seventh rate in a hyperpower structure for an iterative method. The scheme is a matrix product method and employs only five products to reach this rate. Clearly, the efficiency index will hit the bound $7^{1/5} \simeq 1.47577$.

Several computational tests for calculating the Drazin inverses of several randomly generated matrices were provided to show the superiority and stability of the scheme in doing this task. Other computational problems of different sizes for different matrices were also done and showed similar behavior and the superiority of (13) for the Drazin inverse.

Author Contributions: All authors contributed equally in preparing and writing this work.

Acknowledgments: We are grateful to four anonymous referees for several comments which improved the readability of this work.

References

1. Drazin, M.P. Pseudoinverses in associative rings and semigroups. *Am. Math. Mon.* **1958**, *65*, 506–514. [CrossRef]
2. Wilkinson, J.H. *Note on the Practical Significance of the Drazin Inverse*; Campbell, S.L., Ed.; Recent Applications of Generalized Inverses, Pitman Advanced Publishing Program, Research Notes in Mathematics, No. 66, Boston; NASA: Washington, DC, USA, 1982; pp. 82–99.
3. Kyrchei, I. Explicit formulas for determinantal representations of the Drazin inverse solutions of some matrix and differential matrix equations. *Appl. Math. Comput.* **2013**, *219*, 7632–7644. [CrossRef]
4. Liu, X.; Zhu, G.; Zhou, G.; Yu, Y. An analog of the adjugate matrix for the outer inverse $A^{(2)}_{T,S}$. *Math. Prob. Eng.* **2012**, *2012*, 591256.
5. Moghani, Z.N.; Khanehgir, M.; Karizaki, M.M. Explicit solution to the operator equation $AXD + FX^*B = C$ over Hilbert C*-modules. *J. Math. Anal.* **2019**, *10*, 52–64.
6. Ben-Israel, A.; Greville, T.N.E. *Generalized Inverses: Theory and Applications*, 2nd ed.; Springer: New York, NY, USA, 2003.
7. Wei, Y. Index splitting for the Drazin inverse and the singular linear system. *Appl. Math. Comput.* **1998**, *95*, 115–124. [CrossRef]
8. Ma, H.; Li, N.; Stanimirović, P.S.; Katsikis, V.N. Perturbation theory for Moore–Penrose inverse of tensor via Einstein product. *Comput. Appl. Math.* **2019**, *38*, 111. [CrossRef]
9. Soleimani, F.; Soleymani, F.; Shateyi, S. Some iterative methods free from derivatives and their basins of attraction. *Discret. Dyn. Nat. Soc.* **2013**, *2013*, 301718. [CrossRef]

10. Soleymani, F. Efficient optimal eighth-order derivative-free methods for nonlinear equations. *Jpn. J. Ind. Appl. Math.* **2013**, *30*, 287–306. [CrossRef]
11. Pan, V.Y. *Structured Matrices and Polynomials: Unified Superfast Algorithms*; BirkhWauser: Boston, MA, USA; Springer: New York, NY, USA, 2001.
12. Li, X.; Wei, Y. Iterative methods for the Drazin inverse of a matrix with a complex spectrum. *Appl. Math. Comput.* **2004**, *147*, 855–862. [CrossRef]
13. Stanimirović, P.S.; Ciric, M.; Stojanović, I.; Gerontitis, D. Conditions for existence, representations, and computation of matrix generalized inverses. *Complexity* **2017**, *2017*, 6429725. [CrossRef]
14. Schulz, G. Iterative Berechnung der Reziproken matrix. *Z. Angew. Math. Mech.* **1933**, *13*, 57–59. [CrossRef]
15. Li, H.-B.; Huang, T.-Z.; Zhang, Y.; Liu, X.-P.; Gu, T.-X. Chebyshev-type methods and preconditioning techniques. *Appl. Math. Comput.* **2011**, *218*, 260–270. [CrossRef]
16. Krishnamurthy, E.V.; Sen, S.K. *Numerical Algorithms: Computations in Science and Engineering*; Affiliated East-West Press: New Delhi, India, 1986.
17. Ma, J.; Gao, F.; Li, Y. An efficient method to compute different types of generalized inverses based on linear transformation. *Appl. Math. Comput.* **2019**, *349*, 367–380. [CrossRef]
18. Soleymani, F.; Stanimirović, P.S.; Khaksar Haghani, F. On Hyperpower family of iterations for computing outer inverses possessing high efficiencies. *Linear Algebra Appl.* **2015**, *484*, 477–495. [CrossRef]
19. Qin, Y.; Liu, X.; Benítez, J. Some results on the symmetric representation of the generalized Drazin inverse in a Banach algebra. *Symmetry* **2019**, *11*, 105. [CrossRef]
20. Wang, G.; Wei, Y.; Qiao, S. *Generalized Inverses: Theory and Computations*; Science Press: Beijing, China; New York, NY, USA, 2004.
21. Xiong, Z.; Liu, Z. The forward order law for least Squareg-inverse of multiple matrix products. *Mathematics* **2019**, *7*, 277. [CrossRef]
22. Zhao, L. The expression of the Drazin Inverse with rank constraints. *J. Appl. Math.* **2012**, *2012*, 390592. [CrossRef]
23. Sen, S.K.; Prabhu, S.S. Optimal iterative schemes for computing Moore-Penrose matrix inverse. *Int. J. Syst. Sci.* **1976**, *8*, 748–753. [CrossRef]
24. Soleymani, F. An efficient and stable Newton–type iterative method for computing generalized inverse $A_{T,S}^{(2)}$. *Numer. Algorithms* **2015**, *69*, 569–578. [CrossRef]
25. Ostrowski, A.M. Sur quelques transformations de la serie de LiouvilleNewman. *C.R. Acad. Sci. Paris* **1938**, *206*, 1345–1347.
26. Jebreen, H.B.; Chalco-Cano, Y. An improved computationally efficient method for finding the Drazin inverse. *Discret. Dyn. Nat. Soc.* **2018**, *2018*, 6758302. [CrossRef]
27. Sánchez León, J.G. *Mathematica Beyond Mathematics: The Wolfram Language in the Real World*; Taylor & Francis Group: Boca Raton, FL, USA, 2017.
28. Wagon, S. *Mathematica in Action*, 3rd ed.; Springer: Berlin, Germany, 2010.
29. Soleymani, F. Efficient semi-discretization techniques for pricing European and American basket options. *Comput. Econ.* **2019**, *53*, 1487–1508. [CrossRef]
30. Soleymani, F.; Barfeie, M. Pricing options under stochastic volatility jump model: A stable adaptive scheme. *Appl. Numer. Math.* **2019**, *145*, 69–89. [CrossRef]

3

A Higher Order Chebyshev-Halley-Type Family of Iterative Methods for Multiple Roots

Ramandeep Behl [1], Eulalia Martínez [2,*], Fabricio Cevallos [3] and Diego Alarcón [4]

1 Department of Mathematics, King Abdualziz University, Jeddah 21589, Saudi Arabia; ramanbehl87@yahoo.in
2 Instituto Universitario de Matemática Multidisciplinar, Universitat Politècnica de València, Camino de Vera s/n, 46022 Valencia, Spain
3 Fac. de Ciencias Económicas, Universidad Laica "Eloy Alfaro de Manabí", Manabí 130214, Ecuador; alfa2205@gmail.com
4 Departamento de Matemática Aplicada, Universitat Politècnica de València, Camino de Vera s/n, 46022 Valencia, Spain; diealcor@doctor.upv.es
* Correspondence: eumarti@mat.upv.es

Abstract: The aim of this paper is to introduce new high order iterative methods for multiple roots of the nonlinear scalar equation; this is a demanding task in the area of computational mathematics and numerical analysis. Specifically, we present a new Chebyshev–Halley-type iteration function having at least sixth-order convergence and eighth-order convergence for a particular value in the case of multiple roots. With regard to computational cost, each member of our scheme needs four functional evaluations each step. Therefore, the maximum efficiency index of our scheme is 1.6818 for $\alpha = 2$, which corresponds to an optimal method in the sense of Kung and Traub's conjecture. We obtain the theoretical convergence order by using Taylor developments. Finally, we consider some real-life situations for establishing some numerical experiments to corroborate the theoretical results.

Keywords: nonlinear equations; multiple roots; Chebyshev–Halley-type; optimal iterative methods; efficiency index

---

1. Introduction

One important field in the area of computational methods and numerical analysis is to find approximations to the solutions of nonlinear equations of the form:

$$f(x) = 0, \tag{1}$$

where $f : \mathbb{D} \subset \mathbb{C} \to \mathbb{C}$ is the analytic function in the enclosed region $\mathbb{D}$, enclosing the required solution. It is almost impossible to obtain the exact solution in an analytic way for such problems. Therefore, we concentrate on obtaining approximations of the solution up to any specific degree of accuracy by means of an iterative procedure, of course doing it also with the maximum efficiency. In [1], Kung and Traub conjectured that a method without memory that uses $n+1$ functional evaluations per iteration can have at most convergence order $p = 2^n$. If this bound is reached, the method is said to be optimal.

For solving nonlinear Equation (1) by means of iterations, we have the well-known cubically-convergent family of Chebyshev–Halley methods [2], which is given by:

$$x_{n+1} = x_n - \left[1 + \frac{1}{2}\frac{L_f(x_n)}{1 - \alpha L_f(x_n)}\right]\frac{f(x_n)}{f'(x_n)}, \quad \alpha \in \mathbb{R}, \tag{2}$$

where $L_f(x_n) = \frac{f''(x_n)f(x_n)}{\{f'(x_n)\}^2}$. A great variety of iterative methods can be reported in particular cases. For example, the classical Chebyshev's method [1,3], Halley's method [1,3], and the super-Halley method [1,3] can be obtained if $\alpha = 0$, $\alpha = \frac{1}{2}$, and $\alpha = 1$, respectively. Despite the third-order convergence, the scheme (2) is considered less practical from a computational point of view because of the computation of the second-order derivative.

For this reason, several variants of Chebyshev–Halley's methods free from the second-order derivative have been presented in [4–7]. It has been shown that these methods are comparable to the classical third-order methods of the Chebyshev–Halley-type in their performance and can also compete with Newton's method. One family of these methods is given as follows:

$$\begin{aligned} y_n &= x_n - \frac{f(x_n)}{f'(x_n)}, \\ x_{n+1} &= x_n - \left(1 + \frac{f(y_n)}{f(x_n) - \alpha f(y_n)}\right) \frac{f(x_n)}{f'(x_n)}, \quad \alpha \in \mathbb{R} \end{aligned} \tag{3}$$

We can easily obtain some well-known third-order methods proposed by Potra and Pták [4] and Sharma [5] (the Newton-secant method (NSM)) for $\alpha = 0$ and $\alpha = 1$. In addition, we have Ostrowski's method [8] having optimal fourth-order convergence, which is also a special case for $\alpha = 2$. This family is important and interesting not only because of not using a second- or higher order derivative. However, this scheme also converges at least cubically and has better results in comparison to the existing ones. Moreover, we have several higher order modifications of the Chebyshev–Halley methods available in the literature, and some of them can be seen in [9–12].

In this study, we focus on the case of the multiple roots of nonlinear equations. We have some fourth-order optimal and non-optimal modifications or improvements of Newton's iteration function for multiple roots in the research articles [13–17]. Furthermore, we can find some higher order methods for this case, but some of them do not reach maximum efficiency [18–23]; so, this topic is of interest in the current literature.

We propose a new Chebyshev–Halley-type iteration function for multiple roots, which reaches a high order of convergence. Specifically, we get a family of iterative methods with a free parameter α, with sixth-order convergence. Therefore, the efficiency index is $6^{1/4}$, and for $\alpha = 2$, this index is $8^{1/4}$, which is the maximum value that one can get with four functional evaluations, reaching optimality in the sense of Kung and Traub's conjecture. Additionally, an extensive analysis of the convergence order is presented in the main theorem.

We recall that $\xi \in \mathbb{C}$ is a multiple root of the equation $f(x) = 0$, if it is verified that:

$$f(\xi) = 0, f'(\xi) = 0, \cdots, f^{(m-1)}(\xi) = 0 \quad \textit{and} \quad f^{(m)}(\xi) \neq 0,$$

the positive integer ($m \geq 1$) being the multiplicity of the root.

We deal with iterative methods in which the multiplicity must be known in advance, because this value, m, is used in the iterative expression. However, we point out that these methods also work when one uses an estimation of the multiplicity, as was proposed in the classical study carried out in [24].

Finally, we consider some real-life situations that start from some given conditions to investigate and some standard academic test problems for numerical experiments. Our iteration functions here are found to be more comparable and effective than the existing methods for multiple roots in terms of residual errors and errors among two consecutive iterations, and also, we obtain a more stable computational order of convergence. That is, the proposed methods are competitive.

2. Construction of the Higher Order Scheme

In this section, we present the new Chebyshev–Halley-type methods for multiple roots of nonlinear equations, for the first time. In order to construct the new scheme, we consider the following scheme:

$$\begin{aligned} y_n &= x_n - m\frac{f(x_n)}{f'(x_n)}, \\ z_n &= x_n - m\left(1 + \frac{\eta}{1-\alpha\eta}\right)\frac{f(x_n)}{f'(x_n)}, \\ x_{n+1} &= z_n - H(\eta,\ \tau)\frac{f(x_n)}{f'(x_n)}, \end{aligned} \tag{4}$$

where the function:

$$H(\eta,\ \tau) = \frac{\eta\tau\left(\beta - (\alpha-2)^2\eta^2(\eta+1) + \tau^3 + \tau^2\right)}{(\eta+1)(\tau+1)}$$

with:

$$\eta = \left(\frac{f(y_n)}{f(x_n)}\right)^{\frac{1}{m}},$$

$$\tau = \left(\frac{f(z_n)}{f(y_n)}\right)^{\frac{1}{m}},$$

$$\beta = m\left((\alpha(\alpha+2)+9)\eta^3 + \eta^2(\alpha(\alpha+3) - 6\tau - 3) + \eta(\alpha + 8\tau + 1) + 2\tau + 1\right),$$

where $\alpha \in \mathbb{R}$ is a free disposable variable. For $m = 1$, we can easily obtain the scheme (3) from the first two steps of the scheme (4).

In Theorem 1, we illustrate that the constructed scheme attains at least sixth-order convergence and for $\alpha = 2$, it goes to eighth-order without using any extra functional evaluation. It is interesting to observe that $H(\eta,\ \tau)$ plays a significant role in the construction of the presented scheme (for details, please see Theorem 1).

Theorem 1. *Let us consider $x = \xi$ to be a multiple zero with multiplicity $m \geq 1$ of an analytic function $f : \mathbb{C} \to \mathbb{C}$ in the region containing the multiple zero ξ of $f(x)$. Then, the present scheme (4) attains at least sixth-order convergence for each α, but for a particular value of $\alpha = 2$, it reaches the optimal eighth-order convergence.*

Proof. We expand the functions $f(x_n)$ and $f'(x_n)$ about $x = \xi$ with the help of a Taylor's series expansion, which leads us to:

$$f(x_n) = \frac{f^{(m)}(\xi)}{m!}e_n^m\left(1 + c_1e_n + c_2e_n^2 + c_3e_n^3 + c_4e_n^4 + c_5e_n^5 + c_6e_n^6 + c_7e_n^7 + c_8e_n^8 + O(e_n^9)\right), \tag{5}$$

and:

$$\begin{aligned} f'(x_n) =& \frac{f^m(\xi)}{m!}e_n^{m-1}\Big(m + (m+1)c_1e_n + (m+2)c_2e_n^2 + (m+3)c_3e_n^3 + (m+4)c_4e_n^4 + (m+5)c_5e_n^5 \\ &+ (m+6)c_6e_n^6 + (m+7)c_7e_n^7 + (m+8)c_8e_n^8 + O(e_n^9)\Big), \end{aligned} \tag{6}$$

respectively, where $c_k = \frac{m!}{(m-1+k)!}\frac{f^{m-1+k}(\xi)}{f^m(\xi)}$, $k = 2, 3, 4 \ldots, 8$ and $e_n = x_n - \xi$ is the error in the nth iteration.

Inserting the above expressions (5) and (6) into the first substep of scheme (4) yields:

$$y_n - \xi = \frac{c_1}{m}e_n^2 + \sum_{i=0}^{5}\phi_i e_n^{i+3} + O(e_n^9), \tag{7}$$

where $\phi_i = \phi_i(m, c_1, c_2, \ldots, c_8)$ are given in terms of $m, c_2, c_3, \ldots, c_8$, for example $\phi_0 = \frac{1}{m^2}(2mc_2 - (m+1)c_1^2)$ and $\phi_1 = \frac{1}{m^3}\left[3m^2c_3 + (m+1)^2c_1^3 - m(3m+4)c_1c_2\right]$, etc.

Using the Taylor series expansion and the expression (7), we have:

$$\begin{aligned} f(y_n) =& f^{(m)}(\xi)e_n^{2m}\Bigg[\frac{\left(\frac{c_1}{m}\right)^m}{m!} + \frac{(2mc_2 - (m+1)c_1^2)\left(\frac{c_1}{m}\right)^m e_n}{m!c_1} + \left(\frac{c_1}{m}\right)^{1+m}\frac{1}{2m!c_1^3}\{(3+3m+3m^2+m^3)c_1^4 \\ & - 2m(2+3m+2m^2)c_1^2c_2 + 4(m-1)m^2c_2^2 + 6m^2c_1c_3\}e_n^2 + \sum_{i=0}^{5}\bar{\phi}_i e_n^{i+3} + O(e_n^9)\Bigg]. \end{aligned} \tag{8}$$

We obtain the following expression by using (5) and (8):

$$\eta = \frac{c_1e_n}{m} + \frac{2mc_2 - (m+2)c_1^2}{m^2}e_n^2 + \theta_0 e_n^3 + \theta_1 e_n^4 + \theta_2 e_n^5 + O(e_n^6), \tag{9}$$

where $\theta_0 = \frac{(2m^2+7m+7)c_1^3 + 6m^2c_3 - 2m(3m+7)c_1c_2}{2m^3}$, $\theta_1 = -\frac{1}{6m^4}\left[12m^2(2m+5)c_1c_3 + 12m^2((m+3)c_2^2 - 2mc_4) - 6m(4m^2+16m+17)c_1^2c_2 + (6m^3+29m^2+51m+34)c_1^4\right]$ and $\theta_2 = \frac{1}{24m^5}\left[12m^2(10m^2+43m+49)c_1^2c_3 - 24m^3((5m+17)c_2c_3 - 5mc_5) + 12m^2\Big((10m^2+47m+53)c_2^2 - 2m(5m+13)c_4\Big)c_1 - 4m(30m^3+163m^2+306m+209)c_1^3c_2 + (24m^4+146m^3+355m^2+418m+209)c_1^5\right]$.

With the help of Expressions (5)–(9), we obtain:

$$z_n - \xi = -\frac{(\alpha-2)c_1^2}{m^2}e_n^3 + \sum_{i=0}^{4}\psi_i e_n^{i+4} + O(e_n^9), \tag{10}$$

where $\psi_i = \psi_i(\alpha, m, c_1, c_2, \ldots, c_8)$ are given in terms of $\alpha, m, c_2, c_3, \ldots, c_8$ with the first two coefficients explicitly written as $\psi_0 = -\frac{1}{2m^3}\left[(2\alpha^2 - 10\alpha + (7-4\alpha)m + 11)c_1^3 + 2m(4\alpha-7)c_1c_2\right]$ and $\psi_1 = \frac{1}{6m^4}\Big[(-6\alpha^3 + 42\alpha^2 - 96\alpha + (29-18\alpha)m^2 + 6(3\alpha^2 - 14\alpha + 14)m + 67)c_1^4 + 12m^2(5-3\alpha)c_1c_3 + 12m^2(3-2\alpha)c_2^2 + 12m(-3\alpha^2 + 14\alpha + (5\alpha-8)m - 14)c_1^2c_2\Big]$.

By using the Taylor series expansion and (10), we have:

$$f(z_n) = f^{(m)}(\xi)e_n^{3m}\left[\frac{\left(-\frac{(\alpha-2)c_1^2}{m^2}\right)^m}{m!} + \sum_{i=1}^{5}\bar{\psi}_i e_n^i + O(e_n^6)\right]. \tag{11}$$

From Expressions (8) and (11), we further have:

$$\tau = -\frac{(\alpha-2)c_1}{m}e_n + \frac{((-2\alpha^2 + 8\alpha + (2\alpha-3)m - 7)c_1^2 + 2m(3-2\alpha)c_2)}{2m^2}e_n^2 + \gamma_1 e_n^3 + \gamma_2 e_n^4 + O(e_n^5), \tag{12}$$

where $\gamma_1 = \frac{1}{3m^3}\Big[(-3\alpha^3 + 18\alpha^2 - 30\alpha + (4-3\alpha)m^2 + 3(2\alpha^2 - 7\alpha + 5)m + 11)c_1^3 + 3m^2(4-3\alpha)c_3 + 3m(-4\alpha^2 + 14\alpha + 3\alpha m - 4m - 10)c_1c_2\Big]$ and $\gamma_2 = \frac{1}{24m^4}\Big[24m^2(-6\alpha^2 + 20\alpha + (4\alpha-5)m - 14)c_1c_3 + 12m^2((-8\alpha^2 + 24\alpha + 4\alpha m - 5m - 13)c_2^2 + 2m(5-4\alpha)c_4) - 12m(12\alpha^3 - 66\alpha^2 + 100\alpha + 2(4\alpha-5)m^2 + (-20\alpha^2 + 64\alpha - 41)m - 33)c_1^2c_2 + (-24\alpha^4 + 192\alpha^3 - 492\alpha^2 + 392\alpha + 6(4\alpha-5)m^3 + (-72\alpha^2 + 232\alpha - 151)m^2 + 6(12\alpha^3 - 66\alpha^2 + 100\alpha - 33)m + 19)c_1^4\Big]$.

By using Expressions (9) and (12), we obtain:

$$H(\eta, \tau) = -\frac{(\alpha-2)c_1^2}{m^2}e_n^2 + \lambda_1 e_n^3 + \lambda_2 e_n^4 + O(e_n^5) \tag{13}$$

where $\lambda_1 = \frac{c_1}{2m^3}\left[c_1^2\left(-2\alpha^2+8\alpha+(4\alpha-7)m-7\right)+2(7-4\alpha)c_2m\right]$ and $\lambda_2 = \frac{1}{6m^3}\left[c_1^4\left(-6\alpha^3+36\alpha^2-66\alpha+(29-18\alpha)m^2+3(6\alpha^2-22\alpha+17)m+34\right)+12(5-3\alpha)c_3c_1m^2+12(3-2\alpha)c_2^2m^2+6c_2c_1^2m\left(-6\alpha^2+22\alpha+2(5\alpha-8)m-17\right)\right]$.

Now, we use the expressions (5)–(13) in the last substep of Scheme (4), and we get:

$$e_{n+1} = \sum_{i=1}^{3} L_i e_n^{i+5} + O(e_n^9), \tag{14}$$

where $L_1 = \frac{(\alpha-2)^2c_1^3}{m^6}\left[c_1^2\left(\alpha^2-\alpha+m^2-\left(\alpha^2+4\alpha-17\right)m-3\right)-2c_2(m-1)m\right]$, $L_2 = (\alpha-2)c_1^2\big[-12c_2c_1^2m\{10\alpha^3-24\alpha^2-39\alpha+(16\alpha-27)m^2-(10\alpha^3+27\alpha^2-262\alpha+301)m+91\}+12c_3c_1m^2(-4\alpha+(4\alpha-7)m+8)+12c_2^2m^2(-12\alpha+4(3\alpha-5)m+21)+c_1^4\{-24\alpha^4+168\alpha^3-156\alpha^2-662\alpha+(52\alpha-88)m^3-(60\alpha^3+162\alpha^2-1616\alpha+1885)m^2+2(18\alpha^4-12\alpha^3-711\alpha^2+2539\alpha-2089)m+979\}\big]$ and $L_3 = \frac{c_1}{24m^8}\Big[-24c_2c_3c_1m^3((42\alpha^2-146\alpha+125)m-6(7\alpha^2-26\alpha+24))-24c_2^3m^3(-24\alpha^2+84\alpha+(24\alpha^2-80\alpha+66)m-73)+12c_3c_1^3m^2\{2(15\alpha^4-63\alpha^3-5\alpha^2+290\alpha-296)+(54\alpha^2-190\alpha+165)m^2+(-30\alpha^4-28\alpha^3+968\alpha^2-2432\alpha+1697)m\}+12c_1^2m^2\Big\{c_2^2\Big(80\alpha^4-304\alpha^3-226\alpha^2+1920\alpha+2(81\alpha^2-277\alpha+234)m^2+(-80\alpha^4-112\alpha^3+2712\alpha^2-6410\alpha+4209)m-1787\Big)-4(\alpha-2)c_4m(-3\alpha+(3\alpha-5)m+6)\Big\}-2c_2c_1^4m\Big\{-3(96\alpha^5-804\alpha^4+1504\alpha^3+2676\alpha^2-10612\alpha+8283)+4(177\alpha^2-611\alpha+521)m^3-3(220\alpha^4+280\alpha^3-7556\alpha^2+18400\alpha-12463)m^2+4(108\alpha^5-234\alpha^4-4302\alpha^3+22902\alpha^2-38593\alpha+20488)m\Big\}+c_1^6\Big\{48\alpha^6-480\alpha^5+996\alpha^4+5472\alpha^3-29810\alpha^2+50792\alpha+(276\alpha^2-956\alpha+818)m^4+(-360\alpha^4-448\alpha^3+12434\alpha^2-30518\alpha+20837)m^3+(432\alpha^5-1236\alpha^4-16044\alpha^3+92306\alpha^2-161292\alpha+88497)m^2+(-168\alpha^6+888\alpha^5+5352\alpha^4-55580\alpha^3+173290\alpha^2-224554\alpha+97939)m-29771\Big\}\Big]$.

It is noteworthy that we reached at least sixth-order convergence for all α. In addition, we can easily obtain $L_1 = L_2 = 0$ by using $\alpha = 2$.

Now, by adopting $\alpha = 2$ in Expression (14), we obtain:

$$e_{n+1} = \frac{A_0\left(12c_3c_1m^3-12c_2c_1^2m(3m^2+30m-1)+12c_2^2m^2(2m-1)+c_1^4(10m^3+183m^2+650m-3)\right)}{24m^8}e_n^8 + O(e_n^9), \tag{15}$$

where $A_0 = (c_1^3(m+1)-2c_1c_2m)$. The above Expression (15) demonstrates that our proposed Scheme (4) reaches eighth-order convergence for $\alpha = 2$ by using only four functional evaluations per full iteration. Hence, it is an optimal scheme for a particular value of $\alpha = 2$ according to the Kung–Traub conjecture, completing the proof. □

3. Numerical Experiments

In this section, we illustrate the efficiency and convergence behavior of our iteration functions for particular values $\alpha = 0$, $\alpha = 1$, $\alpha = 1.9$, and $\alpha = 2$ in Expression (4), called $OM1, OM2$, $OM3$, and $OM4$, respectively. In this regards, we choose five real problems having multiple and simple zeros. The details are outlined in the examples (1)–(3).

For better comparison of our iterative methods, we consider several existing methods of order six and the optimal order eight. Firstly, we compare our methods with the two-point family of sixth-order methods proposed by Geum et al. in [18], and out of them, we pick Case 4c, which is mentioned as follows:

$$\begin{aligned} y_n &= x_n - m\frac{f(x_n)}{f'(x_n)}, \quad m > 1, \\ x_{n+1} &= y_n - \left[\frac{m+a_1u_n}{1+b_1u_n+b_2{u_n}^2} \times \frac{1}{1+c_1s_n}\right]\frac{f(y_n)}{f'(y_n)}, \end{aligned} \tag{16}$$

where:

$$a_1 = \frac{2m\left(4m^4 - 16m^3 + 31m^2 - 30m + 13\right)}{(m-1)\left(4m^2 - 8m + 7\right)}, \quad b_1 = \frac{4\left(2m^2 - 4m + 3\right)}{(m-1)\left(4m^2 - 8m + 7\right)},$$
$$b_2 = -\frac{4m^2 - 8m + 3}{4m^2 - 8m + 7}, \quad c_1 = 2(m-1),$$
$$u_n = \left(\frac{f(y_n)}{f(x_n)}\right)^{\frac{1}{m}}, \quad s_n = \left(\frac{f'(y_n)}{f'(x_n)}\right)^{\frac{1}{m-1}},$$

called $GM1$.

In addition, we also compare them with one more non-optimal family of sixth-order iteration functions given by the same authors of [19], and out of them, we choose Case 5YD, which is given by:

$$\begin{aligned} y_n &= x_n - m\frac{f(x_n)}{f'(x_n)}, \ m \geq 1, \\ w_n &= x_n - m\left[\frac{(u_n - 2)(2u_n - 1)}{(u_n - 1)(5u_n - 2)}\right]\frac{f(x_n)}{f'(x_n)}, \\ x_{n+1} &= x_n - m\left[\frac{(u_n - 2)(2u_n - 1)}{(5u_n - 2)(u_n + v_n - 1)}\right]\frac{f(x_n)}{f'(x_n)}, \end{aligned} \tag{17}$$

where $u_n = \left(\frac{f(y_n)}{f(x_n)}\right)^{\frac{1}{m}}$ and $v_n = \left(\frac{f(w_n)}{f(x_n)}\right)^{\frac{1}{m}}$, and this method is denoted as $GM2$.

Moreover, we compare our methods with the optimal eighth-order iterative methods proposed by Zafar et al. [21]. We choose the following two schemes out of them:

$$\begin{aligned} y_n &= x_n - m\frac{f(x_n)}{f'(x_n)}, \\ w_n &= y_n - mu_n(6u_n^3 - u_n^2 + 2u_n + 1)\frac{f(x_n)}{f'(x_n)}, \\ x_{n+1} &= w_n - mu_nv_n(1 + 2u_n)(1 + v_n)\left(\frac{2w_n + 1}{A_2P_0}\right)\frac{f(x_n)}{f'(x_n)} \end{aligned} \tag{18}$$

and:

$$\begin{aligned} y_n &= x_n - m\frac{f(x_n)}{f'(x_n)}, \\ w_n &= y_n - mu_n\left(\frac{1 - 5u_n^2 + 8u_n^3}{1 - 2u_n}\right)\frac{f(x_n)}{f'(x_n)}, \\ x_{n+1} &= w_n - mu_nv_n(1 + 2u_n)(1 + v_n)\left(\frac{3w_n + 1}{A_2P_0(1 + w_n)}\right)\frac{f(x_n)}{f'(x_n)}, \end{aligned} \tag{19}$$

where $u_n = \left(\frac{f(y_n)}{f(x_n)}\right)^{\frac{1}{m}}$, $v_n = \left(\frac{f(w_n)}{f(y_n)}\right)^{\frac{1}{m}}$, $w_n = \left(\frac{f(w_n)}{f(x_n)}\right)^{\frac{1}{m}}$, and these iterative methods are denoted in our tables as $ZM1$ and $ZM2$, respectively.

Finally, we demonstrate their comparison with another optimal eighth-order iteration function given by Behl et al. [22]. However, we consider the following the best schemes (which was claimed by them):

$$\begin{aligned} y_n &= x_n - m\frac{f(x_n)}{f'(x_n)}, \\ z_n &= y_n - m\frac{f(x_n)}{f'(x_n)}h_n(1 + 2h_n), \\ x_{n+1} &= z_n + m\frac{f(x_n)}{f'(x_n)}\frac{t_nh_n}{1 - t_n}\left[-1 - 2h_n - h_n^2 + 4h_n^3 - 2k_n\right] \end{aligned} \tag{20}$$

and:

$$\begin{aligned} y_n &= x_n - m\frac{f(x_n)}{f'(x_n)}, \\ z_n &= y_n - m\frac{f(x_n)}{f'(x_n)}h_n(1+2h_n), \\ x_{n+1} &= z_n - m\frac{f(x_n)}{f'(x_n)}\frac{t_n h_n}{1-t_n}\Big[\frac{1+9h_n^2+2k_n+h_n(6+8k_n)}{1+4h_n}\Big], \end{aligned} \tag{21}$$

with $h_n = \left(\frac{f(y_n)}{f(x_n)}\right)^{\frac{1}{m}}$, $k_n = \left(\frac{f(z_n)}{f(x_n)}\right)^{\frac{1}{m}}$ $t_n = \left(\frac{f(z_n)}{f(y_n)}\right)^{\frac{1}{m}}$, which are denoted $BM1$ and $BM2$, respectively.

In order to compare these schemes, we perform a numerical experience, and in Tables 1 and 2, we display the difference between two consecutive iterations $|x_{n+1} - x_n|$, the residual error in the corresponding function $|f(x_n)|$, and the computational order of convergence (ρ) (we used the formula given by Cordero and Torregrosa [25]:

$$\rho \approx \frac{\ln(\mid x_{k+1} - x_k \mid / \mid x_k - x_{k-1} \mid)}{\ln(\mid x_k - x_{k-1} \mid / \mid x_{k-1} - x_{k-2} \mid)} \tag{22}$$

We make our calculations with several significant digits (a minimum of 3000 significant digits) to minimize the round-off error. Moreover, the computational order of convergence is provided up to five significant digits. Finally, we display the initial guess and approximated zeros up to 25 significant digits in the corresponding example where an exact solution is not available.

All computations have been performed using the programming package Mathematica 11 with multiple precision arithmetic. Further, the meaning of $a(\pm b)$ is shorthand for $a \times 10^{(\pm b)}$ in the numerical results.

Example 1. *Population growth problem:*

The law of population growth is defined as follows:

$$\frac{dN(t)}{dt} = \gamma N(t) + \eta,$$

where N(t) = the population at time t, η = the fixed/constant immigration rate, and γ = the fixed/constant birth rate of the population. We can easily obtain the following solution of the above differential equation:

$$N(t) = N_0 e^{\gamma t} + \frac{\eta}{\gamma}(e^{\gamma t - 1}),$$

where N_0 is the initial population.

For a particular case study, the problem is given as follows: Suppose a certain population contains 1,000,000 individuals initially, that 300,000 individuals immigrate into the community in the first year, and that 1,365,000 individuals are present at the end of one year. Find the birth rate (γ) of this population.

To determine the birth rate, we must solve the equation:

$$f_1(x) = 1365 - 1000e^x - \frac{300}{x}(e^x - 1). \tag{23}$$

wherein $x = \gamma$ and our desired zero of the above function f_1 is 0.05504622451335177827483421. The reason for considering the simple zero problem is to confirm that our methods also work for simple zeros. We choose the starting point as $x_0 = 0.5$.

Example 2. *The van der Waals equation of state:*

$$\left(P+\frac{a_1 n^2}{V^2}\right)(V-na_2)=nRT,$$

explains the behavior of a real gas by introducing in the ideal gas equations two parameters, a_1 and a_2, specific for each gas. The determination of the volume V of the gas in terms of the remaining parameters requires the solution of a nonlinear equation in V,

$$PV^3-(na_2P+nRT)V^2+a_1n^2V-a_1a_2n^2=0.$$

Given the constants a_1 and a_2 of a particular gas, one can find values for n, P, and T, such that this equation has three simple roots. By using the particular values, we obtain the following nonlinear function:

$$f_2(x)=x^3-5.22x^2+9.0825x-5.2675. \tag{24}$$

which has three zeros; out of them, one is the multiple zero $\alpha=1.75$ of multiplicity two, and the other is the simple zero $\alpha=1.72$. Our desired root is $\alpha=1.75$, and we chose $x_0=1.8$ as the initial guess.

Example 3. *Eigenvalue problem:*
For this, we choose the following 8×8 matrix:

$$A=\begin{bmatrix}-12 & -12 & 36 & -12 & 0 & 0 & 12 & 8\\ 148 & 129 & -397 & 147 & -12 & 6 & -109 & -74\\ 72 & 62 & -186 & 66 & -8 & 4 & -54 & -36\\ -32 & -24 & 88 & -36 & 0 & 0 & 24 & 16\\ 20 & 13 & -45 & 19 & 8 & 6 & -13 & -10\\ 120 & 98 & -330 & 134 & -8 & 24 & -90 & -60\\ -132 & -109 & 333 & -115 & 12 & -6 & 105 & 66\\ 0 & 0 & 0 & 0 & 0 & 0 & 0 & 4\end{bmatrix}.$$

The corresponding characteristic polynomial of this matrix is as follows:

$$f_3(x)=(x-4)^3(x+4)(x-8)(x-20)(x-12)(x+12).$$

The above function has one multiple zero at $\alpha=4$ of multiplicity three. In addition, we consider $x_0=2.7$ as the starting point.

Example 4. *Let us consider the following polynomial equation:*

$$f_4(z)=\left((x-1)^3-1\right)^{50}. \tag{25}$$

The desired zero of the above function f_4 is $\alpha=2$ with multiplicity of order 50, and we choose initial guess $x_0=2.1$ for this problem.

Table 1. Comparison on the basis of the difference between two consecutive iterations $|x_{n+1} - x_n|$ for the functions f_1–f_4.

f	n	*OM1*	*OM2*	*OM3*	*OM4*	*GM1*	*GM2*	*ZM1*	*ZM2*	*BM1*	*BM2*
f_1	1	2.3 (−3)	8.4 (−4)	9.3 (−5)	3.5 (−5)	*	3.6 (−5)	1.6 (−4)	2.3 (−4)	7.6 (−5)	3.7 (−5)
	2	2.0 (−16)	9.0 (−20)	8.8 (−28)	2.0 (−37)	*	1.4 (−29)	4.2 (−31)	8.9 (−30)	2.6 (−34)	5.0 (−37)
	3	9.7 (−95)	1.3 (−115)	6.4 (−166)	2.5 (−295)	*	5.4 (−173)	1.0 (−243)	5.5 (−233)	5.4 (−270)	5.7 (−292)
	ρ	5.9997	6.0000	6.0001	8.0000	*	6.0000	8.0000	8.0000	8.0000	8.0000
f_2	1	1.3 (−3)	8.2 (−4)	4.0 (−3)	3.5 (−4)	9.5 (−4)	3.9 (−4)	3.9 (−4)	4.1 (−3)	2.7 (−4)	2.6 (−4)
	2	2.5 (−10)	4.2 (−12)	6.4 (−16)	8.7 (−18)	2.7 (−11)	1.0 (−14)	5.2 (−17)	9.8 (−17)	1.1 (−18)	1.4 (−19)
	3	2.0 (−50)	8.7 (−62)	6.5 (−87)	1.5 (−126)	2.0 (−56)	3.9 (−78)	5.9 (−120)	1.2 (−117)	6.3 (−134)	1.0 (−141)
	ρ	5.9757	5.9928	6.0214	7.9963	5.9836	5.9975	7.9945	7.9941	7.9971	8.0026
f_3	1	9.1 (−5)	3.6 (−5)	8.0 (−6)	6.0 (−6)	8.5 (−5)	4.8 (−5)	4.9 (−6)	5.2 (−6)	2.0 (−6)	1.8 (−6)
	2	1.8 (−28)	1.4 (−31)	9.8 (−38)	2.0 (−47)	1.0 (−28)	5.0 (−31)	6.0 (−48)	1.0 (−47)	1.5 (−51)	2.8 (−52)
	3	1.2 (−170)	4.4 (−190)	3.3 (−229)	2.5 (−379)	3.1 (−172)	5.8 (−187)	2.7 (−383)	2.3 (−381)	1.4 (−412)	1.3 (−418)
	ρ	6.0000	6.0000	6.0000	8.0000	6.0000	6.0000	8.0000	8.0000	8.0000	8.0000
f_4	1	2.4 (−5)	7.1 (−6)	4.2 (−7)	1.4 (−7)	1.8 (−5)	2.0 (−7)	4.8 (−7)	6.5 (−7)	1.9 (−7)	6.3 (−8)
	2	1.5 (−26)	1.7 (−30)	3.9 (−40)	6.7 (−54)	1.1 (−26)	1.8 (−41)	5.7 (−49)	8.4 (−48)	8.0 (−53)	4.2 (−57)
	3	7.5 (−154)	3.2 (−178)	2.6 (−438)	1.7 (−424)	6.6 (−154)	1.0 (−245)	2.2 (−384)	6.6 (−375)	9.6 (−416)	5.9 (−169)
	ρ	6.0000	6.0000	6.0000	8.0000	6.0000	6.0000	8.0000	8.0000	8.0000	2.2745

* means that the corresponding method does not work.

Table 2. Comparison on the basis of residual errors $|f(x_n)|$ for the functions f_1–f_4.

f	n	*OM1*	*OM2*	*OM3*	*OM4*	*GM1*	*GM2*	*ZM1*	*ZM2*	*BM1*	*BM2*
f_1	1	2.7	1.0	1.1 (−1)	4.2 (−2)	*	4.4 (−2)	1.9 (−1)	2.7 (−1)	9.2 (−2)	4.4 (−2)
	2	2.4 (−13)	1.1 (−16)	1.1 (−24)	2.4 (−34)	*	1.7 (−26)	5.1 (−28)	1.1 (−26)	3.2 (−31)	6.0 (−34)
	3	1.2 (−91)	1.6 (−112)	7.8 (−163)	3.0 (−292)	*	5.4 (−173)	1.2 (−240)	6.7 (−230)	6.5 (−267)	7.0 (−289)
f_2	1	5.0 (−8)	2.1 (−8)	4.8 (−9)	3.6 (−9)	2.8 (−8)	4.6 (−9)	4.6 (−9)	5.1 (−9)	2.3 (−9)	2.0 (−9)
	2	1.8 (−21)	5.3 (−25)	1.2 (−32)	2.3 (−36)	2.2 (−23)	3.2 (−30)	8.0 (−35)	2.9 (−34)	3.4 (−38)	5.9 (−40)
	3	1.2 (−101)	2.2 (−124)	1.3 (−174)	6.9 (−254)	1.2 (−113)	4.6 (−157)	1.1 (−240)	4.3 (−236)	1.2 (−268)	3.1 (−284)
f_3	1	4.9 (−8)	3.1 (−9)	3.1 (−11)	1.4 (−11)	4.1 (−8)	7.4 (−9)	7.8 (−12)	9.1 (−12)	5.2 (−13)	3.6 (−13)
	2	3.9 (−79)	1.8 (−88)	6.1 (−107)	4.9 (−136)	7.1 (−80)	8.0 (−87)	1.4 (−137)	6.9 (−137)	2.1 (−148)	1.5 (−150)
	3	1.0 (−505)	5.6 (−564)	2.4 (−681)	1.1 (−1131)	1.9 (−510)	1.2 (−554)	1.3 (−1143)	7.5 (−1138)	1.9 (−1231)	1.3 (−1249)
f_4	1	1.2 (−207)	2.7 (−234)	1.1 (−295)	3.3 (−319)	3.5 (−214)	1.0 (−311)	6.6 (−293)	2.3 (−286)	1.8 (−313)	6.2 (−337)
	2	1.9 (−1268)	2.6 (−1465)	3.8 (−1947)	1.6 (−2635)	1.9 (−1274)	9.8 (−2014)	3.4 (−2389)	9.4 (−2331)	9.8 (−2582)	1.1 (−2795)
	3	4.2 (−7633)	2.3 (−8851)	7.5 (−11856)	6.1 (−21166)	6.0 (−7636)	7.3 (−12226)	1.6 (−19159)	7.1 (−18686)	8.8 (−20728)	3.4 (−8388)

4. Conclusions

We presented an eighth-order modification of the Chebyshev–Halley-type iteration scheme having optimal convergence to obtain the multiple solutions of the scalar equation. The proposed scheme is optimal in the sense of the classical Kung–Traub conjecture. Thus, the efficiency index of the present methods is $E = \sqrt[4]{8} \approx 1.682$, which is better than the classical Newton's method $E = \sqrt[2]{2} \approx 1.414$. Finally, the numerical experience corroborates the theoretical results about the convergence order, and moreover, it can be concluded that our proposed methods are highly efficient and competitive.

Author Contributions: First two authors have contribute to the theoretical results and the other two authors have carried out the numerical experience.

References

1. Traub, J.F. *Iterative Methods for the Solution of Equations*; Prentice-Hall: Englewood Cliffs, NJ, USA, 1964.
2. Gutiérrez, J.M.; Hernández, M.A. A family of Chebyshev–Halley type methods in Banach spaces. *Bull. Aust. Math. Soc.* **1997**, *55*, 113–130. [CrossRef]
3. Kanwar, V.; Singh, S.; Bakshi, S. Simple geometric constructions of quadratically and cubically convergent iterative functions to solve nonlinear equations. *Numer. Algorithms* **2008**, *47*, 95–107. [CrossRef]
4. Potra, F.A.; Pták, V. *Nondiscrete Induction and Iterative Processes*; Research Notes in Mathematics; Pitman: Boston, MA, USA, 1984; Volume 103.
5. Sharma, J.R. A composite third order Newton-Steffensen method for solving nonlinear equations. *Appl. Math. Comput.* **2005**, *169*, 242–246.
6. Argyros, I.K.; Ezquerro, J.A.; Gutiérrez, J.M.; Hernández, M.A.; Hilout, S. On the semilocal convergence of efficient Chebyshev-Secant-type methods. *J. Comput. Appl. Math.* **2011**, *235*, 3195–3206. [CrossRef]
7. Xiaojian, Z. Modified Chebyshev–Halley methods free from second derivative. *Appl. Math. Comput.* **2008**, *203*, 824–827. [CrossRef]
8. Ostrowski, A.M. *Solutions of Equations and System of Equations*; Academic Press: New York, NY, USA, 1960.
9. Amat, S.; Hernández, M.A.; Romero, N. A modified Chebyshev's iterative method with at least sixth order of convergence. *Appl. Math. Comput.* **2008**, *206*, 164–174. [CrossRef]
10. Kou, J.; Li, Y. Modified Chebyshev–Halley method with sixth-order convergence. *Appl. Math. Comput.* **2007**, *188*, 681–685. [CrossRef]
11. Li, D.; Liu, P.; Kou, J. An improvement of Chebyshev–Halley methods free from second derivative. *Appl. Math. Comput.* **2014**, *235*, 221–225. [CrossRef]
12. Sharma, J.R. Improved Chebyshev–Halley methods with sixth and eighth order convergence. *Appl. Math. Comput.* **2015**, *256*, 119–124. [CrossRef]
13. Neta, B. Extension of Murakami's high-order non-linear solver to multiple roots. *Int. J. Comput. Math.* **2010**, *87*, 1023–1031. [CrossRef]
14. Zhou, X.; Chen, X.; Song, Y. Constructing higher-order methods for obtaining the multiple roots of nonlinear equations. *J. Comput. Math. Appl.* **2011**, *235*, 4199–4206. [CrossRef]
15. Hueso, J.L.; Martínez, E.; Teruel, C. Determination of multiple roots of nonlinear equations and applications. *J. Math. Chem.* **2015**, *53*, 880-892. [CrossRef]
16. Behl, R.; Cordero, A.; Motsa, S.S.; Torregrosa, J.R. On developing fourth-order optimal families of methods for multiple roots and their dynamics. *Appl. Math. Comput.* **2015**, *265*, 520-532. [CrossRef]
17. Behl, R.; Cordero, A.; Motsa, S.S.; Torregrosa, J.R.; Kanwar, V. An optimal fourth-order family of methods for multiple roots and its dynamics. *Numer. Algorithms* **2016**, *71*, 775–796. [CrossRef]
18. Geum, Y.H.; Kim, Y.I.; Neta, B. A class of two-point sixth-order multiple-zero finders of modified double-Newton type and their dynamics. *Appl. Math. Comput.* **2015**, *270*, 387–400. [CrossRef]
19. Geum, Y.H.; Kim, Y.I.; Neta, B. A sixth-order family of three-point modified Newton-like multiple-root finders and the dynamics behind their extraneous fixed points. *Appl. Math. Comput.* **2016**, 283, 120–140. [CrossRef]
20. Behl, R.; Cordero, A.; Motsa, S.S.; Torregrosa, J.R. An eighth-order family of optimal multiple root finders and its dynamics. *Numer. Algorithms* **2018**, *77*, 1249–1272. [CrossRef]

21. Zafar, F.; Cordero, A.; Motsa, S.S.; Torregrosa, J.R. Optimal iterative methods for finding multiple roots of nonlinear equations using free parameters. *J. Math. Chem.* **2018**, *56*, 1884–1901. [CrossRef]
22. Behl, R.; Alshomrani, A.S.; Motsa, S.S. An optimal scheme for multiple roots of nonlinear equations with eighth-order convergence. *J. Math. Chem.* **2018**, *56*, 2069–2084. [CrossRef]
23. Behl, R.; Zafar, F.; Alshomrani, A.S.; Junjuaz, M.; Yasmin, N. An optimal eighth-order scheme for multiple zeros of univariate functions. *Int. J. Comput. Methods* **2018**, 1843002. [CrossRef]
24. McNamee, J.M. A comparison of methods for accelerating convergence of Newton's method for multiple polynomial roots. *ACM Signum Newsl.* **1998**, *33*, 17–22. [CrossRef]
25. Cordero, A.; Torregrosa, J.R. Variants of Newton's method using fifth-order quadrature formulas. *Appl. Math. Comput.* **2007**, *190*, 686–698. [CrossRef]

4

The Modified Inertial Iterative Algorithm for Solving Split Variational Inclusion Problem for Multi-Valued Quasi Nonexpansive Mappings with Some Applications

Pawicha Phairatchatniyom [1], Poom Kumam [1,2,*], Yeol Je Cho [3,4], Wachirapong Jirakitpuwapat [1] and Kanokwan Sitthithakerngkiet [5]

1 KMUTTFixed Point Research Laboratory, Room SCL 802 Fixed Point Laboratory, Science Laboratory Building, Department of Mathematics, Faculty of Science, King Mongkut's University of Technology Thonburi, 126 Pracha Uthit Rd., Bang Mod, Thung Khru, Bangkok 10140, Thailand; pairatchat@gmail.com (P.P.); wachirapong.jira@hotmail.com (W.J.)
2 Department of Medical Research, China Medical University Hospital, China Medical University, Taichung 40402, Taiwan
3 Department of Mathematics Education, Gyeongsang National University, Jinju 52828, Korea; yjchomath@gmail.com
4 School of Mathematical Sciences, University of Electronic Science and Technology of China, Chengdu 611731, China
5 Intelligent and Nonlinear Dynamic Innovations Research Center, Department of Mathematics, Faculty of Applied Science, King Mongkut's University of Technology North Bangkok (KMUTNB), Wongsawang, Bangsue, Bangkok 10800, Thailand; kanokwan.s@sci.kmutnb.ac.th
* Correspondence: poom.kum@kmutt.ac.th.

Abstract: Based on the very recent work by Shehu and Agbebaku in Comput. Appl. Math. 2017, we introduce an extension of their iterative algorithm by combining it with inertial extrapolation for solving split inclusion problems and fixed point problems. Under suitable conditions, we prove that the proposed algorithm converges strongly to common elements of the solution set of the split inclusion problems and fixed point problems.

Keywords: variational inequality problem; split variational inclusion problem; multi-valued quasi-nonexpasive mappings; Hilbert space

MSC: 47H06; 47H09; 47J05; 47J25

---

1. Introduction

The *split monotone variational inclusion problem* **(SMVIP)** was introduced by Moudafi [1]. This problem is as follows:

$$\text{Find a point } x^* \in H_1 \text{ such that } 0 \in \hat{f}(x^*) + B_1(x^*) \tag{1}$$

and such that

$$y^* = Ax^* \in H_2 \text{ solves } 0 \in \hat{g}(y^*) + B_2(y^*), \tag{2}$$

where 0 is the zero vector, H_1 and H_2 are real Hilbert spaces, $\hat{f}$ and $\hat{g}$ are given single-valued operators defined on H_1 and H_2, respectively, B_1 and B_2 are multi-valued maximal monotone mappings defined on H_1 and H_2, respectively, and A is a bounded linear operator defined on H_1 to H_2.

It is well known (see [1]) that

$$0 \in \hat{f}(x^*) + B_1(x^*) \iff x^* = J_\lambda^{B_1}(x^* - \lambda \hat{f}(x^*)),$$

and that

$$0 \in \hat{g}(y^*) + B_2(y^*) \iff y^* = J_\lambda^{B_2}(y^* - \lambda \hat{g}(y^*)), \ y^* = Ax^*,$$

where $J_\lambda^{B_1} := (I + \lambda B_1)^{-1}$ and $J_\lambda^{B_2} := (I + \lambda B_2)^{-1}$ are the resolvent operators of B_1 and B_2, respectively, with $\lambda > 0$. Note that $J_\lambda^{B_1}$ and $J_\lambda^{B_2}$ are nonexpansive and firmly nonexpansive.

Recently, Shehu and Agbebaku [2] proposed an algorithm involving a step-size selected and proved strong convergence theorem for split inclusion problem and fixed point problem for multi-valued quasi-nonexpansive mappings. In [1], Moudafi pointed out that the problem **(SMVIP)** [3–5] includes, as special cases, the split variational inequality problem [6], the split zero problem, the split common fixed point problem [7–9] and the split feasibility problem [10,11], which have already been studied and used in image processing and recovery [12], sensor networks in computerized tomography and data compression for models of inverse problems [13].

If $\hat{f} \equiv 0$ and $\hat{g} \equiv 0$ in the problem **(SMVIP)**, then the problem reduces to the *split variational inclusion problem* **(SVIP)** as follows:

$$\text{Find a point } x^* \in H_1 \text{ such that } 0 \in B_1(x^*) \tag{3}$$

and such that

$$y^* = Ax^* \in H_2 \text{ solves } 0 \in B_2(y^*). \tag{4}$$

Note that the problem **(SVIP)** is equivalent to the following problem:

$$\text{Find a point } x^* \in H_1 \text{ such that } x^* = J_\lambda^{B_1}(x^*) \text{ and } y^* = J_\lambda^{B_2}(y^*), \ y^* = Ax^*$$

for some $\lambda > 0$.

We denote the solution set of the problem **(SVIP)** by Ω, i.e.,

$$\Omega = \{x^* \in H_1 : 0 \in B_1(x^*) \text{ and } 0 \in B_2(y^*), \ y^* = Ax^*\}.$$

Many works have been developed to solve the split variational inclusion problem **(SVIP)**. In 2002, Byrne et al. [7] introduced the iterative method $\{x_n\}$ as follows: For any $x_0 \in H_1$,

$$x_{n+1} = J_\lambda^{B_1}(x_n + \gamma A^*(J_\lambda^{B_2} - I)Ax_n) \tag{5}$$

for each $n \geq 0$, where A^* is the adjoint of the bounded linear operator A, $\gamma \in (0, 2/L)$, $L = \|A^*A\|$ and $\lambda > 0$. They have shown the weak and strong convergence of the above iterative method for solving the problem **(SVIP)**.

Later, inspired by the above iterative algorithm, many authors have extended the algorithm $\{x_n\}$ generated by (5). In particular, Kazmi and Rizvi [4] proposed an algorithm $\{x_n\}$ for approximating a solution of the problem **(SVIP)** as follows:

$$\begin{cases} u_n = J_\lambda^{B_1}(x_n + \gamma_n A^*(J_\lambda^{B_2} - I)Ax_n), \\ x_{n+1} = \alpha_n f_n(x_n) + (1 - \alpha_n)Su_n \end{cases} \tag{6}$$

for each $n \geq 0$, where $\{\alpha_n\}$ is a sequence in $(0, 1)$, $\lambda > 0$, $\gamma \in (0, 1/L)$, L is the spectral radius of the operator A^*A, $f : H_1 \to H_1$ is a contraction and $S : H_1 \to H_1$ is a nonexpansive mapping. In 2015,

Sitthithakerngkiet et al. [5] proposed an algorithm $\{x_n\}$ for solving the problem **(SVIP)** and the fixed point problem **(FPP)** of a countable family of nonexpansive mappings as follows:

$$\begin{cases} y_n = J_\lambda^{B_1}(x_n + \gamma_n A^*(J_\lambda^{B_2} - I)Ax_n), \\ x_{n+1} = \alpha_n f(x_n) + (1 - \alpha_n D)S_n y_n \end{cases} \tag{7}$$

for each $n \geq 0$, where $\{\alpha_n\}$ is a sequence in $(0,1)$, $\lambda > 0$, $\gamma \in (0, 1/L)$, L is the spectral radius of the operator A^*A, $f : H_1 \to H_1$ is a contraction, $D : H_1 \to H_2$ is strongly positive bounded linear operator and, for each $n \geq 1$, $S_n : H_1 \to H_1$ is a nonexpansive mapping.

In both their works, they obtained some strong convergence results by using their proposed iterative methods (for some more results on algorithms, see [14,15]).

Recall that a point $x^* \in H_1$ is called a fixed point of a given multi-valued mapping $S : H_1 \to 2^{H_1}$ if

$$x^* \in Sx^* \tag{8}$$

and the *fixed point problem* **(FPP)** for a multi-valued mapping $S : H_1 \to 2^{H_1}$ is as follows:

$$\text{Find a point } x^* \in H_1 \text{ such that } x^* \in Sx^*.$$

The set of fixed points of the multi-valued mapping S is denoted by $F(S)$.

As applications, the fixed point theory for multi-valued mappings was applied to various fields, especially mathematical economics and game theory (see [16–18]).

Recently, motivated by the results of Byrne et al. [7], Kazmi and Rizvi [4] and Sitthithakerngkiet [5], Shehu and Agbebaku [2] introduced the *split fixed point inclusion problem* **(SFPIP)** from the problems **(SVIP)** and **(FPP)** for a multi-valued quasi-nonexpansive mapping $S : H_1 \to 2^{H_1}$ as follows:

$$\text{Find a point } x^* \in H_1 \text{ such that } 0 \in B_1(x^*), \ x^* \in Sx^* \tag{9}$$

and such that

$$y^* = Ax^* \in H_2 \text{ solves } 0 \in B_2(y^*), \tag{10}$$

where H_1 and H_2 are real Hilbert spaces, B_1 and B_2 are multi-valued maximal monotone mappings defined on H_1 and H_2, respectively, and A is a bounded linear operator defined on H_1 to H_2.

Note that the problem **(SFPIP)** is equivalent to the following problem: for some $\lambda > 0$,

$$\text{Find a point } x^* \in H_1 \text{ such that } x^* = J_\lambda^{B_1}(x^*), \ x^* \in Sx^* \text{ and } Ax^* = J_\lambda^{B_2}(Ax^*).$$

The solution set of the problem **(SFPIP)** is denoted by $F(S) \bigcap \Omega$, i.e.,

$$F(S) \bigcap \Omega = \{x^* \in H_1 : 0 \in B_1(x^*), \ x^* \in Sx^* \text{ and } 0 \in B_2(Ax^*)\}.$$

Notice that, if S is the identity operator, then the problem **(SFPIP)** reduces to the problem **(SVIP)**. Moreover, if $J_\lambda^{B_1} = J_\lambda^{B_2} = A = I$, then the problem **(SFPIP)** reduces to the problem **(FPP)** for a multi-valued quasi-nonexpansive mapping.

Furthermore, Shehu and Agbebaku [2] introduced an algorithm $\{x_n\}$ for solving the problem **(SFPIP)** for a multi-valued quai-nonexpasive mapping S as follows: For any $x_1 \in H_1$,

$$\begin{cases} u_n = J_\lambda^{B_1}(x_n + \gamma_n A^*(J_\lambda^{B_2} - 1)Ax_n), \\ x_{n+1} = \alpha_n f_n(x_n) + \beta_n x_n + \delta_n(\sigma w_n + (1 - \sigma)u_n), \quad w_n \in Sx_n, \end{cases} \tag{11}$$

for each $n \geq 1$, where $\{\alpha_n\}$, $\{\beta_n\}$ and $\{\delta_n\}$ are the real sequences in $(0,1)$ such that

$$\alpha_n + \beta_n + \delta_n = 1, \quad \sigma \in (0,1), \quad \gamma_n := \tau_n \frac{\|(J_\lambda^{B_2} - I)Ax_n\|^2}{\|A^*(J_\lambda^{B_2} - I)\|^2},$$

where $0 < a \leq \tau_n \leq b < 1$, and $\{f_n(x)\}$ is the uniform convergence sequence for any x in a bounded subset D of H_1, and proved that the sequences $\{u_n\}$ and $\{x_n\}$ generated by (11) both converge strongly to $p \in F(S) \cap \Omega$, where $p = P_{F(S)\cap\Omega} f(p)$.

In optimization theory, the second-order dynamical system, which is called the heavy ball method, is used to accelerate the convergence rate of algorithms. This method is a two-step iterative method for minimizing a smooth convex function which was firstly introduced by Polyak [19].

The following is a modified heavy ball method for the improvement of the convergence rate, which was introduced by Nesterov [20]:

$$\begin{cases} y_n = x_n + \theta_n(x_n - x_{n-1}), \\ x_{n+1} = y_n - \lambda_n \nabla f(y_n) \end{cases}$$

for each $n \geq 1$, where $\lambda_n > 0$, $\theta_n \in [0,1)$ is an extrapolation factor. Here, the term $\theta_n(x_n - x_{n-1})$ is the inertia (for more recent results on the inertial algorithms, see [21,22]).

The following method is called the *inertial proximal point algorithm*, which was introduced by Alvarez and Attouch [23]. This method combined the proximal point algorithm [24] with the inertial extrapolation [25,26]:

$$\begin{cases} y_n = x_n + \theta_n(x_n - x_{n-1}), \\ x_{n+1} = (I + \lambda_n \widehat{T})^{-1}(y_n) \end{cases} \tag{12}$$

for each $n \geq 1$, where I is identity operator and $\widehat{T}$ is a maximal monotone operator. It was proven that, if a positive sequence λ_n is non-decreasing, $\theta_n \in [0,1)$ and the following summability condition holds:

$$\sum_{n=1}^{\infty} \theta_n \|x_n - x_{n-1}\|^2 < \infty, \tag{13}$$

then $\{x_n\}$ generated by (12) converges to a zero point of T.

In fact, recently, some authors have pointed out some problems in this summability condition (13) given in [27], that is, to satisfy this summability condition (13) of the sequence $\{x_n\}$, one needs to calculate $\{\theta_n\}$ at each step. Recently, Bot et al. [28] improved this condition, that is, they got rid of the summability condition (13) and replaced the other conditions.

In this paper, inspired by the results of Shehu and Agbebaku [2], Nesterov [20] and Alvarez and Attouch [23], we proposed a new algorithm by combining the iterative algorithm (11) with the inertial extrapolation for solving the problem **(SFPIP)** and prove some strong convergence theorems of the proposed algorithm to show the existence of a solution of the problem **(SFPIP)**. Furthermore, as applications, we consider our proposed algorithm for solving the variational inequality problem and give some applications in game theory.

2. Preliminaries

In this section, we recall some definitions and results which will be used in the proof of the main results.

Let H_1 and H_2 be two real Hilbert spaces with the inner product $\langle \cdot, \cdot \rangle$ and the norm $\| \cdot \|$. Let C be a nonempty closed and convex subset of H_1 and D be a nonempty bounded subset of H_1. Let $A : H_1 \to H_2$ be a bounded linear operator and $A^* : H_2 \to H_1$ be the adjoint of A.

Let $\{x_n\}$ be a sequence in H, we denote the strong and weak convergence of a sequence $\{x_n\}$ by $x_n \to x$ and $x_n \rightharpoonup x$, respectively.

Recall that a mapping $T : C \to C$ is said to be:

(1) *Lipschitz* if there exists a positive constant α such that, for all $x, y \in C$,

$$\|Tx - Ty\| \leq \alpha \|x - y\|.$$

If $\alpha \in (0, 1)$ and $\alpha = 1$, then the mapping T is contractive and nonexpansive, respectively.

(2) *firmly nonexpansive* if

$$\|Tx - Ty\|^2 \leq \langle Tx - Ty, x - y \rangle$$

for all $x, y \in C$.

A mapping P_C is said to be the *metric projection* of H_1 onto C if, for all point $x \in H_1$, there exists a unique nearest point in C, denoted by $P_C x$, such that

$$\|x - P_C x\| \leq \|x - y\|$$

for all $y \in C$.

It is well known that P_C is nonexpansive mapping and satisfies

$$\langle x - y, P_C x - P_C y \rangle \leq \|P_C x - P_C y\|^2$$

for all $x, y \in H_1$. Moreover, $P_C x$ is characterized by the fact $P_C x \in C$ and

$$\langle x - P_C x, y - P_C x \rangle \leq 0$$

for all $y \in C$ and $x \in H_1$ (see [6,22]).

A multi-valued mapping $B_1 : H_1 \to 2^{H_1}$ is said to be *monotone* if, for all $x, y \in H_1$, $u \in B_1(x)$ and $v \in B_1(y)$,

$$\langle x - y, u - v \rangle \geq 0.$$

A monotone mapping $B_1 : H_1 \to 2^{H_1}$ is said to be *maximal* if the graph $G(B_1)$ of B_1 is not properly contained in the graph of any other monotone mapping. It is known that a monotone mapping B_1 is maximal if and only if, for all $(x, u) \in H_1 \times H_1$,

$$\langle x - y, u - v \rangle \geq 0$$

for all $(y, v) \in G(B_1)$ implies that $u \in B_1(x)$.

Let $B_1 : H_1 \to 2^{H_1}$ be a multi-valued maximal monotone mapping. Then the *resolvent mapping* $J_\lambda^{B_1} : H_1 \to H_1$ associated with B_1 is defined by

$$J_\lambda^{B_1}(x) := (I + \lambda B_1)^{-1}(x)$$

for all $x \in H_1$ and for some $\lambda > 0$, where I is the identity operator on H_1. It is well known that, for any $\lambda > 0$, the resolvent operator $J_\lambda^{B_1}$ is single-valued firmly nonexpansive (see [2,5,6,14]).

Definition 1. *Suppose that $\{f_n(x)\}$ is a sequence of functions defined on a bounded set D. Then $f_n(x)$ converges uniformly to the function $f(x)$ on D if, for all $x \in D$,*

$$f_n(x) \to f(x) \text{ as } n \to \infty.$$

Let $f_n : D \to H_1$ be a uniformly convergent sequence of contraction mappings on D, i.e., there exists $\mu_n \in (0,1)$ such that

$$f_n(x) - f_n(y)\| \leq \mu_n\|x - y\|$$

for all $x, y \in D$.

Let $CB(H_1)$ denote the family of nonempty closed and bounded subsets of H_1. The *Hausdorff metric* on $CB(H_1)$ is defined by

$$\widehat{H}(x,y) = \max\left\{\sup_{x\in A}\inf_{y\in B}\|x-y\|, \sup_{y\in B}\inf_{x\in A}\|x-y\|\right\}$$

for all $A, B \in CB(H_1)$ (see [18]).

Definition 2. [2] *Let $S : H_1 \to CB(H_1)$ be a multi-valued mapping. Assume that $p \in H_1$ is a fixed point of S, that is, $p \in Sp$. The mapping S is said to be:*

(1) *nonexpansive if, for all $x, y \in H_1$,*

$$\widehat{H}(Sx, Sy) \leq \|x - y\|.$$

(2) *quasi-nonexpansive if $F(S) \neq \emptyset$ and, for all $x \in H_1$ and $p \in F(S)$,*

$$\widehat{H}(Sx, Sp) \leq \|x - p\|$$

Definition 3. [2] A single-valued mapping $S : H \to H$ is said to be *demiclosed at the origin* if, for any sequence $\{x_n\} \subset H$ with $x_n \rightharpoonup x$ and $Sx_n \to 0$, we have $Sx = 0$.

Definition 4. [2] A multi-valued mapping $S : H_1 \to CB(H_1)$ is said to be *demiclosed at the origin* if, for any sequence $\{x_n\} \subset H$ with $x_n \rightharpoonup x$ and $d(x_n, Sx_n) \to 0$, we have $x \in Sx$.

Lemma 1. [29,30] *Let H be a Hilbert space. Then, for any $x, y, z \in X$ and $\alpha, \beta, \gamma \in [0,1]$ with $\alpha + \beta + \gamma = 1$, we have*

$$\|\alpha x + \beta y + \gamma z\|^2 = \alpha\|x\|^2 + \beta\|y\|^2 + \gamma\|z\|^2 - \alpha\beta\|x-y\|^2 - \alpha\gamma\|x-z\|^2 - \beta\gamma\|y-z\|^2.$$

Lemma 2. [2,31] *Let H be a real Hilbert space. Then the following results hold:*

(1) $\|x - y\|^2 = \|x\|^2 - 2\langle x, y\rangle + \|y\|^2$.
(2) $\|x + y\|^2 = \|x\|^2 + 2\langle x, y\rangle + \|y\|^2$.
(3) $\|x + y\|^2 \leq \|x\|^2 + 2\langle y, x + y\rangle$ *for all $x, y \in H$.*

Lemma 3. [2,32,33] *Let $\{a_n\}$, $\{c_n\} \subset \mathbb{R}_+$, $\{\sigma_n\} \subset (0,1)$ and $\{b_n\} \subset \mathbb{R}$ be sequences such that*

$$a_{n+1} \leq (1 - \sigma_n)a_n + b_n + c_n \quad \text{for all} \quad n \geq 0.$$

Assume $\sum_{n=0}^{\infty} |c_n| < \infty$. Then the following results hold:

(1) *If $b_n \leq \beta\sigma_n$ for some $\beta \geq 0$, then $\{a_n\}$ is a bounded sequence.*

(2) *If we have*

$$\sum_{n=0}^{\infty} \sigma_n = \infty \text{ and } \limsup_{n\to\infty} \frac{b_n}{\sigma_n} \le 0,$$

then $\lim_{n\to\infty} a_n = 0$.

Lemma 4. [32,33] *Let* $\{s_n\}$ *be a sequence of non-negative real numbers such that*

$$s_{n+1} \le (1-\lambda_n)s_n + \lambda_n t_n + r_n$$

for each $n \ge 1$, *where*

(a) $\{\lambda_n\} \subset [0,1]$ *and* $\sum_{n=1}^{\infty} \lambda_n = \infty$;
(b) $\limsup t_n \le 0$;
(c) $r_n \ge 0$ *and* $\sum_{n=1}^{\infty} r_n < \infty$.

Then $s_n \to 0$ *as* $n \to \infty$.

3. The Main Results

In this section, we prove some strong convergence theorems of the proposed algorithm for solving the problem **(SFPIP)**.

Theorem 1. *Let* H_1, H_2 *be two real Hilbert spaces,* $A : H_1 \to H_2$ *be bounded operator with adjoint operator* A^* *and* $B_1 : H_1 \to 2^{H_1}$, $B_2 : H_2 \to 2^{H_2}$ *be maximal monotone mappings. Let* $S : H_1 \to CB(H_1)$ *be a multi-valued quasi-nonexpansive mapping and* S *be demiclosed at the origin. Let* $\{f_n\}$ *be a sequence of* μ_n*-contractions* $f_n : H_1 \to H_1$ *with* $0 < \mu_* \le \mu_n \le \mu^* < 1$ *and* $\{f_n(x)\}$ *be uniformly convergent for any* x *in a bounded subset* D *of* H_1. *Suppose that* $F(S) \cap \Omega \ne \emptyset$. *For any* $x_0, x_1 \in H_1$, *let the sequences* $\{y_n\}$, $\{u_n\}$, $\{z_n\}$ *and* $\{x_n\}$ *be generated by*

$$\begin{cases} y_n = x_n + \theta_n(x_n - x_{n-1}), \\ u_n = J_\lambda^{B_1}(y_n + \gamma_n A^*(J_\lambda^{B_2} - I)Ay_n), \\ z_n = \xi v_n + (1-\xi)u_n, \ v_n \in Sx_n, \\ x_{n+1} = \alpha_n f_n(x_n) + \beta_n x_n + \delta_n z_n \end{cases} \tag{14}$$

for each $n \ge 1$, *where* $\xi \in (0,1)$, $\gamma_n := \tau_n \frac{\|(J_\lambda^{B_2} - I)Ay_n\|^2}{\|A^*(J_\lambda^{B_2} - I)Ay_n\|^2}$ *with* $0 < \tau_* \le \tau_n \le \tau^* < 1$, $\{\theta_n\} \subset [0, \bar\omega)$ *for some* $\bar\omega > 0$ *and* $\{\alpha_n\}, \{\beta_n\}, \{\delta_n\} \in (0,1)$ *with* $\alpha_n + \beta_n + \delta_n = 1$ *satisfying the following conditions:*

(C1) $\lim_{n\to\infty} \alpha_n = 0$;
(C2) $\sum_{n=1}^{\infty} \alpha_n = \infty$;
(C3) $0 < \epsilon_1 \le \beta_n$ *and* $0 < \epsilon_2 \le \delta_n$;
(C4) $\lim_{n\to\infty} \frac{\theta_n}{\alpha_n} \|x_n - x_{n-1}\| = 0$.

Then $\{x_n\}$ *generated by* (14) *converges strongly to* $p \in F(S) \cap \Omega$, *where* $p = P_{F(S)\cap\Omega} f(p)$.

Proof. First, we show that $\{x_n\}$ is bounded. Let $p = P_{F(S)\cap\Omega} f(p)$. Then $p \in F(S) \cap \Omega$ and so $J_\lambda^{B_1} p = p$ and $J_\lambda^{B_2} Ap = Ap$. By the triangle inequality, we get

$$\begin{aligned} \|y_n - p\| &= \|x_n + \theta_n(x_n - x_{n-1}) - p\| \\ &\le \|x_n - p\| + \theta_n \|x_n - x_{n-1}\|. \end{aligned} \tag{15}$$

By the Cauchy-Schwarz inequality and Lemma 2 (1) and (2), we get

$$\begin{aligned} \|y_n - p\|^2 &= \|x_n + \theta_n(x_n - x_{n-1}) - p\|^2 \\ &= \|x_n - p\|^2 + \theta_n^2\|x_n - x_{n-1}\|^2 + 2\theta_n\langle x_n - p, x_n - x_{n-1}\rangle \\ &\leq \|x_n - p\|^2 + \theta_n^2\|x_n - x_{n-1}\|^2 + 2\theta_n\|x_n - x_{n-1}\|\|x_n - p\|. \end{aligned} \tag{16}$$

By using (15) and the fact that S is quasi-nonexpansive S, we get

$$\begin{aligned} \|z_n - p\| &= \|\xi v_n + (1-\xi)u_n - p\| \\ &= \|\xi(v_n - p) + (1-\xi)(u_n - p)\| \\ &\leq \xi\|v_n - p\| + (1-\xi)\|u_n - p\| \\ &\leq \xi d(v_n, Sp) + (1-\xi)\|y_n - p\| \\ &\leq \xi\widehat{H}(Sx_n, Sp) + (1-\xi)[\|x_n - p\| + \theta_n\|x_n - x_{n-1}\|] \\ &\leq \xi\|x_n - p\| + (1-\xi)\|x_n - p\| + (1-\xi)\theta_n\|x_n - x_{n-1}\| \\ &\leq \|x_n - p\| + \theta_n\|x_n - x_{n-1}\|, \end{aligned} \tag{17}$$

which implies that

$$\begin{aligned} \|z_n - p\|^2 &\leq (\|x_n - p\| + \theta_n\|x_n - x_{n-1}\|)^2 \\ &= \|x_n - p\|^2 + 2\theta_n\|x_n - x_{n-1}\|\|x_n - p\| + \theta_n^2\|x_n - x_{n-1}\|^2. \end{aligned} \tag{18}$$

Since $J_\lambda^{B_1}$ is nonexpansive, by Lemma 2 (2), we get

$$\begin{aligned} \|u_n - p\|^2 &= \|J_\lambda^{B_1}(y_n + \gamma_n A^*(J_\lambda^{B_2} - I)Ay_n) - p\|^2 \\ &= \|J_\lambda^{B_1}(y_n + \gamma_n A^*(J_\lambda^{B_2} - I)Ay_n) - J_\lambda^{B_1}p\|^2 \\ &\leq \|y_n + \gamma_n A^*(J_\lambda^{B_2} - I)Ay_n - p\|^2 \\ &= \|y_n - p\|^2 + \gamma_n^2\|A^*(J_\lambda^{B_2} - I)Ay_n\|^2 + 2\gamma_n\langle y_n - p, A^*(J_\lambda^{B_2} - I)Ay_n\rangle. \end{aligned} \tag{19}$$

Again, by Lemma 2 (2), we get

$$\begin{aligned} &\langle y_n - p, A^*(J_\lambda^{B_2} - I)Ay_n\rangle \\ &= \langle A(y_n - p), (J_\lambda^{B_2} - I)Ay_n\rangle \\ &= \langle J_\lambda^{B_2}Ay_n - Ap - (J_\lambda^{B_2} - I)Ay_n, (J_\lambda^{B_2} - I)Ay_n\rangle \\ &= \langle J_\lambda^{B_2}Ay_n - Ap, (J_\lambda^{B_2} - I)Ay_n\rangle - \langle (J_\lambda^{B_2} - I)Ay_n, (J_\lambda^{B_2} - I)Ay_n\rangle \\ &= \langle J_\lambda^{B_2}Ay_n - Ap, (J_\lambda^{B_2} - I)Ay_n\rangle - \|(J_\lambda^{B_2} - I)Ay_n\|^2 \\ &= \frac{1}{2}(\|J_\lambda^{B_2}Ay_n - Ap\|^2 + \|(J_\lambda^{B_2} - I)Ay_n\|^2 \\ &\quad - \|J_\lambda^{B_2}Ay_n - Ap - (J_\lambda^{B_2} - I)Ay_n\|^2) - \|(J_\lambda^{B_2} - I)Ay_n\|^2 \\ &= \frac{1}{2}(\|J_\lambda^{B_2}Ay_n - Ap\|^2 + \|(J_\lambda^{B_2} - I)Ay_n\|^2 - \|J_\lambda^{B_2}Ay_n - Ap - J_\lambda^{B_2}Ay_n + Ay_n\|^2) \\ &\quad - \|(J_\lambda^{B_2} - I)Ay_n\|^2 \\ &= \frac{1}{2}(\|J_\lambda^{B_2}Ay_n - Ap\|^2 + \|(J_\lambda^{B_2} - I)Ay_n\|^2 - \|Ay_n - Ap\|^2) - \|(J_\lambda^{B_2} - I)Ay_n\|^2 \\ &= \frac{1}{2}(\|J_\lambda^{B_2}Ay_n - Ap\|^2 - \|Ay_n - Ap\|^2 - \|(J_\lambda^{B_2} - I)Ay_n\|^2) \\ &\leq \frac{1}{2}(\|Ay_n - Ap\|^2 - \|Ay_n - Ap\|^2 - \|(J_\lambda^{B_2} - I)Ay_n\|^2) \\ &= -\frac{1}{2}\|(J_\lambda^{B_2} - I)Ay_n\|^2. \end{aligned} \tag{20}$$

Using (20) into (19), we get

$$\begin{aligned}\|u_n - p\|^2 &\le \|y_n - p\|^2 + \gamma_n^2\|A^*(J_\lambda^{B_2} - I)Ay_n\|^2 - \gamma_n\|(J_\lambda^{B_2} - I)Ay_n\|^2 \\ &= \|y_n - p\|^2 - \gamma_n(\|(J_\lambda^{B_2} - I)Ay_n\|^2 - \gamma_n\|A^*(J_\lambda^{B_2} - I)Ay_n\|^2). \end{aligned} \tag{21}$$

By the definition of γ_n, (21) can then be written as follows:

$$\|u_n - p\|^2 \le \|y_n - p\|^2 - \gamma_n(1 - \tau_n)\|(J_\lambda^{B_2} - I)Ay_n\|^2 \le \|y_n - p\|^2.$$

Thus we have

$$\|u_n - p\| \le \|y_n - p\|. \tag{22}$$

Using the condition (C3) and (17), we get

$$\begin{aligned}\|x_{n+1} - p\| &= \|\alpha_n f_n(x_n) + \beta_n x_n + \delta_n z_n - p\| \\ &= \|\alpha_n(f_n(x_n) - f_n(p)) + \alpha_n(f_n(p) - p) + \beta_n(x_n - p) + \delta_n(z_n - p)\| \\ &\le \alpha_n\|f_n(x_n) - f_n(p)\| + \alpha_n\|f_n(p) - p\| + \beta_n\|x_n - p\| + \delta_n\|z_n - p\| \\ &\le \alpha_n\mu_n\|x_n - p\| + \alpha_n\|f_n(p) - p\| + \beta_n\|x_n - p\| + \delta_n(\|x_n - p\| \\ &\quad + (1 - \xi)\theta_n\|x_n - x_{n-1}\|) \\ &\le (\alpha_n\mu^* + (\beta_n + \delta_n))\|x_n - p\| + (1 - \xi)\delta_n\theta_n\|x_n - x_{n-1}\| + \alpha_n\|f_n(p) - p\| \\ &= (1 - \alpha_n(1 - \mu^*)\|x_n - p\| + (1 - \xi)\delta_n\alpha_n\frac{\theta_n}{\alpha_n}\|x_n - x_{n-1}\| + \alpha_n\|f_n(p) - p\|. \end{aligned}$$

Since $\{f_n\}$ is the uniform convergence on D, there exists a constant $M > 0$ such that

$$\|f_n(p) - p\| \le M$$

for each $n \ge 1$. So we can choose $\beta := \dfrac{M}{1 - \mu^*}$ and set

$$a_n := \|x_n - p\|, \quad b_n := \alpha_n\|f_n(p) - p\|,$$

$$c_n := (1 - \xi)\delta_n\alpha_n\frac{\theta_n}{\alpha_n}\|x_n - x_{n-1}\|, \quad \sigma_n := \alpha_n(1 - \mu^*).$$

By Lemma 3 (1) and our assumptions, it follows that $\{x_n\}$ is bounded. Moreover, $\{u_n\}$ and $\{y_n\}$ are also bounded.

Now, by Lemma 2, we get

$$\begin{aligned}
&\|x_{n+1}-p\|^2 \\
&= \|\alpha_n(f_n(x_n)-f_n(p))+\alpha_n(f_n(p)-p)+\beta_n(x_n-p)+\delta_n(z_n-p)\|^2 \\
&\le \|\alpha_n(f_n(x_n)-f_n(p))+\beta_n(x_n-p)+\delta_n(z_n-p)\|^2+2\alpha_n\langle f_n(p)-p, x_{n+1}-p\rangle \\
&= \|\beta_n(x_n-p)+\delta_n(z_n-p)\|^2+\alpha_n^2\|f_n(x_n)-f_n(p)\|^2 \\
&\quad +2\alpha_n\langle f_n(x_n)-f_n(p), \beta_n(x_n-p)+\delta_n(z_n-p)\rangle+2\alpha_n\langle f_n(p)-p, x_{n+1}-p\rangle \\
&\le \beta_n^2\|x_n-p\|^2+\delta_n^2\|z_n-p\|^2+2\beta_n\delta_n\langle x_n-p, z_n-p\rangle+\alpha_n^2\mu_n^2\|x_n-p\|^2 \\
&\quad +2\alpha_n\langle f_n(p)-p, x_{n+1}-p\rangle+2\alpha_n\|f_n(x_n)-f_n(p)\|\|\beta_n(x_n-p)+\delta_n(z_n-p)\| \\
&\le \beta_n^2\|x_n-p\|^2+\delta_n^2\|z_n-p\|^2+\beta_n\delta_n\left(\|x_n-p\|^2+\|z_n-p\|^2-\|x_n-z_n\|^2\right) \\
&\quad +\alpha_n^2\mu^{*2}\|x_n-p\|^2+2\alpha_n\mu_n\|x_n-p\|\left(\beta_n\|x_n-p\|+\delta_n\|z_n-p\|\right) \\
&\quad +2\alpha_n\langle f_n(p)-p, x_{n+1}-p\rangle \\
&\le \beta_n(\beta_n+\delta_n)\|x_n-p\|^2+\delta_n(\beta_n+\delta_n)\|z_n-p\|^2-\beta_n\delta_n\|x_n-z_n\|^2+\alpha_n^2\mu^{*2}\|x_n-p\|^2 \\
&\quad +2\mu^*\alpha_n(\beta_n+\delta_n)\|x_n-p\|^2+2\mu^*\alpha_n(1-\xi)\delta_n\theta_n\|x_n-x_{n-1}\|\|x_n-p\| \\
&\quad +2\alpha_n\langle f_n(p)-p, x_{n+1}-p\rangle \\
&\le \beta_n(\beta_n+\delta_n)\|x_n-p\|^2+\delta_n(\beta_n+\delta_n)(\|x_n-p\|^2+\theta_n^2\|x_n-x_{n-1}\|^2 \\
&\quad +2\theta_n\|x_n-x_{n-1}\|\|x_n-p\|)-\beta_n\delta_n\|x_n-z_n\|^2+\alpha_n^2\mu^{*2}\|x_n-p\|^2 \\
&\quad +2\mu^*\alpha_n(\beta_n+\delta_n)\|x_n-p\|^2+2\mu^*\alpha_n(1-\xi)\delta_n\theta_n\|x_n-x_{n-1}\|\|x_n-p\| \\
&\quad +2\alpha_n\langle f_n(p)-p, x_{n+1}-p\rangle \\
&= ((1-\alpha_n)^2+\alpha_n^2\mu^{*2}+2\mu^*\alpha_n(1-\alpha_n))\|x_n-p\|^2-\beta_n\delta_n\|x_n-z_n\|^2 \\
&\quad +2(1-\alpha_n(1-\mu^*(1-\xi)))\delta_n\theta_n\|x_n-x_{n-1}\|\|x_n-p\|+(1-\alpha_n)\delta_n\theta_n^2\|x_n-x_{n-1}\|^2 \\
&\quad +2\alpha_n\langle f_n(p)-p, x_{n+1}-p\rangle. \qquad (23)
\end{aligned}$$

Now, we consider two steps for the proof as follows:

Case 1. Suppose that there exists $n_0 \in \mathbb{N}$ such that $\{\|x_n-p\|\}_{n=n_0}^{\infty}$ is non-increasing and then $\{\|x_n-p\|\}$ converges. By Lemma 1, we get

$$\begin{aligned}
\|x_{n+1}-p\|^2 &= \|\alpha_n f_n(x_n)+\beta_n x_n+\delta_n z_n-p\|^2 \\
&= \alpha_n\|f_n(x_n)-p\|^2+\beta_n\|x_n-p\|^2+\delta_n\|z_n-p\|^2-\alpha_n\beta_n\|f_n(x_n)-x_n\|^2 \\
&\quad -\alpha_n\gamma_n\|f_n(x_n)-z_n\|^2-\beta_n\gamma_n\|x_n-z_n\|^2 \\
&\le \alpha_n\|f_n(x_n)-p\|^2+\beta_n\|x_n-p\|^2+\delta_n\|z_n-p\|^2 \\
&\le \alpha_n\|f_n(x_n)-p\|^2+\beta_n\|x_n-p\|^2+\delta_n(\xi\|x_n-p\|^2+(1-\xi)\|u_n-p\|^2) \\
&\le \alpha_n\|f_n(x_n)-p\|^2+(\beta_n+\xi\delta_n)\|x_n-p\|^2+(1-\xi)\delta_n\|u_n-p\|^2,
\end{aligned}$$

which implies that

$$-\|u_n-p\|^2 \le \frac{1}{(1-\xi)\delta_n}\left(\alpha_n\|f_n(x_n)-p\|^2+(\beta_n+\xi\delta_n)\|x_n-p\|^2-\|x_{n+1}-p\|^2\right). \qquad (24)$$

Applying (16) and (24) to (21), we get

$$\begin{aligned}
&\gamma_n(\|(J_\lambda^{B_2}-I)Ay_n\|^2-\gamma_n\|A^*(J_\lambda^{B_2}-I)Ay_n\|^2)\\
&\le \|y_n-p\|^2-\|u_n-p\|^2\\
&\le \|x_n-p\|^2+2\theta_n\|x_{n-1}-p\|\|x_n-p\|+\theta_n^2\|x_n-x_{n-1}\|^2\\
&\quad+\frac{1}{(1-\xi)\delta_n}(\alpha_n\|f_n(x_n)-p\|^2+(\beta_n+\xi\delta_n)\|x_n-p\|^2-\|x_{n+1}-p\|^2)\\
&=\frac{\beta_n+\delta_n}{(1-\xi)\delta_n}\|x_n-p\|^2+\frac{\alpha_n}{(1-\xi)\delta_n}\|f_n(x_n)-p\|^2-\frac{1}{(1-\xi)\delta_n}\|x_{n+1}-p\|^2\\
&\quad+\theta_n\|x_n-x_{n-1}\|\,(2\|x_n-p\|+\theta_n\|x_n-x_{n-1}\|)\\
&\le\frac{1}{(1-\xi)\epsilon_2}(\|x_n-p\|^2-\|x_{n+1}-p\|^2)+\frac{\alpha_n}{(1-\xi)\epsilon_2}\Big(\|f_n(x_n)-p\|^2-\|x_n-p\|^2\\
&\quad+\frac{\theta_n}{\alpha_n}\|x_n-x_{n-1}\|\Big(2\|x_n-p\|+\alpha_n\frac{\theta_n}{\alpha_n}\|x_n-x_{n-1}\|\Big)\Big).
\end{aligned}$$

Since $\{\|x_n-p\|\}$ is convergent, we have $\|x_n-p\|-\|x_{n+1}-p\|\to 0$ as $n\to\infty$. By the conditions (C2) and (C4), we get

$$\gamma_n(\|(J_\lambda^{B_2}-I)Ay_n\|^2-\gamma_n\|A^*(J_\lambda^{B_2}-I)Ay_n\|^2)\to 0 \text{ as } n\to\infty.$$

From the definition of γ_n, we get

$$\frac{\tau_n(1-\tau_n)\|(J_\lambda^{B_2}-I)Ay_n\|^4}{\|A^*(J_\lambda^{B_2}-I)Ay_n\|^2}\to 0 \text{ as } n\to\infty$$

or

$$\frac{\|(J_\lambda^{B_2}-I)Ay_n\|^2}{\|A^*(J_\lambda^{B_2}-I)Ay_n\|}\to 0 \text{ as } n\to\infty.$$

Since

$$\|A^*(J_\lambda^{B_2}-I)Ay_n\|\le\|A^*\|\|(J_\lambda^{B_2}-I)Ay_n\|=\|A\|\|(J_\lambda^{B_2}-I)Ay_n\|,$$

it is easy to see that

$$\|(J_\lambda^{B_2}-I)Ay_n\|\le\|A\|\frac{\|(J_\lambda^{B_2}-I)Ay_n\|^2}{\|A^*(J_\lambda^{B_2}-I)Ay_n\|}.$$

Consequently, we get

$$\|(J_\lambda^{B_2}-I)Ay_n\|\to 0 \text{ as } n\to\infty \tag{25}$$

and also

$$\|A^*(J_\lambda^{B_2}-I)Ay_n\|\to 0 \text{ as } n\to\infty. \tag{26}$$

Similarly, from (23) and our assumptions, we get

$$\begin{aligned}
&\|x_n - z_n\|^2 \\
&= \frac{1}{\beta_n\delta_n}\{\|x_n - p\|^2 - \|x_{n+1} - p\|^2 + (1-\alpha_n)\delta_n\theta_n^2\|x_n - x_{n-1}\|^2 \\
&\quad + 2(1-\alpha_n(1-\mu^*(1-\xi)))\delta_n\theta_n\|x_n - x_{n-1}\|\|x_n - p\| \\
&\quad + \alpha_n[(\alpha_n(1+\mu^{*2}) - 2(1-\mu^*(1-\alpha_n)))\|x_n - p\|^2 + 2\langle f_n(p) - p, x_{n+1} - p\rangle]\} \\
&\leq \frac{1}{\epsilon_1\epsilon_2}\{\|x_n - p\|^2 - \|x_{n+1} - p\|^2 + \frac{\theta_n}{\alpha_n}\|x_n - x_{n-1}\|[\delta_n(1-\alpha_n)\alpha_n^2\frac{\theta_n}{\alpha_n}\|x_n - x_{n-1}\| \\
&\quad + 2\delta_n(1-\alpha_n(1-\mu^*(1-\xi)))\theta_n\|x_n - p\|] + \alpha_n[2\langle f_n(p) - p, x_{n+1} - p\rangle \\
&\quad + (\alpha_n(1+\mu^{*2}) - 2(1-\mu^*(1-\alpha_n)))\|x_n - p\|^2]\} \to 0 \text{ as } n \to \infty.
\end{aligned}$$

Therefore, we have

$$\|x_n - z_n\| \to 0 \text{ as } n \to \infty. \tag{27}$$

By the condition (C2) and (27), we get

$$\begin{aligned}
\|x_{n+1} - x_n\| &= \|\alpha_n f_n(x_n) + \beta_n x_n + \delta_n z_n - x_n\| \\
&\leq \alpha_n\|f_n(x_n) - x_n\| + \delta_n\|x_n - z_n\| \to 0 \text{ as } n \to \infty.
\end{aligned}$$

Thus we have

$$\|x_{n+1} - z_n\| \leq \|x_{n+1} - x_n\| + \|x_n - z_n\| \to 0 \text{ as } n \to \infty.$$

Since $J_\lambda^{B_1}$ is firmly nonexpansive, we have

$$\begin{aligned}
&\|u_n - p\|^2 \\
&= \|J_\lambda^{B_1}(y_n + \gamma_n A^*(J_\lambda^{B_2} - I)Ay_n) - J_\lambda^{B_1}p\|^2 \\
&\leq \langle u_n - p, y_n + \gamma_n A^*(J_\lambda^{B_2} - I)Ay_n - p\rangle \\
&= \frac{1}{2}(\|u_n - p\|^2 + \|y_n + \gamma_n A^*(J_\lambda^{B_2} - I)Ay_n - p\|^2 - \|u_n - y_n - \gamma_n A^*(J_\lambda^{B_2} - I)Ay_n\|^2) \\
&= \frac{1}{2}(\|u_n - p\|^2 + \|y_n - p\|^2 + \gamma_n^2\|A^*(J_\lambda^{B_2} - I)Ay_n\|^2 + 2\langle y_n - p, \gamma_n A^*(J_\lambda^{B_2} - I)Ay_n\rangle \\
&\quad - \|u_n - y_n\|^2 - \gamma_n^2\|A^*(J_\lambda^{B_2} - I)Ay_n\|^2 + 2\langle u_n - y_n, \gamma_n A^*(J_\lambda^{B_2} - I)Ay_n\rangle) \\
&\leq \frac{1}{2}(\|y_n - p\|^2 + \|y_n - p\|^2 - \|u_n - y_n\|^2 + 2\langle u_n - p, \gamma_n A^*(J_\lambda^{B_2} - I)Ay_n\rangle) \\
&\leq \frac{1}{2}(2\|y_n - p\|^2 - \|u_n - y_n\|^2 + 2\gamma_n\|u_n - p\|\|A^*(J_\lambda^{B_2} - I)Ay_n\|) \\
&\leq \|y_n - p\|^2 - \frac{1}{2}\|u_n - y_n\|^2 + \gamma_n\|u_n - p\|\|A^*(J_\lambda^{B_2} - I)Ay_n\|
\end{aligned}$$

or

$$\|u_n - y_n\|^2 \leq 2(\|y_n - p\|^2 - \|u_n - p\|^2 + \gamma_n\|u_n - p\|\|A^*(J_\lambda^{B_2} - 1)Ay_n\|). \tag{28}$$

From (28), (16), (24) and (26) and our assumptions, it follows that

$$\begin{aligned}\|u_n - y_n\|^2 &\leq 2[\|x_n - p\|^2 + 2\theta_n\|x_n - x_{n-1}\|\|x_n - p\| + \theta_n^2\|x_n - x_{n-1}\|^2 \\ &+ \frac{1}{(1-\xi)\delta_n}(\alpha_n\|f_n(x_n) - p\|^2 + (\beta_n + \xi\delta_n)\|x_n - p\|^2 - \|x_{n+1} - p\|^2) \\ &+ \gamma_n\|u_n - p\|\|A^*(J_\lambda^{B_2} - 1)Ay_n\|] \\ &= 2[\frac{1}{(1-\xi)\epsilon_2}(\|x_n - p\|^2 - \|x_{n+1} - p\|^2) + \gamma_n\|u_n - p\|\|A^*(J_\lambda^{B_2} - 1)Ay_n\| \\ &+ \frac{\alpha_n}{(1-\xi)\epsilon_2}(\|f_n(x_n) - p\|^2 - \|x_n - p\|^2 \\ &+ \frac{\theta_n}{\alpha_n}\|x_n - x_{n-1}\|(2\|x_n - p\| + \alpha_n\frac{\theta_n}{\alpha_n}\|x_n - x_{n-1}\|))] \to 0 \text{ as } n \to \infty,\end{aligned}$$

that is, we have

$$\|u_n - y_n\| \to 0 \text{ as } n \to \infty. \tag{29}$$

From $y_n := x_n + \theta_n(x_n - x_{n-1})$, we get

$$\|y_n - x_n\| = \|x_n + \theta_n(x_n - x_{n-1}) - x_n\| = \alpha_n\frac{\theta_n}{\alpha_n}\|x_n - x_{n-1}\|,$$

which, with the condition (C4), implies that

$$\|y_n - x_n\| \to 0 \text{ as } n \to \infty. \tag{30}$$

In addition, using (27), (29) and (30), we obtain

$$\begin{aligned}\|z_n - u_n\| &\leq \|u_n - y_n\| + \|y_n - z_n\| \\ &\leq \|u_n - y_n\| + \|y_n - x_n\| + \|x_n - z_n\| \to 0 \text{ as } n \to \infty.\end{aligned}$$

From $z_n := \xi v_n + (1-\xi)u_n$, we get

$$\|v_n - u_n\| = \frac{1}{\xi}\|z_n - u_n\| \to 0 \text{ as } n \to \infty. \tag{31}$$

Thus, by (29)–(31), we also get

$$\begin{aligned}\|x_n - v_n\| &\leq \|x_n - u_n\| + \|u_n - v_n\| \\ &\leq \|x_n - y_n\| + \|y_n - u_n\| + \|u_n - v_n\| \to 0 \text{ as } n \to \infty.\end{aligned}$$

Therefore, we have

$$d(x_n, Sx_n) \leq \|x_n - v_n\| \to 0 \text{ as } n \to \infty. \tag{32}$$

Since $\{x_n\}$ is bounded, there exists a subsequence $\{x_{n_k}\}$ of $\{x_n\}$ such that $x_{n_k} \rightharpoonup x^* \in H_1$ and, consequently, $\{u_{n_k}\}$ and $\{y_{n_k}\}$ converge weakly to the point x^*.

From (32), Lemma 4 and the demiclosedness principle for a multi-valued mapping S at the origin, we get $x^* \in Sx^*$, which implies that

$$x^* \in F(S).$$

Next, we show that $x^* \in \Omega$. Let $(v, z) \in G(B_1)$, that is, $z \in B_1(v)$. On the other hand, $u_{n_k} = J_\lambda^{B_1}(y_{n_k} + \gamma_{n_k}A^*(J_\lambda^{B_2} - I)Ay_{n_k})$ can be written as

$$y_{n_k} + \gamma_{n_k}A^*(J_\lambda^{B_1} - I)Ay_{n_k} \in u_{n_k} + \lambda B_1(u_{n_k}),$$

or, equivalently,

$$\frac{(y_{n_k} - u_{n_K}) + \gamma_{n_k} A^*(J_\lambda^{B_1} - I)Ay_{n_k}}{\lambda} \in B_1(u_{n_k}).$$

Since B_1 is maximal monotone, we get

$$\left\langle v - u_{n_k}, z - \frac{(y_{n_k} - u_{n_k}) + \gamma_{n_k} A^*(J_\lambda^{B_2} - I)Ay_{n_k}}{\lambda} \right\rangle \geq 0.$$

Therefore, we have

$$\begin{aligned} \langle v - u_{n_k}, z\rangle &\geq \left\langle v - u_{n_k}, \frac{(y_{n_k} - u_{n_k}) + \gamma_{n_k} A^*(J_\lambda^{B_2} - I)Ay_{n_k}}{\lambda} \right\rangle \\ &= \left\langle v - u_{n_k}, \frac{y_{n_k} - u_{n_k}}{\lambda} \right\rangle + \left\langle v - u_{n_k}, \frac{\gamma_{n_k} A^*(J_\lambda^{B_2} - I)Ay_{n_k}}{\lambda} \right\rangle. \end{aligned} \tag{33}$$

Since $u_{n_k} \rightharpoonup x^*$, we have

$$\lim_{k\to\infty} \langle v - u_{n_k}, z\rangle = \langle v - x^*, z\rangle.$$

By (26) and (29), it follows that (33) becomes $\langle v - x^*, z\rangle \geq 0$, which implies that

$$0 \in B_1(x^*).$$

Moreover, from (29), we know that $\{Ay_{n_k}\}$ converges weakly to Ax^* and, by (25), the fact that $J_\lambda^{B_2}$ is nonexpansive and the demiclosedness principle for a multi-valued mapping, we have

$$0 \in B_2(Ax^*),$$

which implies that $x^* \in \Omega$. Thus $x^* \in F(S) \cap \Omega$. Since $\{f_n(x)\}$ is uniformly convergent on D, we get

$$\begin{aligned} \limsup_{n\to\infty} \langle f_n(p) - p, x_{n+1} - p\rangle &= \limsup_{j\to\infty} \langle f_{n_j}(p) - p, x_{n_j+1} - p\rangle \\ &= \langle f(p) - p, x^* - p\rangle \leq 0. \end{aligned}$$

From (23), we get

$$\begin{aligned} \|x_{n+1} - p\|^2 &\leq (1 - 2\alpha_n(1 - \mu^*(1 - \alpha_n)) + \alpha_n^2(1 + \mu^{*2}))\|x_n - p\|^2 - \beta_n\delta_n\|x_n - z_n\|^2 \\ &\quad + 2(1 - \alpha_n(1 - \mu^*(1 - \xi)))\delta_n\theta_n\|x_n - x_{n-1}\|\|x_n - p\| \\ &\quad + (1 - \alpha_n)\delta_n\theta_n^2\|x_n - x_{n-1}\|^2 + 2\alpha_n\langle f_n(p) - p, x_{n+1} - p\rangle \\ &\leq (1 - 2\alpha_n(1 - \mu^*))\|x_n - p\|^2 + 2\alpha_n(1 - \mu^*)\frac{\langle f_n(p) - p, x_{n+1} - p\rangle}{1 - \mu^*} \\ &\quad + \alpha_n\big[\delta_n\frac{\theta_n}{\alpha_n}\|x_n - x_{n-1}\|(2(1 - \alpha_n(1 - \mu^*(1 - \xi)))\|x_n - p\| \\ &\quad + ((1 - \alpha_n)\alpha_n\frac{\theta_n}{\alpha_n}\|x_n - x_{n-1}\|) + \alpha_n(1 + \mu^{*2})\|x_n - p\|^2\big]. \end{aligned}$$

By Lemma 4, we obtain

$$\lim_{n\to\infty} x_n = p.$$

Case 2. Suppose that $\{\|x_n - p\|\}_{n=n_0}^{\infty}$ is not a monotonically decreasing sequence for some n_0 large enough. Set $\Gamma_n = \|x_n - p\|^2$ and let $\tau : \mathbb{B} \to \mathbb{N}$ be a mapping defined by

$$\tau(n) := \max\{k \in \mathbb{N} : k \leq n, \Gamma_k \leq \Gamma_{k+1}\}$$

for all $n \geq n_0$. Obviously, τ is a non-decreasing sequence. Thus we have

$$0 \leq \Gamma_{\tau(n)} \leq \Gamma_{\tau(n)+1}$$

for all $n \geq n_0$. That is, $\|x_{\tau(n)} - p\| \leq \|x_{\tau(n)+1} - p\|$ for all $n \geq n_0$. Thus $\lim_{n\to\infty} \|x_{\tau(n)} - p\|$ exists. As in Case 1, we can show that

$$\lim_{n\to\infty} \|(J_\lambda^{B_2} - I)Ay_{\tau(n)}\| = 0, \quad \lim_{n\to\infty} \|A^*(J_\lambda^{B_2} - I)Ay_{\tau(n)}\| = 0, \tag{34}$$

$$\lim_{n\to\infty} \|x_{\tau(n)+1} - x_{\tau(n)}\| = 0, \quad \lim_{n\to\infty} \|u_{\tau(n)} - x_{\tau(n)}\| = 0, \tag{35}$$

$$\lim_{n\to\infty} \|v_{\tau(n)} - u_{\tau(n)}\| = 0, \quad \lim_{n\to\infty} \|x_{\tau(n)} - v_{\tau(n)}\| = 0. \tag{36}$$

Therefore, we have

$$d(x_{\tau(n)}, Sx_{\tau(n)}) \leq \|x_{\tau(n)} - v_{\tau(n)}\| \to 0 \text{ as } n \to \infty. \tag{37}$$

Since $\{x_{\tau(n)}\}$ is bounded, there exists a subsequence $\{u_{\tau(n)}\}$ of $\{x_{\tau(n)}\}$ that converges weakly to a point $x^* \in H_1$. From $\|u_{\tau(n)} - x_{\tau(n)}\| \to 0$, it follows that $u_{\tau(n)} \rightharpoonup x^* \in H_1$.

Moreover, as in Case 1, we show that $x^* \in F(S) \cap \Omega$. Furthermore, since $\{f_n(x)\}$ is uniformly convergent on $D \subset H_1$, we obtain that

$$\limsup_{n\to\infty} \langle f_{\tau(n)}(p) - p, x_{\tau(n)+1} - p\rangle \leq 0.$$

From (23), we get

$$\begin{aligned}
\|x_{\tau(n)+1} - p\|^2 \leq\ & (1 - 2\alpha_{\tau(n)}(1 - \mu^*(1 - \alpha_{\tau(n)})) + \alpha_{\tau(n)}^2(1 + \mu^{*2}))\|x_{\tau(n)} - p\|^2 \\
& - \beta_{\tau(n)}\delta_{\tau(n)}\|x_{\tau(n)} - z_{\tau(n)}\|^2 + 2\alpha_{\tau(n)}\langle f_{\tau(n)}(p) - p, x_{\tau(n)+1} - p\rangle \\
& + 2(1 - \alpha_{\tau(n)}(1 - \mu^*(1 - \xi)))\delta_{\tau(n)}\theta_{\tau(n)}\|x_{\tau(n)} - x_{\tau(n)-1}\|\|x_{\tau(n)} - p\| \\
& + (1 - \alpha_{\tau(n)})\delta_{\tau(n)}\theta_{\tau(n)}^2\|x_{\tau(n)} - x_{\tau(n)-1}\|^2 \\
\leq\ & (1 - 2\alpha_{\tau(n)}(1 - \mu^*))\|x_{\tau(n)} - p\|^2 + \alpha_{\tau(n)}^2(1 + \mu^{*2})\|x_{\tau(n)} - p\|^2 \\
& + \delta_{\tau(n)}\theta_n\|x_{\tau(n)} - x_{\tau(n)-1}\|(2(1 - \alpha_{\tau(n)}(1 - \mu^*))\|x_{\tau(n)} - p\| \\
& + (1 - \alpha_{\tau(n)})\theta_{\tau(n)}\|x_{\tau(n)} - x_{\tau(n)-1}\|) + 2\alpha_{\tau(n)}\langle f_{\tau(n)}(p) - p, x_{\tau(n)+1} - p\rangle,
\end{aligned}$$

which implies that

$$\begin{aligned}
2\alpha_{\tau(n)}(1 - \mu^*)\|x_{\tau(n)} - p\|^2 \leq\ & \|x_{\tau(n)} - p\|^2 - \|x_{\tau(n)+1} - p\|^2 + \alpha_{\tau(n)}^2(1 + \mu^{*2})\|x_{\tau(n)} - p\|^2 \\
& + \delta_{\tau(n)}\theta_n\|x_{\tau(n)} - x_{\tau(n)-1}\|(2(1 - \alpha_{\tau(n)}(1 - \mu^*))\|x_{\tau(n)} - p\| \\
& + (1 - \alpha_{\tau(n)})\theta_{\tau(n)}\|x_{\tau(n)} - x_{\tau(n)-1}\|) \\
& + 2\alpha_{\tau(n)}\langle f_{\tau(n)}(p) - p, x_{\tau(n)+1} - p\rangle,
\end{aligned}$$

or

$$\begin{aligned}
2(1 - \mu^*)\|x_{\tau(n)} - p\|^2 \leq\ & \alpha_{\tau(n)}(1 + \mu^{*2})\|x_{\tau(n)} - p\|^2 + 2\langle f_{\tau(n)}(p) - p, x_{\tau(n)+1} - p\rangle \\
& + \delta_{\tau(n)}\frac{\theta_{\tau(n)}}{\alpha_{\tau(n)}}\|x_{\tau(n)} - x_{\tau(n)-1}\|\Big(2(1 - \alpha_{\tau(n)}(1 - \mu^*))\|x_{\tau(n)} - p\| \\
& + (1 - \alpha_{\tau(n)})\alpha_{\tau(n)}\frac{\theta_{\tau(n)}}{\alpha_{\tau(n)}}\|x_{\tau(n)} - x_{\tau(n)-1}\|\Big).
\end{aligned}$$

Thus we have

$$\limsup_{n\to\infty} \|x_{\tau(n)} - p\| \le 0$$

and so

$$\lim_{n\to\infty} \|x_{\tau(n)} - p\| = 0. \tag{38}$$

By (35) and (38), we get

$$\|x_{\tau(n)+1} - p\| \le \|x_{\tau(n)+1} - x_{\tau(n)}\| + \|x_{\tau(n)} - p\| \to 0,\ n \to \infty.$$

Furthermore, for all $n \ge n_0$, it is easy to see that $\Gamma_{\tau(n)} \le \Gamma_{\tau(n)+1}$ if $n \ne \tau(n)$ (that is, $\tau(n) < n$) because of $\Gamma_j \ge \Gamma_{j+1}$ for $\tau(n)+1 \le j \le n$. Consequently, it follows that, for all $n \ge n_0$,

$$0 \le \Gamma_n \le \max\{\Gamma_{\tau(n)}, \Gamma_{\tau(n)+1}\} = \Gamma_{\tau(n)+1}.$$

Therefore, $\lim \Gamma_n = 0$, that is, $\{x_n\}$ converges strongly to the point x^*. This completes the proof. □

Remark 1. [22] *The condition (C4) is easily implemented in numerical results because the value of $\|x_n - x_{n-1}\|$ is known before choosing θ_n. Indeed, we can choose the parameter θ_n such as*

$$\theta_n = \begin{cases} \min\left\{\bar{\omega}, \frac{\omega_n}{\|x_n - x_{n-1}\|}\right\}, & \text{if } \|x_n - x_{n-1}\| \ne 0, \\ \bar{\omega}, & \text{otherwise}, \end{cases}$$

where $\{\omega_n\}$ is a positive sequence such that $\omega_n = o(\alpha_n)$. Moreover, in the condition (C4), we can take $\alpha_n = \dfrac{1}{n+1}$, $\bar{\omega} = \dfrac{4}{5}$ and

$$\theta_n = \begin{cases} \min\left\{\bar{\omega}, \frac{\alpha_n^2}{\|x_n - x_{n-1}\|}\right\}, & \text{if } \|x_n - x_{n-1}\| \ne 0, \\ \bar{\omega}, & \text{otherwise}, \end{cases}$$

or

$$\theta_n = \begin{cases} \min\left\{\frac{4}{5}, \frac{1}{(n+1)^2\|x_n - x_{n-1}\|}\right\}, & \text{if } \|x_n - x_{n-1}\| \ne 0, \\ \frac{4}{5}, & \text{otherwise}. \end{cases}$$

If the multi-valued quasi-nonexpansive mapping S in Theorem 1 is a single-valued quasi-nonexpansive mapping, then we obtain the following:

Corollary 1. *Let H_1 and H_2 be two real Hilbert spaces. Suppose that $A : H_1 \to H_2$ is a bounded linear operator with adjoint operator A^*. Let $\{f_n\}$ be a sequence of μ_n-contractions $f_n : H_1 \to H_1$ with $0 < \mu_* \le \mu_n \le \mu^* < 1$ and $\{f_n(x)\}$ be uniformly convergent for any x in a bounded subset D of H_1. Suppose that $S : H_1 \to H_1$ is a single-valued quasi-nonexpansive mapping, $I - S$ is demiclosed at the origin and $F(S) \cap \Omega \ne \emptyset$. For any $x_0, x_1 \in H_1$, let the sequences $\{y_n\}$, $\{u_n\}$, $\{z_n\}$ and $\{x_n\}$ be generated by*

$$\begin{cases} y_n = x_n + \theta_n(x_n - x_{n-1}), \\ u_n = J_\lambda^{B_1}(y_n + \gamma_n A^*(J_\lambda^{B_2} - I)Ay_n), \\ z_n = \xi S x_n + (1 - \xi)u_n, \\ x_{n+1} = \alpha_n f_n(x_n) + \beta_n x_n + \delta_n z_n \end{cases} \tag{39}$$

for each $n \geq 1$, *where* $\xi \in (0,1)$, $\gamma_n := \tau_n \frac{\|(J_\lambda^{B_2}-I)Ay_n\|^2}{\|A^*(J_\lambda^{B_2}-I)Ay_n\|^2}$ *with* $0 < \tau_* \leq \tau_n \leq \tau^* < 1$, $\{\theta_n\} \subset [0,\bar{\omega})$ *for some* $\bar{\omega} > 0$ *and* $\{\alpha_n\}, \{\beta_n\}, \{\delta_n\} \in (0,1)$ *with* $\alpha_n + \beta_n + \delta_n = 1$ *satisfying the following conditions:*

(C1) $\lim_{n\to\infty} \alpha_n = 0$;

(C2) $\sum_{n=1}^{\infty} \alpha_n = \infty$;

(C3) $0 < \epsilon_1 \leq \beta_n$ *and* $0 < \epsilon_2 \leq \delta_n$;

(C4) $\lim_{n\to\infty} \frac{\theta_n}{\alpha_n}\|x_n - x_{n-1}\| = 0$.

Then the sequence $\{x_n\}$ *generated by* (39) *converges strongly to a point* $p \in F(S) \cap \Omega$, *where* $p = P_{F(S)\cap\Omega} f(p)$.

Remark 2. *If* $\theta_n = 0$, *then the iterative scheme (14) in Theorem 1 reduces to the iterative (11).*

4. Applications

In this section, we give some applications of the problem **(SFPIP)** in the variational inequality problem and game theory. First, we introduce variational inequality problem in [34] and game theory (see [35]).

4.1. The Variational Inequality Problem

Let C be a nonempty closed and convex subset of a real Hilbert space H_1. Suppose that an operator $F : H_1 \to H_1$ is monotone.

Now, we consider the following variational inequality problem **(VIP)**:

$$\text{Find a point } x^* \in C \text{ such that } \langle Fx^*, y - x^*\rangle \geq 0 \text{ for all } y \in C. \tag{40}$$

The solution set of the problem **(VIP)** is denoted by Γ.

Moreover, it is well-known that x^* is a solution of the problem **(VIP)** if and only if x^* is a solution of the problem **(FPP)** [34], that is, for any $\gamma > 0$,

$$x^* = P_C(x^* - \gamma Fx^*).$$

The following lemma is extracted from [2,36]. This lemma is used for finding a solution of the split inclusion problem and the variational inequality problem:

Lemma 5. *Let* H_1 *be a real Hilbert space,* $F : H_1 \to H_1$ *be a monotone and L-Lipschitz operator on a nonempty closed and convex subset* C *of* H_1. *For any* $\gamma > 0$, *let* $T = P_C(I - \gamma F(P_C(I - \gamma F)))$. *Then, for any* $y \in \Gamma$ *and* $L\gamma < 1$, *we have*

$$\|Tx - Ty\| \leq \|x - y\|,$$

$I - T$ *is demiclosed at the origin and* $F(T) = \Gamma$.

Now, we apply our Theorem 1, by combining with Lemma 5, to find a solution of the problem **(VIP)**, that is, a point in the set Γ.

let $B_1 : H_1 \to 2^{H_1}$ and $B_2 : H_2 \to 2^{H_2}$ be maximal monotone mappings defined on H_1 and H_2, respectively, and $A : H_1 \to H_2$ be a bounded linear operator with its adjoint A^*.

Now, we consider the *split fixed point variational inclusion problem* **(SFPVIP)** as follows:

$$\text{Find a point } x^* \in H_1 \text{ such that } 0 \in B_1(x^*), \; x^* \in \Gamma \tag{41}$$

and

$$y^* = Ax^* \in H_2 \text{ such that } 0 \in B_2(y^*). \tag{42}$$

Theorem 2. *Let H_1 and H_2 be two real Hilbert spaces, $A : H_1 \to H_2$ be a bounded linear operator with its adjoint A^*. Let $\{f_n\}$ be a sequence of μ_n-contractions $f_n : H_1 \to H_1$ with $0 < \mu_* \leq \mu_n \leq \mu^* < 1$ and $\{f_n(x)\}$ be uniformly convergent for any x in a bounded subset D of H_1. For any $\lambda > 0$, let $T = P_C(I - \gamma F(P_C(I - \gamma F)))$ with $L\gamma < 1$, where $F : H_1 \to H_1$ is a L-Lipschitz and monotone operator on $C \subset H_1$ and $F(T) \cap \Omega \neq \emptyset$. For any $x_0, x_1 \in H_1$, let the sequences $\{y_n\}$, $\{u_n\}$, $\{z_n\}$ and $\{x_n\}$ be generated by*

$$\begin{cases} y_n = x_n + \theta_n(x_n - x_{n-1}), \\ u_n = J_\lambda^{B_1}(y_n + \gamma_n A^*(J_\lambda^{B_2} - I)Ay_n), \\ z_n = \xi T x_n + (1 - \xi)u_n, \\ x_{n+1} = \alpha_n f_n(x_n) + \beta_n x_n + \delta_n z_n \end{cases} \tag{43}$$

for each $n \geq 1$, where $\xi \in (0,1)$, $\gamma_n := \tau_n \frac{\|(J_\lambda^{B_2} - I)Ay_n\|^2}{\|A^*(J_\lambda^{B_2} - I)Ay_n\|^2}$ with $0 < \tau_* \leq \tau_n \leq \tau^* < 1$, $\{\theta_n\} \subset [0, \bar{\omega})$ for some $\bar{\omega} > 0$ and $\{\alpha_n\}, \{\beta_n\}, \{\delta_n\} \in (0,1)$ with $\alpha_n + \beta_n + \delta_n = 1$ satisfying the following conditions:

(C1) $\lim_{n\to\infty} \alpha_n = 0$;
(C2) $\sum_{n=1}^{\infty} \alpha_n = \infty$;
(C3) $0 < \epsilon_1 \leq \beta_n, 0 < \epsilon_2 \leq \delta_n$;
(C4) $\lim_{n\to\infty} \frac{\theta_n}{\alpha_n} \|x_n - x_{n-1}\| = 0$.

Then the sequence $\{x_n\}$ generated by (43) converges strongly to a point $p \in F(T) \cap \Omega = \Gamma \cap \Omega$, where $p = P_{\Gamma\cap\Omega} f(p)$.

Proof. Since $I - T$ is demiclosed at the origin and $F(T) = \Gamma$, by using Lemma (5) and Corollary (1), the sequence $\{x_n\}$ converges strongly to a point $p \in F(T) \cap \Omega$, that is, the sequence $\{x_n\}$ converges strongly to a point $p \in \Gamma$. □

4.2. Game Theory

Now, we consider a game of N players in strategic form

$$G = (p_i, S_i),$$

where $i = 1, \cdots, N$, $p_i : S = S_1 \times S_2 \times \cdots \times S_N \to \mathbb{R}$ is the pay-off function (continuous) of the ith player and $S_i \in \mathbb{R}^{M_i}$ is the set of strategy of the ith player such that $M_i = |S_i|$.

Let S_i be nonempty compact and convex set, $s_i \in S_i$ be the strategy of the ith player and $s = (s_1, s_2, \cdots, s_N)$ be the collective strategy of all players. For any $s \in S$ and $z_i \in S_i$ of the ith player for each i, the symbols S_{-i}, s_{-i} and (z_i, s_{-i}) are defined by

- $S_{-i} := (S_1 \times \cdots \times S_{i-1} \times S_{i+1} \times \cdots \times S_N)$ is the set of strategies of the remaining players when s_i was chosen by ith player,
- $s_{-i} := (s_1, \cdots, s_{i-1}, s_{i+1}, \cdots, s_N)$ is the strategies of the remaining players when ith player has s_i and
- $(z_i, s_{-i}) := (s_1, \cdots, s_{i-1}, z_i, s_{i+1}, \cdots, s_N)$ is the strategies of the situation that z_i was chosen by ith player when the rest of the remaining players have chosen s_{-i}.

Moreover, $\bar{s}_i$ is a special strategy of the ith player, supporting the player to maximize his pay-off, which equivalent to the following:

$$p_i(\bar{s}_i, s_{-i}) = \max_{z_i \in S_i} p_i(z_i, s_{-i}).$$

Definition 5. [37,38] *Given a game of N players in strategic form, the collective strategies $s^* \in S$ is said to be a Nash equilibrium point if*

$$p_i(s^*) = \max_{z_i \in S_i} p_i(z_i, s_i^*)$$

for all $i = 1, \cdots, N$ and $s_i^* \in S_{-i}$.

If no player can change his strategy to bring advantages, then the collective strategies $s^* = (s_i^*, s_{-i}^*)$ is a Nash equilibrium point. Furthermore, a Nash equilibrium point is the collective strategies of all players, i.e., s_i^* (for each $i \geq 1$) is the best response of ith player. There is a multi-valued mapping $T_i : S_{-i} \to 2^{S_i}$ such that

$$\begin{aligned} T_i(s_{-i}) &= \arg\max p_i(z_i, s_{-i}) \\ &= \{s_i \in S_i : p_i(s_i, s_{-i}) = \max_{z_i \in S_i} p_i(z_i, s_{-i})\} \end{aligned}$$

for all $s_{-i} \in S_{-i}$. Therefore, we can define the mapping $T : S \to 2^S$ by

$$T := T_1 \times T_2 \times \cdots \times T_N$$

such that the Nash equilibrium point is the collective strategies s^*, where $s^* \in F(T)$. Note that $s^* \in F(T)$ is equivalent to $s_i^* \in T(s_{-i}^*)$.

Let H_1 and H_2 be two real Hilbert spaces, $B_1 : H_1 \to 2^{H_1}$ and $B_2 : H_2 \to 2^{H_2}$ be multi-valued mappings. Suppose S is nonempty compact and convex subset of $H_1 = \mathbb{R}^{M_N}$, $H_2 = \mathbb{R}$ and the rest of the players have made their best responses s_{-i}^*. For each $s \in S$, define a mapping $A : S \to H_2$ by

$$As = p_i(s) - p_i(z_i, s_{-i}^*),$$

where p_i is linear, bounded and convex. Indeed, A is also linear, bounded and convex.

The *Nash equilibrium problem* **(NEP)** is the following:

$$\text{Find a point } \ s^* \in S \ \text{ such that } \ As^* > 0, \ \ 0 \in H_2. \tag{44}$$

However, the solution to the problem **(NEP)** may not be single-valued. Then the problem **(NEP)** reduces to finding the fixed point problem **(FPP)** of a multi-valued mapping, i.e.,

$$\text{Find a point } \ s^* \in S \ \text{ such that } \ s^* \in Ts^*, \tag{45}$$

where T is multi-valued pay-off function.

Now, we apply our Theorem 1 to find a solution to the problem **(FPP)**.

Let $B_1 : H_1 \to 2^{H_1}$ and $B_2 : H_2 \to 2^{H_2}$ be maximal monotone mappings defined on H_1 and H_2, respectively, and $A : H_1 \to H_2$ be a bounded linear operator with its adjoint A^*.

Now, we consider the following problem:

$$\text{Find a point } \ s^* \in H_1 \ \text{ such that } \ 0 \in B_1(s^*), \ \ s^* \in Ts^* \tag{46}$$

and

$$y^* = As^* \in H_2 \ \text{ such that } \ 0 \in B_2(y^*). \tag{47}$$

Theorem 3. *Assume that B_1 and B_2 are maximal monotone mappings defined on Hilbert spaces H_1 and H_2, respectively. Let $T : S \to CB(S)$ be a multi-valued quasi-nonexpansive mapping such that T is demiclosed at the origin. Let $\{f_n\}$ be a sequence of μ_n-contractions $f_n : H_1 \to H_1$ with $0 < \mu_* \leq \mu_n \leq \mu^* < 1$ and $\{f_n(x)\}$ be uniformly convergent for any x in a bounded subset D of H_1. Suppose that the problem* **(NEP)** *has* *a nonempty solution and* $F(T) \cap \Omega \neq \emptyset$. *For arbitrarily chosen* $x_0, x_1 \in H_1$, *let the sequences* $\{y_n\}$, $\{u_n\}$, $\{z_n\}$ *and* $\{x_n\}$ *be generated by*

$$\begin{cases} y_n = x_n + \theta_n(x_n - x_{n-1}), \\ u_n = J_\lambda^{B_1}(y_n + \gamma_n A^*(J_\lambda^{B_2} - I)Ay_n), \\ z_n = \xi v_n + (1-\xi)u_n, \quad v_n \in Tx_n, \\ x_{n+1} = \alpha_n f_n(x_n) + \beta_n x_n + \delta_n z_n \end{cases} \tag{48}$$

for each $n \geq 1$, *where* $\xi \in (0,1)$, $\gamma_n := \tau_n \frac{\|(J_\lambda^{B_2} - I)Ay_n\|^2}{\|A^*(J_\lambda^{B_2} - I)Ay_n\|^2}$ *with* $0 < \tau_* \leq \tau_n \leq \tau^* < 1$, $\{\theta_n\} \subset [0, \bar{\omega})$ *for some* $\bar{\omega} > 0$ *and* $\{\alpha_n\}, \{\beta_n\}, \{\delta_n\} \in (0,1)$ *with* $\alpha_n + \beta_n + \delta_n = 1$ *satisfying the following conditions:*

(C1) $\lim_{n\to\infty} \alpha_n = 0$;
(C2) $\sum_{n=1}^{\infty} \alpha_n = \infty$;
(C3) $0 < \epsilon_1 \leq \beta_n$ *and* $0 < \epsilon_2 \leq \delta_n$;
(C4) $\lim_{n\to\infty} \frac{\theta_n}{\alpha_n}\|x_n - x_{n-1}\| = 0$.

Then the sequence $\{x_n\}$ *generated by Equation* (48) *converges strongly to Nash equilibrium point.*

Proof. By Theorem 1, the sequence $\{x_n\}$ converges strongly to a point $p \in F(T) \cap \Omega$, then the sequence $\{x_n\}$ converges strongly to a Nash equilibrium point. □

Author Contributions: All five authors contributed equally to work. All authors read and approved the final manuscript. P.K. and K.S. conceived and designed the experiments. P.P., W.J. and Y.J.C. analyzed the data. P.P. and W.J. wrote the paper. Authorship must be limited to those who have contributed substantially to the work reported.

Acknowledgments: The authors acknowledge the financial support provided by King Mongkut's University of Technology Thonburi through the "KMUTT 55th Anniversary Commemorative Fund". Pawicha Phairatchatniyom would like to thank the *"Science Graduate Scholarship"*, Faculty of Science, King Mongkut's University of Technology Thonburi (KMUTT) (Grant No. 11/2560). Wachirapong Jirakitpuwapat would like to thank the Petchra Pra Jom Klao Ph.D. Research Scholarship and the King Mongkut's University of Technology Thonburi (KMUTT) for financial support. Moreover, Kanokwan Sitthithakerngkiet was funded by King Mongkut's University of Technology North Bangkok, Contract No. KMUTNB-KNOW-61-035.

References

1. Moudafi. Split monotone variational inclusions. *J. Opt. Theory Appl.* **2011**, *150*, 275–283. [CrossRef]
2. Shehu, Y.; Agbebaku, D. On split inclusion problem and fixed point problem for multi-valued mappings. *Comput. Appl. Math.* **2017**, *37*. [CrossRef]
3. Shehu, Y.; Ogbuisi, F.U. An iterative method for solving split monotone variational inclusion and fixed point problems. *Rev. Real Acad. Cienc. Exact. Fís. Nat. Serie A. Mat.* **2015**, *110*, 503–518. [CrossRef]
4. Kazmi, K.; Rizvi, S. An iterative method for split variational inclusion problem and fixed point problem for a nonexpansive mapping. *Opt. Lett.* **2013**, *8*. [CrossRef]
5. Sitthithakerngkiet, K.; Deepho, J.; Kumam, P. A hybrid viscosity algorithm via modify the hybrid steepest descent method for solving the split variational inclusion in image reconstruction and fixed point problems. *Appl. Math. Comput.* **2015**, *250*, 986–1001. [CrossRef]
6. Censor, Y.; Gibali, A.; Reich, S. Algorithms for the split variational inequality problem. *Num. Algorithms* **2012**, *59*, 301–323. [CrossRef]
7. Byrne, C.; Censor, Y.; Gibali, A.; Reich, S. Weak and strong convergence of algorithms for the split common null point problem. Technical Report. *arXiv* **2011**, arXiv:1108.5953.

8. Moudafi, A. The split common fixed-point problem for demicontractive mappings. *Inverse Prob.* **2010**, *26*, 055007. [CrossRef]
9. Yao, Y.; Liou, Y.C.; Postolache, M. Self-adaptive algorithms for the split problem of the demicontractive operators. *Optimization* **2017**, *67*, 1309–1319. [CrossRef]
10. Dang, Y.; Gao, Y. The strong convergence of a KM-CQ-like algorithm for a split feasibility problem. *Inverse Prob.* **2011**, *27*, 015007. [CrossRef]
11. Sahu, D.R.; Pitea, A.; Verma, M. A new iteration technique for nonlinear operators as concerns convex programming and feasibility problems. *Numer. Algorithms* **2019**. [CrossRef]
12. Censor, Y.; Elfving, T. A multiprojection algorithm using Bregman projections in a product space. *Numer. Algorithms* **1994**, *8*, 221–239. [CrossRef]
13. Combettes, P. The convex feasibility problem in image recovery. *Adv. Imag. Electron. Phys.* **1996**, *95*, 155–270. [CrossRef]
14. Kazmi, K.R.; Rizvi, S.H. Iterative approximation of a common solution of a split equilibrium problem, a variational inequality problem and a fixed point problem. *J. Egypt. Math. Soc.* **2013**, *21*, 44–51. [CrossRef]
15. Peng, J.W.; Wang, Y.; Shyu, D.S.; Yao, J.C. Common solutions of an iterative scheme for variational inclusions, equilibrium problems, and fixed point problems. *J. Inequal. Appl.* **2008**, 15, 720371. [CrossRef]
16. Jung, J.S. Strong convergence theorems for multivalued nonexpansive nonself-mappings in Banach spaces. *Nonlinear Anal. Theory, Meth. Appl.* **2007**, *66*, 2345–2354. [CrossRef]
17. Panyanak, B. Mann and Ishikawa iterative processes for multivalued mappings in Banach spaces. *Comput. Math. Appl.* **2007**, *54*, 872–877. [CrossRef]
18. Shahzad, N.; Zegeye, H. On Mann and Ishikawa iteration schemes for multi-valued maps in Banach spaces. *Nonlinear Anal.* **2009**, *71*, 838–844. [CrossRef]
19. Polyak, B. Some methods of speeding up the convergence of iteration methods. *USSR Comput. Math. Math. Phys.* **1964**, *4*, 1–17. [CrossRef]
20. Nesterov, Y. A method of solving a convex programming problem with convergence rate O(1/sqr(k)). *Sov. Math. Dokl.* **1983**, *27*, 372–376.
21. Dang, Y.; Sun, J.; Xu, H. Inertial accelerated algorithms for solving a split feasibility problem. *J. Ind. Manag. Optim.* **2017**, *13*, 1383–1394. [CrossRef]
22. Suantai, S.; Pholasa, N.; Cholamjiak, P. The modified inertial relaxed CQ algorithm for solving the split feasibility problems. *J. Ind. Manag. Opt.* **2017**, *13*, 1–21. [CrossRef]
23. Alvarez, F.; Attouch, H. An inertial proximal method for maximal monotone operators via discretization of a nonlinear oscillator with damping. Wellposedness in optimization and related topics (Gargnano, 1999). *Set-Valued Anal.* **2001**, *9*, 3–11. [CrossRef]
24. Rockafellar, R.T. Monotone operators and the proximal point algorithm. *SIAM J. Control Opt.* **1976**, *14*, 877–898. [CrossRef]
25. Attouch, H.; Peypouquet, J.; Redont, P. A dynamical approach to an inertial forward-backward algorithm for convex minimization. *SIAM J. Opt.* **2014**, *24*, 232–256. [CrossRef]
26. Boţ, R.I.; Csetnek, E.R. An inertial alternating direction method of multipliers. *Min. Theory Appl.* **2016**, *1*, 29–49.
27. Maingé, P.E. Convergence theorems for inertial KM-type algorithms. *J. Comput. Appl. Math.* **2008**, *219*, 223–236. [CrossRef]
28. Bot, R.I., Csetnek, E.R., Hendrich, C. Inertial Douglas-Rachford splitting for monotone inclusion problems. *Appl. Math. Comput.* **2015**, *256*, 472—487.
29. Chuang, C.S. Strong convergence theorems for the split variational inclusion problem in Hilbert spaces. *Fix. Point Theory Appl.* **2013**, *2013*. [CrossRef]
30. Che, H.; Li, M. Solving split variational inclusion problem and fixed point problem for nonexpansive semigroup without prior knowledge of operator norms. *Math. Prob. Eng.* **2015**, *2015*, 1–9. [CrossRef]
31. Ansari, Q.H.; Rehan, A.; Wen, C.F. Split hierarchical variational inequality problems and fixed point problems for nonexpansive mappings. *J. Inequal. Appl.* **2015**, *16*, 274. [CrossRef]
32. Xu, H.K. Iterative algorithms for nonlinear operators. *J. Lond. Math. Soc.* **2002**, *66*, 240–256. [CrossRef]
33. Maingé, P.E. Approximation methods for common fixed points of nonexpansive mappings in Hilbert spaces. *J. Math. Anal. Appl.* **2007**, *325*, 469–479. [CrossRef]

34. Glowinski, R.; Tallec, P. *Augmented Lagrangian and Operator-Splitting Methods in Nonlinear Mechanics*; SIAM Studies in Applied Mathematics; Society for Industrial and Applied Mathematics: Philadelphia, PA, USA, 1989.
35. Von Neumann, J.; Morgenstern, O. *Theory of Games and Economic Behavior*; Princeton University Press: Princeton, NJ, USA, 1947.
36. Kraikaew, R.; Saejung, S. Strong convergence of the Halpern subgradient extragradient method for solving variational inequalities in Hilbert spaces. *J. Opt. Theory Appl.* **2014**, *163*, 399–412. [CrossRef]
37. Nash, J.F., Jr. Equilibrium points in *n*-person games. *Proc. Nat. Acad. Sci. USA* **1950**, *36*, 48–49. [CrossRef] [PubMed]
38. Nash, J. Non-cooperative games. *Ann. Math.* **1951**, *54*, 286–295. [CrossRef]

5

A Few Iterative Methods by Using $[1, n]$-Order Padé Approximation of Function and the Improvements

Shengfeng Li [1,*], Xiaobin Liu [2] and Xiaofang Zhang [1]

[1] Institute of Applied Mathematics, Bengbu University, Bengbu 233030, China; zhangxiaofangah@126.com
[2] School of Computer Engineering, Bengbu University, Bengbu 233030, China; 18807925677@139.com
* Correspondence: lsf@bbc.edu.cn.

Abstract: In this paper, a few single-step iterative methods, including classical Newton's method and Halley's method, are suggested by applying $[1, n]$-order Padé approximation of function for finding the roots of nonlinear equations at first. In order to avoid the operation of high-order derivatives of function, we modify the presented methods with fourth-order convergence by using the approximants of the second derivative and third derivative, respectively. Thus, several modified two-step iterative methods are obtained for solving nonlinear equations, and the convergence of the variants is then analyzed that they are of the fourth-order convergence. Finally, numerical experiments are given to illustrate the practicability of the suggested variants. Henceforth, the variants with fourth-order convergence have been considered as the imperative improvements to find the roots of nonlinear equations.

Keywords: nonlinear equations; Padé approximation; iterative method; order of convergence; numerical experiment

---

1. Introduction

It is well known that a variety of problems in different fields of science and engineering require to find the solution of the nonlinear equation $f(x) = 0$ where $f : I \to D$, for an interval $I \subseteq \mathbb{R}$ and $D \subseteq \mathbb{R}$, is a scalar function. In general, iterative methods, such as Newton's method, Halley's method, Cauchy's method, and so on, are the most used techniques. Hence, iterative algorithms based on these iterative methods for finding the roots of nonlinear equations are becoming one of the most important aspects in current researches. We can see the works, for example, [1–22] and references therein. In the last few years, some iterative methods with high-order convergence have been introduced to solve a single nonlinear equation. By using various techniques, such as Taylor series, quadrature formulae, decomposition techniques, continued fraction, Padé approximation, homotopy methods, Hermite interpolation, and clipping techniques, these iterative methods can be constructed. For instance, there are many ways of introducing Newton's method. Among these ways, using Taylor polynomials to derive Newton's method is probably the most widely known technique [1,2]. By considering different quadrature formulae for the computation of the integral, Weerakoon and Fernando derive an implicit iterative scheme with cubic convergence by the trapezoidal quadrature formulae [4], while Cordero and Torregrosa develope some variants of Newton's method based in rules of quadrature of fifth order [5]. In 2005, Chun [6] have presented a sequence of iterative methods improving Newton's method for solving nonlinear equations by applying the Adomian decomposition method. Based on Thiele's continued fraction of the function, Li et al. [7] give a fourth-order convergent iterative method. Using Padé approximation of the function, Li et al. [8] rederive the Halley's method and by the divided differences to approximate the derivatives, they arrive at some modifications with third-order convergence. In [9], Abbasbandy et al. present an efficient numerical algorithm for solving nonlinear algebraic equations based on Newton–Raphson method and homotopy analysis method. Noor and Khan suggest and analyze a new class of iterative methods by using the homotopy perturbation method in [10]. In 2015, Wang et al. [11] deduce a general family of n-point Newton type iterative methods for solving nonlinear equations by using direct Hermite interpolation. Moreover, for a particular class of functions, for instance, if f is a polynomial, there exist some efficient univariate root-finding algorithms to compute all solutions of the polynomial equation (see [12,13]). In the literature [13], Bartoň et al. present an algorithm for computing all roots of univariate polynomial based on degree reduction, which has the higher convergence rate than Newton's method. In this article, we will mainly solve more general nonlinear algebraic equations.

Newton's method is probably the best known and most widely used iterative algorithm for root-finding problems. By applying Taylor's formula for the function $f(x)$, let us recall briefly how to derive Newton iterative method. Suppose that $f(x) \in C^n[I], n = 1, 2, 3, \ldots$, and $\eta \in I$ is a single root of the nonlinear equation $f(x) = 0$. For a given guess value $x_0 \in I$ and a $\delta \in \mathbb{R}$, assume that $f'(x) \neq 0$ for each x belongs to the neighborhood $(x_0 - \delta, x_0 + \delta)$. For any $x \in (x_0 - \delta, x_0 + \delta)$, we expand $f(x)$ into the following Taylor's formula about x_0:

$$f(x) = f(x_0) + f'(x_0)(x - x_0) + \frac{1}{2!} f''(x_0)(x - x_0)^2 + \cdots + \frac{1}{k!}(x - x_0)^k f^{(k)}(x_0) + \cdots,$$

where $k = 0, 1, 2, \cdots$. Let $|\eta - x_0|$ be sufficiently small. Then the terms involving $(\eta - x_0)^k, k = 2, 3, \ldots$, are much smaller. Hence, we think the fact that the first Taylor polynomial is a good approximation to the function near the point x_0 and give that

$$f(x_0) + f'(x_0)(\eta - x_0) \approx 0.$$

Notice the fact $f'(x_0) \neq 0$, and solving the above equation for η yields

$$\eta \approx x_0 - \frac{f(x_0)}{f'(x_0)},$$

which follows that we can construct the Newton iterative scheme as below

$$x_{k+1} = x_k - \frac{f(x_k)}{f'(x_k)}, k = 0, 1, 2, \ldots.$$

It has been known that Newton iterative method is a celebrated one-step iterative method. The order of convergence of Newton's method is quadratic for a simple zero and linear for multiple root.

Motivated by the idea of the above technique, in this paper, we start with using Padé approximation of a function to construct a few one-step iterative schemes which includes classical Newton's method and Halley's method to find roots of nonlinear equations. In order to avoid calculating the high-order derivatives of the function, then we employ the approximants of the higher derivatives to improve the presented iterative method. As a result, we build several two-step iterative formulae, and some of them do not require the operation of high-order derivatives. Furthermore, it is shown that these modified iterative methods are all fouth-order convergent for a simple root of the equation. Finally, we give some numerical experiments and comparison to illustrate the efficiency and performance of the presented methods.

The rest of this paper is organized as follows. we introduce some basic preliminaries about Padé approximation and iteration theory for root-finding problem in Section 2. In Section 3, we firstly construct several one-step iterative schemes based on Padé approximation. Then, we modify the presented iterative method to obtain a few iterative formulae without calculating the high-order derivatives. In Section 4, we show that the modified methods have fourth-order convergence at least for a simple root of the equation. In Section 5 we give numerical examples to show the performance of the presented methods and compare them with other high-order methods. At last, we draw conclusions from the experiment results in Section 6.

2. Preliminaries

In this section, we briefly review some basic definitions and results for Padé approximation of function and iteration theory for root-finding problem. Some surveys and complete literatures about iteration theory and Padé approximation could be found in Alfio [1], Burden et al. [2], Wuytack [23], and Xu et al. [24].

Definition 1. *Assume that $f(x)$ is a function whose $(n+1)$-st derivative $f^{(n+1)}(x), n = 0,1,2,\ldots,$ exists for any x in an interval I. Then for each $x \in I$, we have*

$$f(x) = f(x_0) + f'(x_0)(x - x_0) + \frac{f''(x_0)}{2!}(x - x_0)^2 + \cdots + \frac{f^{(n)}(x_0)}{n!}(x - x_0)^n + o[(x - x_0)^n], \tag{1}$$

which is called the Taylor's formula with Peano remainder term of order n based at x_0, and the error $o[(x - x_0)^n]$ is called the Peano remainder term or the Peano truncation error.

Definition 2. *If $P(x)$ is a polynomial, the accurate degree of the polynomial is $\partial(P)$, and the order of the polynomial is $\omega(P)$, which is the degree of the first non-zero term of the polynomial.*

Definition 3. *If it can be found two ploynomials*

$$P(x) = \sum_{i=0}^{m} a_i(x - x_0)^i \qquad \text{and} \qquad Q(x) = \sum_{i=0}^{n} b_i(x - x_0)^i$$

such that

$$\partial(P(x)) \le m, \quad \partial(Q(x)) \le n, \quad \omega(f(x)Q(x) - P(x)) \ge m + n + 1,$$

then we have the following incommensurable form of the rational fraction $\frac{P(x)}{Q(x)}$:

$$R_{m,n}(x) = \frac{P_0(x)}{Q_0(x)} = \frac{P(x)}{Q(x)},$$

which is called $[m,n]$-order Padé approximation of function $f(x)$.

We give the computational formula of Padé approximation of function $f(x)$ by the use of determinant, as shown in the following lemma [23,24].

Lemma 1. *Assume that $R_{m,n}(x) = \frac{P_0(x)}{Q_0(x)}$ is Padé approximation of function $f(x)$. If the matrix*

$$A_{m,n} = \begin{pmatrix} a_m & a_{m-1} & \cdots & a_{m+1-n} \\ a_{m+1} & a_m & \cdots & a_{m+2-n} \\ \vdots & \vdots & \ddots & \vdots \\ a_{m+n-1} & a_{m+n-2} & \cdots & a_m \end{pmatrix}$$

is nonsingular, that is the determinant $|A_{m,n}| = d \neq 0$, then $P_0(x), Q_0(x)$ can be written by the following determinants

$$P_0(x) = \frac{1}{d}\begin{vmatrix} T_m(x) & (x-x_0)T_{m-1}(x) & \cdots & (x-x_0)^n T_{m+1-n}(x) \\ a_{m+1} & a_m & \cdots & a_{m+1-n} \\ \vdots & \vdots & \ddots & \vdots \\ a_{m+n} & a_{m+n-1} & \cdots & a_m \end{vmatrix}$$

and

$$Q_0(x) = \frac{1}{d}\begin{vmatrix} 1 & (x-x_0) & \cdots & (x-x_0)^n \\ a_{m+1} & a_m & \cdots & a_{m+1-n} \\ \vdots & \vdots & \ddots & \vdots \\ a_{m+n} & a_{m+n-1} & \cdots & a_m \end{vmatrix},$$

where $a_n = \frac{f^{(n)}(x_0)}{n!}, n = 0, 1, 2, \ldots$, *and we appoint that*

$$T_k(x) = \begin{cases} \sum_{i=0}^{k} a_i(x-x_0)^i, & for \quad k \geq 0, \\ 0, & for \quad k < 0. \end{cases}$$

Next, we recall the speed of convergence of an iterative scheme. Thus, we give the following definition and lemma.

Definition 4. *Assume that a sequence* $\{x_i\}_{i=0}^{\infty}$ *converges to* η, *with* $x_i \neq \eta$ *for all* $i, i = 0, 1, 2, \ldots$. *Let the error be* $e_i = x_i - \eta$. *If there exist two positive constants* α *and* β *such that*

$$\lim_{i\to\infty} \frac{|e_{i+1}|}{|e_i|^{\alpha}} = \beta,$$

then $\{x_i\}_{i=0}^{\infty}$ *converges to the constant* η *of order* α. *When* $\alpha = 1$, *the sequence* $\{x_i\}_{i=0}^{\infty}$ *is linearly convergent. When* $\alpha > 1$, *the sequence* $\{x_i\}_{i=0}^{\infty}$ *is said to be of higher-order convergence.*

For a single-step iterative method, sometimes it is convenient to use the following lemma to judge the order of convergence of the iterative method.

Lemma 2. *Assume that the equation* $f(x) = 0, x \in I$, *can be rewritten as* $x = \varphi(x)$, *where* $f(x) \in C[I]$ *and* $\varphi(x) \in C^{\gamma}[I], \gamma \in \mathbb{N}^+$. *Let* η *be a root of the equation* $f(x) = 0$. *If the iterative function* $\varphi(x)$ *satisfies*

$$\varphi^{(j)}(\eta) = 0, j = 1, 2, \ldots, \gamma - 1, \varphi^{(\gamma)}(\eta) \neq 0,$$

then the order of convergence of the iterative scheme $x_{i+1} = \varphi(x_i), i = 0, 1, 2, \ldots$, *is* γ.

3. Some Iterative Methods

Let η be a simple real root of the equation $f(x) = 0$, where $f : I \to D, I \subseteq \mathbb{R}, D \subseteq \mathbb{R}$. Suppose that $x_0 \in I$ is an initial guess value sufficiently close to η, and the function $f(x)$ has n-th derivative $f^{(n)}(x), n = 1, 2, 3, \ldots$, in the interval I. According to Lemma 1, $[m, n]$-order Padé approximation of function $f(x)$ is denoted by the following rational fraction:

$$f(x) \approx R_{m,n}(x) = \frac{\begin{vmatrix} T_m(x) & (x-x_0)T_{m-1}(x) & \cdots & (x-x_0)^n T_{m+1-n}(x) \\ a_{m+1} & a_m & \cdots & a_{m+1-n} \\ \vdots & \vdots & \ddots & \vdots \\ a_{m+n} & a_{m+n-1} & \cdots & a_m \end{vmatrix}}{\begin{vmatrix} 1 & (x-x_0) & \cdots & (x-x_0)^n \\ a_{m+1} & a_m & \cdots & a_{m+1-n} \\ \vdots & \vdots & \ddots & \vdots \\ a_{m+n} & a_{m+n-1} & \cdots & a_m \end{vmatrix}}. \tag{2}$$

Recall Newton iterative method derived by Taylor's series in Section 1. The first Taylor polynomial is regarded as a good approximation to the function $f(x)$ near the point x_0. Solving the linear equation denoted by $f(x_0) + f'(x_0)(\eta - x_0) \approx 0$ for η gives us the stage for Newton's method. Then, we think whether or not a novel or better linear function is selected to approximate the function $f(x)$ near the point x_0. Maybe Padé approximation can solve this question. In the process of obtaining new iterative methods based on Padé approximation of function, on the one hand, we consider that the degree of the numerator of Equation (2) is always taken as 1, which guarantees to obtain the different linear function. On the other hand, we discuss the equations are mainly nonlinear algebraic equations, which differ rational equations and have not the poles. Clearly, as n grows, the poles of the denominator of Equation (2) do not affect the linear functions that we need. These novel linear functions may be able to set the stage for new methods. Next, let us start to introduce a few iterative methods by using $[1,n]$-order Padé approximation of function.

3.1. *Iterative Method Based on* $[1,0]$*-Order Padé Approximation*

Firstly, when $m = 1, n = 0$, we consider $[1,0]$-order Padé approximation of function $f(x)$. It follows from the expression (2) that

$$f(x) \approx R_{1,0}(x) = T_1(x) = a_0 + a_1(x - x_0).$$

Let $R_{1,0}(x) = 0$, then we have

$$a_0 + a_1(x - x_0) = 0. \tag{3}$$

Due to the determinant $|A_{1,0}| \neq 0$, i.e., $f'(x_0) \neq 0$, we obtain the following equation from Equation (3).

$$x = x_0 - \frac{a_0}{a_1}.$$

In view of $a_0 = f(x_0), a_1 = f'(x_0)$, we reconstruct the Newton iterative method as below.

Method 1. *Assume that the function $f : I \to D$ has its first derivative at the point $x_0 \in I$. Then we obtain the following iterative method based on $[1,0]$-order Padé approximation of function $f(x)$:*

$$x_{k+1} = x_k - \frac{f(x_k)}{f'(x_k)}, k = 0, 1, 2, \ldots. \tag{4}$$

Starting with an initial approximation x_0 that is sufficiently close to the root η and using the above scheme (4), we can get the iterative sequence $\{x_i\}_{i=0}^{\infty}$.

Remark 1. *Method 1 is the well-known Newton's method for solving nonlinear equation [1,2].*

3.2. Iterative Method Based on $[1,1]$-Order Padé Approximation

Secondly, when $m=1, n=1$, we think about $[1,1]$-order Padé approximation of function $f(x)$. Similarly, it follows from the expression (2) that

$$f(x) \approx R_{1,1}(x) = \frac{\begin{vmatrix} T_1(x) & (x-x_0)T_0(x) \\ a_2 & a_1 \end{vmatrix}}{\begin{vmatrix} 1 & (x-x_0) \\ a_2 & a_1 \end{vmatrix}}.$$

Let $R_{1,1}(x)=0$, then we get

$$a_0a_1 + a_1^2(x-x_0) - a_0a_2(x-x_0) = 0. \tag{5}$$

Due to the determinant $|A_{1,1}| \neq 0$, that is,

$$\begin{vmatrix} a_1 & a_0 \\ a_2 & a_1 \end{vmatrix} = \begin{vmatrix} f'(x_0) & f(x_0) \\ \frac{f''(x_0)}{2} & f'(x_0) \end{vmatrix} = f'^2(x_0) - f(x_0)\frac{f''(x_0)}{2} \neq 0.$$

Thus, we obtain the following equality from Equation (5):

$$x = x_0 - \frac{a_0a_1}{a_1^2 - a_0a_2}.$$

Combining $a_0 = f(x_0), a_1 = f'(x_0), a_2 = \frac{1}{2}f''(x_0)$, gives Halley iterative method as follows.

Method 2. *Assume that the function $f: I \to D$ has its second derivative at the point $x_0 \in I$. Then we obtain the following iterative method based on $[1,1]$-order Padé approximation of function $f(x)$:*

$$x_{k+1} = x_k - \frac{2f(x_k)f'(x_k)}{2f'^2(x_k) - f(x_k)f''(x_k)}, k = 0,1,2,\ldots. \tag{6}$$

Starting with an initial approximation x_0 that is sufficiently close to the root η and applying the above scheme (6), we can obtain the iterative sequence $\{x_i\}_{i=0}^{\infty}$.

Remark 2. *Method 2 is the classical Halley's method for finding roots of nonlinear equation [1,2], which converges cubically.*

3.3. Iterative Method Based on $[1,2]$-Order Padé Approximation

Thirdly, when $m=1, n=2$, we take into account $[1,2]$-order Padé approximation of function $f(x)$. By the same manner, it follows from the expression (2) that

$$f(x) \approx R_{1,2}(x) = \frac{\begin{vmatrix} T_1(x) & (x-x_0)T_0(x) & 0 \\ a_2 & a_1 & a_0 \\ a_3 & a_2 & a_1 \end{vmatrix}}{\begin{vmatrix} 1 & x-x_0 & (x-x_0)^2 \\ a_2 & a_1 & a_0 \\ a_3 & a_2 & a_1 \end{vmatrix}}.$$

Let $R_{1,2}(x)=0$, then one has

$$a_0a_1^2 + a_0^2a_2 + (a_1^3 - 2a_0a_1a_2 + a_0^2a_3)(x-x_0) = 0. \tag{7}$$

Due to the determinant $|A_{1,2}| \neq 0$, that is,

$$\begin{vmatrix} a_1 & a_0 & 0 \\ a_2 & a_1 & a_0 \\ a_3 & a_2 & a_1 \end{vmatrix} = \begin{vmatrix} f'(x_0) & f(x_0) & 0 \\ \frac{f''(x_0)}{2} & f'(x_0) & f(x_0) \\ \frac{f'''(x_0)}{6} & \frac{f''(x_0)}{2} & f'(x_0) \end{vmatrix} = f'^3(x_0) - f(x_0)f'(x_0)f''(x_0) + \frac{f^2(x_0)f'''(x_0)}{6} \neq 0.$$

Thus, we gain the following equality from Equation (7):

$$x = x_0 - \frac{a_0a_1^2 - a_0^2a_2}{a_1^3 - 2a_0a_1a_2 + a_0^2a_3}.$$

Substituting $a_0 = f(x_0), a_1 = f'(x_0), a_2 = \frac{1}{2}f''(x_0)$, and $a_3 = \frac{1}{6}f'''(x_0)$ into the above equation gives a single-step iterative method as follows.

Method 3. *Assume that the function $f : I \to D$ has its third derivative at the point $x_0 \in I$. Then we obtain the following iterative method based on $[1,2]$-order Padé approximation of function $f(x)$:*

$$x_{k+1} = x_k - \frac{3f(x_k)\left(2f'^2(x_k) - f(x_k)f''(x_k)\right)}{6f'^3(x_k) - 6f(x_k)f'(x_k)f''(x_k) + f^2(x_k)f'''(x_k)}, k = 0, 1, 2, \ldots. \quad (8)$$

Starting with an initial approximation x_0 that is sufficiently close to the root η and applying the above scheme (8), we can receive the iterative sequence $\{x_i\}_{i=0}^{\infty}$.

Remark 3. *Method 3 could be used to find roots of nonlinear equation. Clearly, for the sake of applying this iterative method, we must compute the second derivative and the third derivative of the function $f(x)$, which may generate inconvenience. In order to overcome the drawback, we suggest approximants of the second derivative and the third derivative, which is a very important idea and plays a significant part in developing some iterative methods free from calculating the higher derivatives.*

3.4. Modified Iterative Method Based on Approximant of the Third Derivative

In fact, we let $z_k = x_k - \frac{f(x_k)}{f'(x_k)}$. Then expanding $f(z_k)$ into third Taylor's series about the point x_k yields

$$f(z_k) \approx f(x_k) + f'(x_k)(z_k - x_k) + \frac{1}{2!}f''(x_k)(z_k - x_k)^2 + \frac{1}{3!}f'''(x_k)(z_k - x_k)^3,$$

which follows that

$$f'''(x_k) \approx \frac{3f^2(x_k)f'(x_k)f''(x_k) - 6f(z_k)f'^3(x_k)}{f^3(x_k)}. \quad (9)$$

Substituting (9) into (8), we can have the following iterative method.

Method 4. *Assume that the function $f : I \to D$ has its second derivative about the point $x_0 \in I$. Then we possess a modified iterative method as below:*

$$\begin{cases} z_k = x_k - \frac{f(x_k)}{f'(x_k)}, \\ x_{k+1} = x_k - \frac{x_k - z_k}{1 + 2f(z_k)f'^2(x_k)L^{-1}(x_k)}, \quad k = 0, 1, 2, \ldots, \end{cases} \quad (10)$$

where $L(x_k) = f(x_k)\left(f(x_k)f''(x_k) - 2f'^2(x_k)\right)$. Starting with an initial approximation x_0 that is sufficiently close to the root η and using the above scheme (10), we can have the iterative sequence $\{x_i\}_{i=0}^{\infty}$.

Remark 4. *Methods 4 is a two-step iterative method free from third derivative of the function.*

3.5. Modified Iterative Method Based on Approximant of the Second Derivative

It is obvious that the iterative method (10) requires the operation of the second derivative of the function $f(x)$. In order to avoid computing the second derivative, we introduce an approximant of the second derivative by using Taylor's series.

Similarly, expanding $f(z_k)$ into second Taylor's series about the point x_k yields

$$f(z_k) \approx f(x_k) + (z_k - x_k)f'(x_k) + \frac{1}{2!}(z_k - x_k)^2 f''(x_k),$$

which means

$$f''(x_k) \approx \frac{2f(z_k)f'^2(x_k)}{f^2(x_k)}. \tag{11}$$

Using (11) in (10), we can get the following modified iterative method without computing second derivative.

Method 5. *Assume that the function $f : I \to D$ has its first derivative about the point $x_0 \in I$. Then we have a modified iterative method as below:*

$$\begin{cases} z_k = x_k - \frac{f(x_k)}{f'(x_k)}, \\ x_{k+1} = x_k - \frac{f(x_k)-f(z_k)}{f(x_k)-2f(z_k)}(x_k - z_k), \quad k = 0, 1, 2, \ldots. \end{cases} \tag{12}$$

Starting with an initial approximation x_0 that is sufficiently close to the root η and using the above scheme (12), we can obtain the iterative sequence $\{x_i\}_{i=0}^{\infty}$.

Remark 5. *Method 5 is another two-step iterative method. It is clear that Method 5 does not require to calculate the high-order derivative. But more importantly, the characteristic of Method 5 is that per iteration it requires two evaluations of the function and one of its first-order derivative. The efficiency of this method is better than that of the well-known other methods involving the second-order derivative of the function.*

4. Convergence Analysis of Iterative Methods

Theorem 1. *Suppose that $f(x)$ is a function whose n-th derivative $f^{(n)}(x), n = 1, 2, 3, \ldots$, exists in a neighborhood of its root η with $f'(\eta) \neq 0$. If the initial approximation x_0 is sufficiently close to η, then the Method 3 defined by (8) is fourth-order convergent.*

Proof of Theorem 1. By the hypothesis $f(\eta) = 0$ and $f'(\eta) \neq 0$, we know that η is an unique single root of the equation $f(x) = 0$. So, for each positive integer $n \geq 1$, we have that the derivatives of high orders $f^{(n)}(\eta) \neq 0$. Considering the iterative scheme (8) in Method 3, we denote its corresponding iterative function as shown below:

$$\varphi(x) = x - \frac{3f(x)\left(2f'^2(x) - f(x)f''(x)\right)}{6f'^3(x) - 6f(x)f'(x)f''(x) + f^2(x)f'''(x)}. \tag{13}$$

By calculating the first and high-order derivatives of the iterative function $\varphi(x)$ with respect to x at the point η, we verify that

$$\varphi'(\eta) = 0, \quad \varphi''(\eta) = 0, \quad \varphi'''(\eta) = 0$$

and

$$\varphi^{(4)}(\eta) = \frac{3f''^3(\eta) - 4f'(\eta)f''(\eta)f'''(\eta) + f'^2(\eta)f^{(4)}(\eta)}{f'^3(\eta)} \neq 0.$$

Thus, it follows from Lemma 2 that Method 3 defined by (8) is fourth-order convergent. This completes the proof. □

Theorem 2. *Suppose that $f(x)$ is a function whose n-th derivative $f^{(n)}(x), n = 1,2,3,\ldots$, exists in a neighborhood of its root η with $f'(\eta) \neq 0$. If the initial approximation x_0 is sufficiently close to η, then the Method 4 defined by (10) is at least fourth-order convergent with the following error equation*

$$e_{k+1} = (b_2^3 - 2b_2b_3)e_k^4 + O(e_k^5),$$

where $e_k = x_k - \eta, k = 1,2,3,\ldots$, and the constants $b_n = \frac{a_n}{f'(\eta)}, a_n = \frac{f^{(n)}(\eta)}{n!}, n = 1,2,3,\ldots$.

Proof of Theorem 2. By the hypothesis, it is clear to see that η is an unique single root of the equation $f(x) = 0$. By expanding $f(x_k)$, $f'(x_k)$ and $f''(x_k)$ into Taylor's series about η, we obtain

$$\begin{aligned} f(x_k) &= e_k f'(\eta) + \tfrac{e_k^2}{2!} f''(\eta) + \tfrac{e_k^3}{3!} f'''(\eta) + \tfrac{e_k^4}{4!} f^{(4)}(\eta) + \tfrac{e_k^5}{5!} f^{(5)}(\eta) + \tfrac{e_k^6}{6!} f^{(6)}(\eta) + O(e_k^7) \\ &= f'(\eta)\left(b_1 e_k + b_2 e_k^2 + b_3 e_k^3 + b_4 e_k^4 + b_5 e_k^5 + b_6 e_k^6 + O(e_k^7)\right), \end{aligned} \tag{14}$$

$$f'(x_k) = f'(\eta)\left(b_1 + 2b_2e_k + 3b_3e_k^2 + 4b_4e_k^3 + 5b_5e_k^4 + 6b_6e_k^5 + O(e_k^6)\right) \tag{15}$$

and

$$f''(x_k) = f'(\eta)\left(2b_2 + 6b_3e_k + 12b_4e_k^2 + 20b_5e_k^3 + 30b_6e_k^4 + O(e_k^5)\right), \tag{16}$$

where $b_n = \frac{1}{n!}\frac{f^{(n)}(\eta)}{f'(\eta)}, n = 1,2,\cdots$. Clearly, $b_1 = 1$. Dividing (14) by (15) directly, gives us

$$\begin{aligned} \tfrac{f(x_k)}{f'(x_k)} = \; & x_k - z_k = e_k - b_2e_k^2 - 2(b_3 - b_2^2)e_k^3 - (4b_2^3 + 3b_4 - 7b_2b_3)e_k^4 \\ & -2(10b_2^2b_3 - 2b_5 + 5b_2b_4 + 4b_2^4 + 3b_3^2)e_k^5 \\ & -(16b_2^5 + 28b_2^2b_4 + 33b_2b_3^2 + 5b_6 - 52b_2^3b_3 - 17b_3b_4 - 13b_2b_5)e_k^6 + O(e_k^7). \end{aligned} \tag{17}$$

By substituting (17) into (10) in Method 4, one has

$$\begin{aligned} z_k = \; & \eta + b_2e_k^2 + 2(b_3 - b_2^2)e_k^3 + (4b_2^3 + 3b_4 - 7b_2b_3)e_k^4 \\ & +2(10b_2^2b_3 - 2b_5 + 5b_2b_4 + 4b_2^4 + 3b_3^2)e_k^5 \\ & +(16b_2^5 + 28b_2^2b_4 + 33b_2b_3^2 + 5b_6 - 52b_2^3b_3 - 17b_3b_4 - 13b_2b_5)e_k^6 + O(e_k^7). \end{aligned} \tag{18}$$

Again, expanding $f(z_k)$ by Taylor's series about η, we have

$$\begin{aligned} f(z_k) = \; & f'(\eta)\left(b_2e_k^2 - 2(b_2^2 - b_3)e_k^3 - (7b_2b_3 - 5b_2^3 - 3b_4)e_k^4\right. \\ & -2(5b_2b_4 + 6b_2^4 + 3b_3^2 - 12b_2^2b_3 - 2b_5)e_k^5 \\ & \left.+ (28b_2^5 + 34b_2^2b_4 + 37b_2b_3^2 + 5b_6 - 73b_2^3b_3 - 17b_3b_4 - 13b_2b_5)e_k^6 + O(e_k^7)\right). \end{aligned} \tag{19}$$

Hence, from (15) and (19), we have

$$\begin{aligned} f(z_k)f'^2(x_k) = \; & f'^3(\eta)\left(b_2e_k^2 + 2(b_2^2 + b_3)e_k^3 + (7b_2b_3 + b_2^3 + 3b_4)e_k^4 + 2(5b_2b_4 + 2b_2^2b_3 + 3b_3^2 + 2b_5)e_k^5\right. \\ & \left.+ (4b_2b_3^2 + 6b_2^2b_4 + b_2^3b_3 + 13b_2b_5 + 17b_3b_4 + 5b_6)e_k^6 + O(e_k^7)\right). \end{aligned} \tag{20}$$

Also, from (14), (15), and (16), one has

$$\begin{aligned} L(x_k) = & -2f'^3(\eta)\left(e_k + 4b_2e_k^2 + 2(3b_2^2+2b_3)e_k^3 + (14b_2b_3+3b_2^3+3b_4)e_k^4\right. \\ & +(14b_2b_4+11b_2^2b_3+9b_3^2+b_5)e_k^5 \\ & \left. + 2(7b_2b_3^2+6b_2^2b_4+6b_2b_5+10b_3b_4+b_6)e_k^6+O(e_k^7)\right). \end{aligned} \tag{21}$$

Therefore, combining (20) and (21), one can have

$$\begin{aligned} \frac{2f(z_k)f'^2(x_k)}{L(x_k)} = & -b_2e_k - 2(b_3-b_2^2)e_k^2 - (3b_2^3+3b_4-5b_2b_3)e_k^3 \\ & -(6b_2^2b_3+4b_5-5b_2b_4-3b_2^4-2b_3^2)e_k^4+O(e_k^5). \end{aligned} \tag{22}$$

Furthermore, from (17) and (22), we get

$$\frac{x_k-y_k}{1+2f(z_k)f'^2(x_k)L^{-1}(x_k)} = e_k - (b_2^3-2b_2b_3)e_k^4 - (12b_2^2b_3-5b_2b_4-4b_2^4-4b_3^2)e_k^5+O(e_k^6). \tag{23}$$

So, substituting (23) into (10) in Method 4, one obtains

$$x_{k+1} = \eta + (b_2^3-2b_2b_3)e_k^4+O(e_k^5). \tag{24}$$

Noticing that the $(k+1)$-st error $e_{k+1}=x_{k+1}-\eta$, from (24) we have the following error equation

$$e_{k+1} = (b_2^3-2b_2b_3)e_k^4+O(e_k^5), \tag{25}$$

which shows that Method 4 defined by (10) is at least fourth-order convergent according to Definition 4. We have shown Theorem 2. □

Theorem 3. *Suppose that $f(x)$ is a function whose n-th derivative $f^{(n)}(x), n = 1,2,3,\ldots$, exists in a neighborhood of its root η with $f'(\eta)\neq 0$. If the initial approximation x_0 is sufficiently close to η, then the Method 5 defined by (12) is also at least fourth-order convergent with the following error equation*

$$e_{k+1} = (b_2^3-b_2b_3)e_k^4+O(e_k^5),$$

where $e_k = x_k-\eta, k=1,2,3,\ldots$, and the constants $b_n = \frac{a_n}{f'(\eta)}, a_n = \frac{f^{(n)}(\eta)}{n!}, n=1,2,3,\ldots$.

Proof of Theorem 3. Referring to (14) and (19) in the proof of Theorem 2, then, dividing $f(z_k)$ by $f(x_k)-f(z_k)$ we see that

$$\begin{aligned} \frac{f(z_k)}{f(x_k)-f(z_k)} = & \; b_2e_k - 2(b_2^2-b_3)e_k^2 - 3(2b_2b_3-b_2^3-b_4)e_k^3 - (3b_2^4+4b_3^2+8b_2b_4-11b_2^2b_3-4b_5)e_k^4 \\ & -(10b_2^3b_3+10b_3b_4+10b_2b_5-11b_2b_3^2-14b_2^2b_4-5b_6)e_k^5 \\ & -(221b_2^4b_3+16b_3^3+78b_2b_3b_4+27b_2^2b_5-73b_2^6 \\ & -158b_2^2b_3^2-91b_2^3b_4-6b_4^2-10b_3b_5-4b_2b_6)e_k^6+O(e_k^7). \end{aligned} \tag{26}$$

From (26), we obtain

$$\begin{aligned} \frac{f(x_k)-f(z_k)}{f(x_k)-2f(z_k)} = & \frac{1}{1-\frac{f(z_k)}{f(x_k)-f(z_k)}} = 1+b_2e_k+(2b_3-b_2^2)e_k^2+(3b_4-2b_2b_3)e_k^3 \\ & +(2b_2^4+4b_5-3b_2^2b_3-2b_2b_4)e_k^4+(14b_2^3b_3+2b_3b_4+5b_6-5b_2^5 \\ & -9b_2b_3^2-5b_2^2b_4-2b_2b_5)e_k^5+(77b_2^6+192b_2^2b_3^2+121b_2^3b_4+15b_4^2 \\ & +26b_3b_5+14b_2b_6-240b_2^4b_3-24b_3^3-130b_2b_3b_4-51b_2^2b_5)e_k^6+O(e_k^7). \end{aligned} \tag{27}$$

Multiplying (27) by (17) yields that

$$\frac{f(x_k)-f(y_k)}{f(x_k)-2f(y_k)}(x_k-z_k)=e_k-(b_2^3-b_2b_3)e_k^4+2(2b_2^4-4b_2^2b_3+b_3^2+b_2b_4)e_k^5+O(e_k^6). \tag{28}$$

Consequently, from (12) and the above Equation (28), and noticing the error $e_{k+1}=x_{k+1}-\eta$, we get the error equation as below:

$$e_{k+1}=(b_2^3-b_2b_3)e_k^4+O(e_k^5). \tag{29}$$

Thus, according to Definition 4, we have shown that Method 5 defined by (12) has fourth-order convergence at least. This completes the proof of Theorem 3. □

Remark 6. *Per iteration, Method 5 requires two evaluations of the function and one of its first-order derivative. If we consider the definition of efficiency index [3] as* $\sqrt[\tau]{\lambda}$, *where* λ *is the order of convergence of the method and* τ *is the total number of new function evaluations (i.e., the values of* f *and its derivatives) per iteration, then Method 5 has the efficiency index equal to* $\sqrt[3]{4}\approx 1.5874$, *which is better than the ones of Halley iterative method* $\sqrt[3]{3}\approx 1.4423$ *and Newton iterative method* $\sqrt{2}\approx 1.4142$.

5. Numerical Results

In this section, we present the results of numerical calculations to compare the efficiency of the proposed iterative methods (Methods 3–5) with Newton iterative method (Method 1, NIM for short), Halley iterative method (Method 2, HIM for short) and a few classical variants defined in literatures [19–22], such as the next iterative schemes with fourth-order convergence:

(i) Kou iterative method (KIM for short) [19].

$$x_{k+1}=x_k-\frac{2}{1+\sqrt{1-2\bar{L}_f(x_k)}}\frac{f(x_k)}{f'(x_k)},\qquad k=0,1,2,\ldots,$$

where $\bar{L}_f(x_k)$ is defined by the equation as follows:

$$\bar{L}_f(x_k)=\frac{f''\left(x_k-f(x_k)/\left(3f'(x_k)\right)\right)f(x_k)}{f'^2(x_k)}.$$

(ii) Double-Newton iterative method (DNIM for short) [20].

$$\begin{cases} z_k=x_k-\frac{f(x_k)}{f'(x_k)}, \\ x_{k+1}=x_k-\frac{f(x_k)}{f'(x_k)}-\frac{f(z_k)}{f'(z_k)}, \end{cases}\qquad k=0,1,2,\ldots.$$

(iii) Chun iterative method (CIM for short) [21].

$$\begin{cases} z_k=x_k-\frac{f(x_k)}{f'(x_k)}, \\ x_{k+1}=x_k-\frac{f(x_k)}{f'(x_k)}-\left(1+2\frac{f(z_k)}{f(x_k)}+\frac{f^2(z_k)}{f^2(x_k)}\right)\frac{f(z_k)}{f'(x_k)}, \end{cases}\qquad k=0,1,2,\ldots.$$

(iv) Jarratt-type iterative method (JIM for short) [22].

$$\begin{cases} z_k=x_k-\frac{2}{3}\frac{f(x_k)}{f'(x_k)}, \\ x_{k+1}=x_k-\frac{4f(x_k)}{f'(x_k)+3f'(z_k)}\left(1+\frac{9}{16}\left(\frac{f'(z_k)}{f'(x_k)}-1\right)^2\right), \end{cases}\qquad k=0,1,2,\ldots.$$

In iterative process, we use the following stopping criteria for computer programs:

$$|x_{k+1} - x_k| < \varepsilon \qquad \textit{and} \qquad |f(x_{k+1})| < \varepsilon,$$

where the fixed tolerance ε is taken as 10^{-14}. When the stopping criteria are satisfied, x_{k+1} can be regarded as the exact root η of the equation. Numerical experiments are performed in Mathematica 10 environment with 64 digit floating point arithmetics (Digits: =64). Different test equations $f_i = 0$, $i = 1, 2, \ldots, 5$, the initial guess value x_0, the number of iterations $k+1$, the approximate root x_{k+1}, the values of $|x_{k+1} - x_k|$ and $|f(x_{k+1})|$ are given in Table 1. The following test equations are used in the numerical results:

$$\begin{aligned}
f_1(x) &= x^3 - 11 = 0,\\
f_2(x) &= \cos x - x = 0,\\
f_3(x) &= x^3 + 4x^2 - 25 = 0,\\
f_4(x) &= x^2 - e^x - 3x + 2 = 0,\\
f_5(x) &= (x+2)e^x - 1 = 0.
\end{aligned}$$

Table 1. Numerical results and comparison of various iterative methods.

Methods	**Equation**	x_0	$k+1$	x_{k+1}	$\|x_{k+1} - x_k\|$	$\|f(x_{k+1})\|$
NIM	$f_1 = 0$	1.5	7	2.22398009056931552116536337672215719652	1.1×10^{-25}	4.1×10^{-47}
HIM	$f_1 = 0$	1.5	5	2.22398009056931552116536337672215719652	1.7×10^{-41}	1.0×10^{-46}
Method 3	$f_1 = 0$	1.5	4	2.22398009056931552116536337672215719652	8.3×10^{-40}	1.6×10^{-48}
Method 4	$f_1 = 0$	1.5	4	2.22398009056931552116536337672215719652	8.3×10^{-22}	1.9×10^{-47}
Method 5	$f_1 = 0$	1.5	4	2.22398009056931552116536337672215719652	7.5×10^{-30}	7.4×10^{-45}
KIM	$f_1 = 0$	1.5	4	2.22398009056931552116536337672215719652	8.5×10^{-38}	3.9×10^{-48}
DNIM	$f_1 = 0$	1.5	4	2.22398009056931552116536337672215719652	1.1×10^{-25}	1.1×10^{-47}
CIM	$f_1 = 0$	1.5	5	2.22398009056931552116536337672215719652	1.5×10^{-41}	6.6×10^{-45}
JIM	$f_1 = 0$	1.5	5	2.22398009056931552116536337672215719652	1.2×10^{-45}	4.3×10^{-47}
NIM	$f_2 = 0$	1	5	0.73908513321516064165531208767387340401	6.4×10^{-21}	1.5×10^{-41}
HIM	$f_2 = 0$	1	4	0.73908513321516064165531208767387340401	3.4×10^{-29}	5.1×10^{-49}
Method 3	$f_2 = 0$	1	3	0.73908513321516064165531208767387340401	8.2×10^{-19}	7.5×10^{-49}
Method 4	$f_2 = 0$	1	3	0.73908513321516064165531208767387340401	1.4×10^{-17}	9.4×10^{-48}
Method 5	$f_2 = 0$	1	3	0.73908513321516064165531208767387340401	1.1×10^{-18}	7.5×10^{-47}
KIM	$f_2 = 0$	1	3	0.73908513321516064165531208767387340401	1.5×10^{-20}	8.3×10^{-49}
DNIM	$f_2 = 0$	1	3	0.73908513321516064165531208767387340401	6.4×10^{-21}	9.5×10^{-48}
CIM	$f_2 = 0$	1	3	0.73908513321516064165531208767387340401	2.2×10^{-17}	9.4×10^{-48}
JIM	$f_2 = 0$	1	3	0.73908513321516064165531208767387340401	7.4×10^{-18}	8.3×10^{-49}
NIM	$f_3 = 0$	3.5	7	2.03526848118195915354755041547361249916	6.4×10^{-28}	2.9×10^{-47}
HIM	$f_3 = 0$	3.5	5	2.03526848118195915354755041547361249916	2.0×10^{-39}	5.8×10^{-47}
Method 3	$f_3 = 0$	3.5	4	2.03526848118195915354755041547361249916	2.0×10^{-33}	6.0×10^{-47}
Method 4	$f_3 = 0$	3.5	4	2.03526848118195915354755041547361249916	2.0×10^{-33}	8.0×10^{-46}
Method 5	$f_3 = 0$	3.5	4	2.03526848118195915354755041547361249916	3.4×10^{-30}	3.1×10^{-45}
KIM	$f_3 = 0$	3.5	4	2.03526848118195915354755041547361249916	4.3×10^{-33}	8.6×10^{-47}
DNIM	$f_3 = 0$	3.5	4	2.03526848118195915354755041547361249916	6.4×10^{-28}	9.9×10^{-46}
CIM	$f_3 = 0$	3.5	4	2.03526848118195915354755041547361249916	1.1×10^{-20}	9.6×10^{-46}
JIM	$f_3 = 0$	3.5	4	2.03526848118195915354755041547361249916	1.9×10^{-22}	1.1×10^{-49}

Table 1. *Cont.*

Methods	Equation	x_0	$k+1$	x_{k+1}	$\lvert x_{k+1}-x_k \rvert$	$\lvert f(x_{k+1}) \rvert$
NIM	$f_4=0$	3.6	8	0.25753028543986076045536730493724178138	6.5×10^{-29}	3.5×10^{-46}
HIM	$f_4=0$	3.6	6	0.25753028543986076045536730493724178138	4.8×10^{-37}	1.6×10^{-46}
Method 3	$f_4=0$	3.6	4	0.25753028543986076045536730493724178138	9.6×10^{-14}	2.7×10^{-46}
Method 4	$f_4=0$	3.6	5	0.25753028543986076045536730493724178138	1.1×10^{-36}	3.3×10^{-44}
Method 5	$f_4=0$	3.6	4	0.25753028543986076045536730493724178138	2.5×10^{-19}	4.9×10^{-44}
KIM	$f_4=0$	3.6	5	0.25753028543986076045536730493724178138	2.1×10^{-14}	8.9×10^{-44}
DNIM	$f_4=0$	3.6	4	0.25753028543986076045536730493724178138	2.6×10^{-14}	3.5×10^{-46}
CIM	$f_4=0$	3.6	4	0.25753028543986076045536730493724178138	2.8×10^{-12}	2.8×10^{-46}
JIM	$f_4=0$	3.6	5	0.25753028543986076045536730493724178138	9.7×10^{-38}	9.7×10^{-46}
NIM	$f_5=0$	3.5	11	−0.44285440100238858314132799999933681972	8.2×10^{-22}	7.7×10^{-43}
HIM	$f_5=0$	3.5	7	−0.44285440100238858314132799999933681972	2.2×10^{-37}	6.1×10^{-45}
Method 3	$f_5=0$	3.5	5	−0.44285440100238858314132799999933681972	1.8×10^{-24}	3.4×10^{-45}
Method 4	$f_5=0$	3.5	5	−0.44285440100238858314132799999933681972	5.3×10^{-37}	7.9×10^{-44}
Method 5	$f_5=0$	3.5	6	−0.44285440100238858314132799999933681972	2.0×10^{-42}	3.9×10^{-42}
KIM	$f_5=0$	3.5	7	−0.44285440100238858314132799999933681972	3.6×10^{-23}	2.7×10^{-42}
DNIM	$f_5=0$	3.5	6	−0.44285440100238858314132799999933681972	8.2×10^{-22}	4.9×10^{-45}
CIM	$f_5=0$	3.5	7	−0.44285440100238858314132799999933681972	3.3×10^{-37}	8.6×10^{-44}
JIM	$f_5=0$	3.5	6	−0.44285440100238858314132799999933681972	9.3×10^{-13}	6.6×10^{-46}

6. Conclusions

In Section 3 of the paper, it is evident that we have obtained a few single-step iterative methods including classical Newton's method and Halley's method, based on $[1,n]$-order Padé approximation of a function for finding a simple root of nonlinear equations. In order to avoid calculating the higher derivatives of the function, we have tried to improve the proposed iterative method by applying approximants of the second derivative and the third derivative. Hence, we have gotten a few modified two-step iterative methods free from the higher derivatives of the function. In Section 4, we have given theoretical proofs of the several methods. It is seen that any modified iterative method reaches the convergence order 4. However, it is worth mentioning that Method 5 is free from second order derivative and its efficiency index is 1.5874. Furthermore, in Section 5, numerical examples are employed to illustrate the practicability of the suggested variants for finding the approximate roots of some nonlinear scalar equations. The computational results presented in Table 1 show that in almost all of the cases the presented variants converge more rapidly than Newton iterative method and Halley iterative method, so that they can compete with Newton iterative method and Halley iterative method. Finally, for more nonlinear equations we tested, the presented variants have at least equal performance compared to the other existing iterative methods that are of the same order.

Author Contributions: The contributions of all of the authors have been similar. All of them have worked together to develop the present manuscript.

Acknowledgments: The authors are thankful to the anonymous reviewers for their valuable comments.

References

1. Alfio, Q.; Riccardo, S.; Fausto, S. Rootfinding for nonlinear equations. In *Numerical Mathematics*; Springer: New York, NY, USA, 2000; pp. 251–285, ISBN 0-387-98959-5.
2. Burden, A.M.; Faires, J.D.; Burden, R.L. Solutions of equations in one variable. In *Numerical Analysis*, 10th ed.; Cengage Learning: Boston, MA, USA, 2014; pp. 48–101, ISBN 978-1-305-25366-7.
3. Gautschi, W. Nonlinear equations. In *Numerical Analysis*, 2nd ed.; Birkhäuser: Boston, MA, USA, 2011; pp. 253–323, ISBN 978-0-817-68258-3.

4. Weerakoon, S.; Fernando, T.G.I. A variant of Newton's method with accelerated third-order convergence. *Appl. Math. Lett.* **2000**, *13*, 87–93. [CrossRef]
5. Cordero, A.; Torregrosa, I.R. Variants of Newton's method using fifth-order quadrature formulas. *Appl. Math. Comput.* **2007**, *190*, 686–698. [CrossRef]
6. Chun, C. Iterative methods improving Newton's method by the decomposition method. *Comput. Math. Appl.* **2005**, *50*, 1559–1568. [CrossRef]
7. Li, S.; Tan, J.; Xie, J.; Dong, Y. A new fourth-order convergent iterative method based on Thiele's continued fraction for solving equations. *J. Inform. Comput. Sci.* **2011**, *8*, 139–145.
8. Li, S.; Wang, R.; Zhang, X.; Xie, J.; Dong, Y. Halley's iterative formula based on Padé approximation and its modifications. *J. Inform. Comput. Sci.* **2012**, *9*, 997–1004.
9. Abbasbandy, S.; Tan, Y.; Liao, S.J. Newton-homotopy analysis method for nonlinear equations. *Appl. Math. Comput.* **2007**, *188*, 1794–1800. [CrossRef]
10. Noor, M.A.; Khan, W.A. New iterative methods for solving nonlinear equation by using homotopy perturbation method. *Appl. Math. Comput.* **2012**, *219*, 3565–3574. [CrossRef]
11. Wang, X.; Qin, Y.; Qian, W.; Zhang, S.; Fan, X. A family of Newton type iterative methods for solving nonlinear equations. *Algorithms* **2015**, *8*, 786–798. [CrossRef]
12. Sederberg, T.W.; Nishita, T. Curve intersection using Bézier clipping. *Comput.-Aided Des.* **1990**, *22*, 538–549. [CrossRef]
13. Bartoň, M.; Jüttler, B. Computing roots of polynomials by quadratic clipping. *Comput. Aided Geom. Des.* **2007**, *24*, 125–141. [CrossRef]
14. Morlando, F. A class of two-step Newton's methods with accelerated third-order convergence. *Gen. Math. Notes* **2015**, *29*, 17–26.
15. Thukral, R. New modification of Newton method with third-order convergence for solving nonlinear equations of type $f(0) = 0$. *Am. J. Comput. Appl. Math.* **2016**, *6*, 14–18. [CrossRef]
16. Rafiq, A.; Rafiullah, M. Some multi-step iterative methods for solving nonlinear equations. *Comput. Math. Appl.* **2009**, *58*, 1589–1597. [CrossRef]
17. Ali, F.; Aslam, W.; Ali, K.; Anwar, M.A.; Nadeem, A. New family of iterative methods for solving nonlinear models. *Disc. Dyn. Nat. Soc.* **2018**, 1–12. [CrossRef]
18. Qureshi, U.K. A new accelerated third-order two-step iterative method for solving nonlinear equations. *Math. Theory Model.* **2018**, *8*, 64–68.
19. Kou, J.S. Some variants of Cauchy's method with accelerated fourth-order convergence. *J. Comput. Appl. Math.* **2008**, *213*, 71–78. [CrossRef]
20. Traub, J.F. *Iterative Methods for the Solution of Equations*; Chelsea Publishing Company: New York, NY, USA, 1977; pp. 1–49.
21. Chun, C. Some fourth-order iterative methods for solving nonlinear equations. *Appl. Math. Comput.* **2008**, *195*, 454–459. [CrossRef]
22. Sharifi, M.; Babajee, D.K.R.; Soleymani, F. Finding the solution of nonlinear equations by a class of optimal methods. *Comput. Math. Appl.* **2012**, *63*, 764–774. [CrossRef]
23. Wuytack, L. Padé approximants and rational functions as tools for finding poles and zeros of analytical functions measured experimentally. In *Padé Approximation and its Applications*; Springer: New York, NY, USA, 1979; pp. 338–351, ISBN 0-387-09717-1.
24. Xu, X.Y.; Li, J.K.; Xu, G.L. Definitions of Padé approximation. In *Introduction to Padé Approximation*; Shanghai Science and Technology Press: Shanghai, China, 1990; pp. 1–8, ISBN 978-7-532-31887-2.

6

Ball Comparison for Some Efficient Fourth Order Iterative Methods Under Weak Conditions

Ioannis K. Argyros [1,†] and Ramandeep Behl [2,*,†]

1 Department of Mathematics Sciences, Cameron University, Lawton, OK 73505, USA; iargyros@cameron.edu
2 Department of Mathematics, Faculty of Science, King Abdulaziz University, Jeddah 21589, Saudi Arabia
* Correspondence: ramanbehl87@yahoo.in.
† These authors contributed equally to this work.

Abstract: We provide a ball comparison between some 4-order methods to solve nonlinear equations involving Banach space valued operators. We only use hypotheses on the first derivative, as compared to the earlier works where they considered conditions reaching up to 5-order derivative, although these derivatives do not appear in the methods. Hence, we expand the applicability of them. Numerical experiments are used to compare the radii of convergence of these methods.

Keywords: fourth order iterative methods; local convergence; banach space; radius of convergence

MSC: 65G99; 65H10; 47H17; 49M15

1. Introduction

Let $\mathbb{E}_1$, $\mathbb{E}_2$ be Banach spaces and $\mathbb{D} \subset \mathbb{E}_1$ be a nonempty and open set. Set $\mathbb{LB}(\mathbb{E}_1, \mathbb{E}_2) = \{M : \mathbb{E}_1 \to \mathbb{E}_2\}$, bounded and linear operators. A plethora of works from numerous disciplines can be phrased in the following way:

$$\lambda(x) = 0, \tag{1}$$

using mathematical modelling, where $\lambda : \mathbb{D} \to \mathbb{E}_2$ is a continuously differentiable operator in the Fréchet sense. Introducing better iterative methods for approximating a solution s_* of expression (1) is a very challenging and difficult task in general. Notice that this task is extremely important, since exact solutions of Equation (1) are available in some occasions.

We are motivated by four iterative methods given as

$$\begin{cases} y_j = x_j - \frac{2}{3}\lambda'(x_j)^{-1}\lambda(x_j) \\ x_{n+1} = x_j - \frac{1}{2}\Big[\Big(3\lambda'(y_j) - \lambda'(x_j)\Big)^{-1}\Big(3\lambda'(y_j) + \lambda'(x_j)\Big)\Big]\lambda'(x_j)^{-1}\lambda(x_j), \end{cases} \tag{2}$$

$$\begin{cases} y_j = x_j - \frac{2}{3}\lambda'(x_j)^{-1}\lambda(x_j) \\ x_{n+1} = x_j - \Big[-\frac{1}{2}I + \frac{9}{8}B_j + \frac{3}{8}A_j\Big]\lambda'(x_j)^{-1}\lambda(x_j), \end{cases} \tag{3}$$

$$\begin{cases} y_j = x_j - \frac{2}{3}\lambda'(x_j)^{-1}\lambda(x_j) \\ x_{n+1} = x_j - \Big[I + \frac{1}{4}(A_j - I) + \frac{3}{8}(A_j - I)^2\Big]\lambda'(y_j)^{-1}\lambda(x_j), \end{cases} \tag{4}$$

and

$$\begin{cases} y_j = x_j - H_j\lambda'(x_j)^{-1}\lambda(x_j) \\ x_{n+1} = z_j - \Big[3I - H_j\lambda'(x_j)^{-1}[x_j, z_j; \lambda]\Big]\lambda'(x_j)^{-1}\lambda(z_j), \end{cases} \tag{5}$$

where $H_j^0 = H^0(x_j)$, $x_0, y_0 \in \mathbb{D}$ are initial points, $H(x) = 2I + H^0(x)$, $H_j = H(X_j) \in \mathcal{LB}(\mathbb{E}_1, \mathbb{E}_1)$, $A_j = \lambda'(x_j)^{-1}\lambda'(y_j)$, $z_j = \frac{x_j+y_j}{2}$, $B_j = \lambda'(y_j)^{-1}\lambda'(x_j)$, and $[\cdot,\cdot;\lambda] : \mathbb{D}\times\mathbb{D} \to \mathcal{LB}(\mathbb{E}_1, \mathbb{E}_1)$ is a first order divided difference. These methods specialize to the corresponding ones (when $\mathbb{E}_1 = \mathbb{E}_2 = \mathbb{R}^i$, i is a natural number) studied by Nedzhibov [1], Hueso et al. [2], Junjua et al. [3], and Behl et al. [4], respectively. The 4-order convergence of them was established by Taylor series and conditions on the derivatives up to order five. Even though these derivatives of higher-order do not appear in the methods (2)–(5). Hence, the usage of methods (2)–(5) is very restricted. Let us start with a simple problem. Set $\mathbb{E}_1 = \mathbb{E}_2 = \mathbb{R}$ and $\mathbb{D} = [-\frac{5}{2}, \frac{3}{2}]$. We suggest a function $\lambda : \mathbb{A} \to \mathbb{R}$ as

$$\lambda(t) = \begin{cases} 0, & t = 0 \\ t^5 + t^3 \ln t^2 - t^4, & t \neq 0 \end{cases}.$$

Then, $s_* = 1$ is a zero of the above function and we have

$$\lambda'(t) = 5t^4 + 3t^2 \ln t^2 - 4t^3 + 2t^2,$$

$$\lambda''(t) = 20t^3 + 6t \ln t^2 - 12t^2 + 10t,$$

and

$$\lambda'''(t) = 60t^2 + 6 \ln t^2 - 24t + 22.$$

Then, the third-order derivative of function $\lambda'''(x)$ is not bounded on $\mathbb{D}$. The methods (2)–(5) cannot be applicable to such problems or their special cases that require the hypotheses on the third or higher-order derivatives of λ. Moreover, these works do not give a radius of convergence, estimations on $\|x_j - s_*\|$, or knowledge about the location of s_*. The novelty of our work is that we provide this information, but requiring only the derivative of order one, for these methods. This expands the scope of utilization of them and similar methods. It is vital to note that the local convergence results are very fruitful, since they give insight into the difficult operational task for choosing the starting points/guesses.

Otherwise with the earlier approaches: (i) We use the Taylor series and high order derivative, (ii) we do not have any clue for the choice of the starting point x_0, (iii) we have no estimate in advance about the number of iterations needed to obtain a predetermined accuracy, and (iv) we have no knowledge of the uniqueness of the solution.

The work lays out as follows: We give the convergence of these iterative schemes (2)–(5) with some main theorems in Section 2. Some numerical problems are discussed in the Section 3. The final conclusions are summarized in Section 4.

2. Local Convergence Analysis

Let us consider that $I = [0, \infty)$ and $\varphi_0 : I \to I$ be a non-decreasing and continuous function with $\varphi_0(0) = 0$.

Assume that the following equation

$$\varphi_0(t) = 1 \tag{6}$$

has a minimal positive solution ρ_0. Let $I_0 = [0, \rho_0)$. Let $\varphi : I_0 \to I$ and $\varphi_1 : I_0 \to I$ be continuous and non-decreasing functions with $\varphi(0) = 0$. We consider functions on the interval I_0 as

$$\psi_1(t) = \frac{\int_0^1 \varphi((1-\tau)t)d\tau + \frac{1}{3}\int_0^1 \varphi_1(\tau t)d\tau}{1 - \varphi_0(t)}$$

and

$$\bar{\psi}_1(t) = \psi_1(t) - 1.$$

Suppose that

$$\varphi_0(t) < 3. \tag{7}$$

Then, by (7), $\bar{\psi}_1(0) < 0$ and $\bar{\psi}_1(t) \to \infty$, as $t \to \rho_0^-$. On the basis of the classical intermediate value theorem, the function $\bar{\psi}_1(t)$ has a minimal solution R_1 in $(0, \rho_0)$. In addition, we assume

$$q(t) = 1 \tag{8}$$

has a minimal positive solution ρ_q, where

$$q(t) = \frac{1}{2}\Big(3\varphi_0(\psi_1(t)t) + \varphi_0(t)\Big).$$

Set $\rho = \min\{\rho_0, \rho_q\}$.

Moreover, we consider two functions ψ_2 and $\bar{\psi}_2$ on $I_1 = [0, \rho)$ as

$$\psi_2(t) = \frac{\int_0^1 \varphi((1-\tau)t)d\tau}{1-\varphi_0(t)} + \frac{3}{2}\frac{\Big(\varphi_0(\psi_1(t)t) + \varphi_0(t)\Big)\int_0^1 \varphi_1(\tau t)d\tau}{(1-q(t))(1-\varphi_0(t))}$$

and

$$\bar{\psi}_2(t) = \psi_2(t) - 1.$$

Then, $\bar{\psi}_2(0) = -1$, and $\bar{\psi}_2(t) \to \infty$, with $t \to \rho^-$. We recall R_2 as the minimal solution of $\bar{\psi}_2(t) = 0$. Set

$$R = \min\{R_1, R_2\}. \tag{9}$$

It follows from (9) that for every $t \in [0, R)$

$$0 \le \varphi_0(t) < 1, \tag{10}$$

$$0 \le \psi_1(t) < 1, \tag{11}$$

$$0 \le q(t) < 1 \tag{12}$$

and

$$0 \le \psi_2(t) < 1 \tag{13}$$

Define by $S(s_*, r) = \{y \in \mathbb{E}_1 : \|s_* - y\| < r, \}$ and denote by $\bar{S}(s_*, r)$ the closure of $S(s_*, r)$.

The local convergence of method (2) uses the conditions (A):

(a_1) $\lambda : \mathbb{D} \to \mathbb{E}_2$ is a continuously differentiable operator in the Fréchet sense, and there exists $s_* \in \mathbb{D}$.

(a_2) There exists a function $\varphi_0 : I \to I$ non-decreasing and continuous with $\varphi_0(0) = 0$ for all $x \in \mathbb{D}$

$$\Big\|\lambda'(s_*)^{-1}\Big(\lambda'(x) - \lambda'(s_*)\Big)\Big\| \le \varphi_0(\|x - s_*\|).$$

Set $\mathbb{D}_0 = \mathbb{D} \cap S(s_*, \rho_0)$, where ρ_0 is given in (6).

(a_3) There exist functions $\varphi : I_0 \to I$, $\varphi_1 : I_0 \to I$ non-decreasing and continuous with $\varphi(0) = 0$ so that for all $x, y \in \mathbb{D}_0$

$$\Big\|\lambda'(s_*)^{-1}\Big(\lambda'(y) - \lambda'(x)\Big)\Big\| \le \varphi(\|y - x\|)$$

and

$$\Big\|\lambda'(s_*)^{-1}\lambda'(x)\Big\| \le \varphi_1(\|y - x\|)$$

(a_4) $S(s_*, R) \subset \mathbb{D}$, radii ρ_0, ρ_q as given, respectively by (6), (8) exist; the condition (7) holds, where R is defined in (9).

(a_5)

$$\int_0^1 \varphi_0(\tau R^*)d\tau < 1, \text{ for some } R^* \geq R.$$

Set $\mathbb{D}_1 = D \cap S(s_*, R^*)$.

We can now proceed with the local convergence study of Equation (2) adopting the preceding notations and the conditions (A).

Theorem 1. *Under the conditions (A) sequence $\{x_j\}$ starting at $x_0 \in S(s_*, R) - \{s_*\}$ converges to s_*, $\{x_j\} \subset S(x, R)$ so that*

$$\|y_j - s_*\| \leq \psi_1(\|x_j - s_*\|)\|x_j - s_*\| \leq \|x_j - s_*\| < R \tag{14}$$

and

$$\|x_{n+1} - s_*\| \leq \psi_2(\|x_j - s_*\|)\|x_j - s_*\| \leq \|x_j - s_*\|, \tag{15}$$

with ψ_1 and ψ_2 functions considered previously and R is given in (9)*. Moreover, s_* is a unique solution in the set $\mathbb{D}_1$.*

Proof. We proof the estimates (14) and (15) by adopting mathematical induction. Therefore, we consider $x \in S(s_*, R) - \{s_*\}$. By (a_1), (a_2), (9), and (10), we have

$$\|\lambda'(s_*)^{-1}(\lambda'(s_*) - \lambda'(x))\| \leq \varphi_0(\|s_* - x_0\|) < \varphi_0(R) < 1, \tag{16}$$

hence $\lambda'(x)^{-1} \in \mathcal{LB}(\mathbb{E}_2, \mathbb{E}_1)$ and

$$\|\lambda'(x)^{-1}\lambda'(s_*)\| \leq \frac{1}{1 - \varphi_0(\|s_* - x_0\|)}. \tag{17}$$

The point y_0 is also exists by (17) for $n = 0$. Now, by using (a_1), we have

$$\lambda(x) = \lambda(x) - \lambda(s_*) = \int_0^1 \lambda'(s_* + \tau(x - s_*))d\tau(x - s_*). \tag{18}$$

From (a_3) and (18), we yield

$$\left\|\lambda'(s_*)^{-1}\lambda(x)\right\| \leq \int_0^1 \varphi_1(\tau\|x - s_*\|)d\tau\|x - s_*\|. \tag{19}$$

We can also write by method (2) for $n = 0$

$$y_0 - s_* = \left(x_0 - s_* - \lambda'(x_0)^{-1}\lambda(x_0)\right) + \frac{1}{3}\lambda'(x_0)^{-1}\lambda(x_0). \tag{20}$$

By expressions (9), (11), (17), (19), and (20), we obtain in turn that

$$\begin{aligned}
\|y_0 - s_*\| \leq{}& \left\|\lambda'(x_0)^{-1}\lambda'(s_*)\right\| \left\|\int_0^1 \lambda'(s_*)^{-1}\left(\lambda'(s_* + \tau(x_0 - s_*)) - \lambda'(x_0)\right)(x_0 - s_*)d\tau\right\| \\
&+ \frac{1}{3}\left\|\lambda'(x_0)^{-1}\lambda'(s_*)\right\| \left\|\lambda'(s_*)^{-1}\lambda(x_0)\right\| \\
\leq{}& \frac{\int_0^1 \varphi((1-\tau)\|x_0 - s_*\|)d\tau + \frac{1}{3}\int_0^1 \varphi(\tau\|x_0 - s_*\|)d\tau}{1 - \varphi_0(\|x_0 - s_*\|)} \\
={}& \psi_1(\|x_0 - s_*\|)\|x_0 - s_*\| \leq \|x_0 - s_*\| < R,
\end{aligned} \tag{21}$$

which confirms $y_0 \in S(s_*,\ R)$ and (14) for $n = 0$. We need to show that $\Big(3\lambda'(y_0) - 3\lambda'(x_0)\Big)^{-1} \in \mathcal{LB}(\mathbb{E}_2, \mathbb{E}_1)$.

In view of (a_2), (12), and (21), we have

$$\begin{aligned} &\left\| (2\lambda'(s_*))^{-1}\Big[3\lambda'(y_0) - \lambda'(x_0) - 3\lambda'(s_*) + \lambda'(s_*)\Big] \right\| \\ &\le \frac{1}{2}\Big[3\left\|\lambda'(s_*)^{-1}(\lambda'(y_0) - \lambda'(s_*))\right\| + \left\|\lambda'(s_*)^{-1}(\lambda'(x_0) - \lambda'(s_*))\right\|\Big] \\ &\le \frac{1}{2}\Big[\varphi_0(\|y_0 - s_*\|) + \varphi_0(\|x_0 - s_*\|)\Big] \\ &\le \frac{1}{2}\Big[\varphi_0(\psi_1(\|x_0 - s_*\|)\|x_0 - s_*\|) + \varphi_0(\|x_0 - s_*\|)\Big] \\ &= q(\|x_0 - s_*\|) < 1, \end{aligned} \tag{22}$$

so

$$\left\| \Big(3\lambda'(y_0) - \lambda'(x_0)\Big)^{-1}\lambda'(s_*) \right\| \le \frac{1}{1 - q(\|x_0 - s_*\|)}. \tag{23}$$

Using (9), (13), (17), (a_3), (21), (23), and the second substep of method (2) (since x_1 exists by (23)), we can first write

$$\begin{aligned} x_1 - s_* &= x_0 - s_* - \lambda'(x_0)^{-1}\lambda(x_0) \\ &\quad + \Big[I - \frac{1}{2}(3\lambda'(y_0) - \lambda'(x_0))^{-1}(3\lambda'(y_0) + \lambda'(x_0))\Big]\lambda'(x_0)^{-1}\lambda(x_0) \end{aligned} \tag{24}$$

so

$$\begin{aligned} \|x_1 - s_*\| &\le \|x_0 - s_* - \lambda'(x_0)^{-1}\lambda(x_0)\| + \frac{3}{2}\|(3\lambda'(y_0) - \lambda'(x_0))^{-1}\lambda'(s_*)\| \\ &\quad \times \Big[\|\lambda'(s_*)^{-1}(\lambda'(y_0) - \lambda'(x_0))\| + \|\lambda'(s_*)^{-1}(\lambda'(x_0) - \lambda'(s_*))^{-1}\|\Big]\|\lambda'(x_0)^{-1}\lambda(s_*)\|\|\lambda'(x_0)^{-1}\lambda(x_0)\| \\ &\le \left[\frac{\int_0^1 \varphi((1-\tau)t)d\tau}{1 - \varphi_0(t)} + \frac{3}{2}\frac{\Big(\varphi_0(\|y_0 - s_*\|) + \varphi_0(\|x_0 - s_*\|)\Big)\int_0^1 \varphi_1(\tau\|x_0 - s_*\|)d\tau}{(1 - q(\|x_0 - s_*\|))(1 - \varphi_0(\|x_0 - s_*\|))}\right]\|x_0 - s_*\| \\ &\le \psi_2(\|x_0 - s_*\|)\|x_0 - s_*\| \le \|x_0 - s_*\|. \end{aligned} \tag{25}$$

So, (15) holds and $x_1 \in S(s_*, R)$.

To obtain estimate (25), we also used the estimate

$$\begin{aligned} &I - \frac{1}{2}(3\lambda'(y_0) - \lambda'(x_0))^{-1}(3\lambda'(y_0) + \lambda'(x_0)) \\ &= \frac{1}{2}(3\lambda'(y_0) - \lambda'(x_0))^{-1}\Big[2(3\lambda'(y_0) - \lambda'(x_0)) - (3\lambda'(y_0) + \lambda'(x_0))\Big] \\ &= \frac{3}{2}(3\lambda'(y_0) - \lambda'(x_0))^{-1}\Big[(\lambda'(y_0) - \lambda'(s_*)) + (\lambda'(s_*) - \lambda'(x_0))\Big] \end{aligned} \tag{26}$$

The induction for (14) and (15) can be finished, if x_m, y_m, x_{m+1} replace $x_0, y_0,\ x_1$ in the preceding estimations. Then, from the estimate

$$\|x_{m+1} - s_*\| \le \mu\|x_m - s_*\| < R,\ \mu = \varphi_2(\|x_0 - s_*\|) \in [0,\ 1), \tag{27}$$

we arrive at $\lim_{m\to\infty} x_m = s_*$ and $x_{m+1} \in S(s_*,\ R)$. Let us consider that $K = \int_0^1 \lambda'(y_* + \tau(s_* - y_*))d\tau$ for $y^* \in \mathbb{D}_1$ with $K(y_*) = 0$. From (a_1) and (a_5), we obtain

$$\begin{aligned} &\|\lambda'(s_*)^{-1}(\lambda'(s_*) - K)\| \le \int_0^1 \varphi_0(\tau\|s_* - y_*\|)d\tau \\ &\le \int_0^1 \varphi_0(\tau R)d\tau < 1. \end{aligned} \tag{28}$$

So, $K^{-1} \in \mathcal{LB}(\mathbb{E}_1, \mathbb{E}_2)$, and $s_* = y_*$ by the identity

$$0 = K(s_*) - K(y_*) = K(s_* - y_*). \tag{29}$$

□

Proof. Next, we deal with method (3) in an analogous way. We shall use the same notation as previously. Let $\varphi_0, \varphi, \varphi_1, \rho_0, \psi_1, R_1$, and $\bar{\psi}_1$, be as previously.

We assume

$$\varphi_0(\psi_1(t)t) = 1 \tag{30}$$

has a minimal solution ρ_1. Set $\rho = \min\{\rho_0, \rho_1\}$. Define functions ψ_2 and $\bar{\psi}_2$ on interval $I_2 = [0, \rho)$ by

$$\psi_2(t) = \frac{\int_0^1 \varphi((1-\tau)t)d\tau}{1-\varphi_0(t)} + \left[2 + \frac{3\Big(\varphi_0(\psi_1(t)t) + \varphi_0(t)\Big)}{8(1-\varphi_0(t))} + \frac{9\Big(\varphi_0(\psi_1(t)t) + \varphi_0(t)\Big)}{8(1-\varphi_0(\psi_1(t)t))}\right] \frac{\int_0^1 \varphi_1(\tau t)d\tau}{1-\varphi_0(t)}$$

and

$$\bar{\psi}_2(t) = \psi_2(t) - 1.$$

Then, $\bar{\psi}_2(0) = -1$ and $\bar{\psi}_2(t) \to \infty$, with $t \to \rho^-$. R_2 is known as the minimal solution of equation $\bar{\psi}_2(t) = 0$ in $(0, \rho)$, and set

$$R = \min\{R_1, R_2\}. \tag{31}$$

Replace ρ_q by ρ_1 in the conditions (A) and call the resulting conditions $(A)'$.
Moreover, we use the estimate obtained for the second substep of method (3)

$$\begin{aligned} x_1 - s_* &= x_0 - s_* - \lambda'(x_0)^{-1}\lambda(x_0) + \Big[\frac{3}{2}I - \frac{9}{8}B_0 - \frac{9}{16}A_0\Big]\lambda'(x_0)^{-1}\lambda(x_0) \\ &= x_0 - s_* - \lambda'(x_0)^{-1}\lambda(x_0) + \Big[-2I + \frac{3}{8}(I - A_0) + \frac{9}{8}(I - B_0)\Big]\lambda'(x_0)^{-1}\lambda(x_0) \\ &= x_0 - s_* - \lambda'(x_0)^{-1}\lambda(x_0) + \Big[-2I + \frac{3}{8}\lambda'(x_0)^{-1}\Big(\lambda'(x_0) - \lambda'(y_0)\Big) \\ &\quad + \frac{9}{8}\lambda'(y_0)^{-1}\Big(\lambda'(y_0) - \lambda'(x_0)\Big)\Big]\lambda'(x_0)^{-1}\lambda(x_0). \end{aligned} \tag{32}$$

Then, by replacing (24) by (32) in the proof of Theorem 1, we have instead of (25)

$$\begin{aligned} \|x_1 - s_*\| &= \left[\frac{\int_0^1 \varphi((1-\tau)\|s_* - x_0\|)d\tau}{1 - \varphi(\|s_* - x_0\|)} + \left\{2 + \frac{3\Big(\varphi(\|y_0 - s_*\|) + \varphi_0(\|s_* - x_0\|)\Big)}{8(1 - \varphi_0(\|s_* - x_0\|))}\right.\right. \\ &\quad \left.\left. + \frac{9\Big(\varphi_0(\|y_0 - s_*\|) + \varphi_0(\|s_* - x_0\|)\Big)}{8(1 - \varphi_0(\|y_0 - s_*\|))}\right\} \frac{\int_0^1 \varphi_1(\|s_* - x_0\|)d\tau}{1 - \varphi_0(\|s_* - x_0\|)}\right] \|s_* - x_0\| \\ &\le \psi_2(\|s_* - x_0\|)\|s_* - x_0\| \le \|s_* - x_0\|. \end{aligned} \tag{33}$$

The rest follows as in Theorem 1. □

Hence, we arrived at the next Theorem.

Theorem 2. *Under the conditions $(A)'$, the conclusions of Theorem 1 hold for method* (3).

Proof. Next, we deal with method (4) in the similar way. Let $\varphi_0, \varphi, \varphi_1, \rho_0, \rho_1, \rho, \psi_1, R_1$, and $\bar{\psi}_1$, be as in the case of method (3). We consider functions ψ_2 and $\bar{\psi}_2$ on I_1 as

$$\psi_2(t) = \frac{\int_0^1 \varphi((1-\tau)t)d\tau}{1-\varphi_0(t)} + \frac{\varphi_0(\psi_1(t)t) + \varphi_0(t)}{(1-\varphi_0(t))(1-\varphi_0(\psi_1(t)t))} + \frac{1}{4}\frac{(\varphi_0(\psi_1(t)t) + \varphi_0(t))}{(1-\varphi_0(t))}$$
$$+ \frac{3}{8}\left(\frac{\varphi_0(\psi_1(t)t) + \varphi_0(t)}{1-\varphi_0(t)}\right)^2$$

and

$$\bar{\psi}_2(t) = \psi_2(t) - 1.$$

The minimal zero of $\bar{\psi}_2(t) = 0$ is denoted by R_2 in $(0, \rho)$, and set

$$R = \min\{R_1, R_2\}. \tag{34}$$

Notice again that from the second substep of method (4), we have

$$\begin{aligned} x_1 - s_* &= x_0 - s_* - \lambda'(x_0)^{-1}\lambda(x_0) + \Big[\lambda'(x_0)^{-1} - \lambda'(y_0)^{-1} - \frac{1}{4}(A_0 - I) - \frac{3}{8}(I - A_0)^2\Big]\lambda(x_0) \\ &= x_0 - s_* - \lambda'(x_0)^{-1}\lambda(x_0) + \Big\{\lambda'(x_0)^{-1}\Big[(\lambda'(y_0) - \lambda'(s_*)) + (\lambda'(s_*) - \lambda'(x_0))\Big] \\ &- \frac{1}{4}\lambda'(x_0)^{-1}\Big[(\lambda'(y_0) - \lambda'(s_*)) + (\lambda'(s_*) - \lambda'(x_0))\Big] \\ &- \frac{3}{8}\lambda'(x_0)^{-1}\Big[(\lambda'(y_0) - \lambda'(s_*)) + (\lambda'(s_*) - \lambda'(x_0))\Big]^2\Big\}\lambda(x_0), \end{aligned} \tag{35}$$

so

$$\begin{aligned} \|x_1 - s_*\| &\le \Bigg[\frac{\int_0^1 \varphi((1-\tau)\|s_* - x_0\|)d\tau}{1 - \varphi(\|s_* - x_0\|)} + \frac{\varphi_0(\psi(\|s_* - x_0\|)\|s_* - x_0\|) + \varphi_0(\|s_* - x_0\|)}{(1 - \varphi_0(\|s_* - x_0\|))(1 - \varphi_0(\psi(\|s_* - x_0\|)\|s_* - x_0\|))} \\ &+ \frac{1}{4}\frac{(\varphi(\psi_1(\|s_* - x_0\|)\|s_* - x_0\|) + \varphi_0(\|s_* - x_0\|))}{(1 - \varphi_0(\|s_* - x_0\|))} \\ &+ \frac{3}{8}\left(\frac{\varphi(\psi_1(\|s_* - x_0\|)\|s_* - x_0\|) + \varphi_0(\|s_* - x_0\|)}{(1 - \varphi_0(\|s_* - x_0\|))}\right)^2\Bigg]\|s_* - x_0\| \\ &\le \psi_2(\|s_* - x_0\|)\|s_* - x_0\| \le \|s_* - x_0\|. \end{aligned} \tag{36}$$

The rest follows as in Theorem 1. □

Hence, we arrived at the next following Theorem.

Theorem 3. *Under the conditions* $(A)'$, *conclusions of Theorem 1 hold for scheme* (4).

Proof. Finally, we deal with method (5). Let $\varphi_0, \varphi, \varphi_1, \rho_0, I_0$ be as in method (2). Let also $\varphi_2 : I_0 \to I$, $\varphi_3 : I_0 \to I$, $\varphi_4 : I_0 \to I$ and $\varphi_5 : I_0 \times I_0 \to I$ be continuous and increasing functions with $\varphi_3(0) = 0$. We consider functions ψ_1 and $\bar{\psi}_1$ on I_0 as

$$\psi_1(t) = \frac{\int_0^1 \varphi((1-\tau)t)d\tau + \varphi_2(t)\int_0^1 \varphi_1(\tau t)d\tau}{1 - \varphi_0(t)}$$

and

$$\bar{\psi}_1(t) = \psi_1(t) - 1.$$

Suppose that

$$\varphi_1(0)\varphi_2(0) < 1. \tag{37}$$

Then, by (6) and (37), we yield $\bar{\psi}_1(0) < 0$ and $\bar{\psi}_1(t) \to \infty$ with $t \to \rho_0^-$. R_1 is known as the minimal zero of $\bar{\psi}_1(t) = 0$ in $(0, \rho_0)$. We assume

$$\varphi_0(g(t)t) = 1, \tag{38}$$

where $g(t) = \frac{1}{2}(1 + \psi_1(t))$, has a minimal positive solution ρ_1. Set $I_1 = [0, \rho)$, where $\rho = \min\{\rho_0, \rho_1\}$. We suggest functions ψ_2 and $\bar{\psi}_2$ on I_1 as

$$\psi_2(t) = \left[\frac{\int_0^1 \varphi((1-\tau)g(t)t)d\tau}{1 - \varphi_0(g(t)t)} + \frac{(\varphi_0(g(t)t) + \varphi_0(t)) \int_0^1 \varphi_1(\tau g(t)t)d\tau}{(1 - \varphi_0(g(t)))(1 - \varphi_0(t))} \right.$$
$$\left. + 2\frac{\varphi_3\left(\frac{t}{2}(1 + \psi_1(t))\right) \int_0^1 \varphi_1(\tau g(t)t)d\tau}{(1 - \varphi_0(t))^2} + \frac{\varphi_4(t)\varphi_5(t, \psi_1(t)t) \int_0^1 \varphi_1(\tau g(t)t)d\tau}{(1 - \varphi_0(t))^2} \right] g(t)$$

and

$$\bar{\psi}_2(t) = \psi_2(t) - 1.$$

Suppose that

$$\Big(2\varphi_3(0) + \varphi_4(0)\varphi_5(0,0)\Big)\varphi_1(0) < 1. \tag{39}$$

By (39) and the definition of I_1, we have $\bar{\psi}_2(0) < 0$, $\bar{\psi}_2(t) \to \infty$ with $t \to \rho^-$. We assume R_2 as the minimal solution of $\bar{\psi}_2(t) = 0$. Set

$$R = \min\{R_1, R_2\}. \tag{40}$$

The study of local convergence of scheme (5) is depend on the conditions (C):

(c_1) = (a_1).
(c_2) = (a_2).
(c_3) There exist functions $\varphi : I_1 \to I$, $\varphi_1 : I_0 \to I$, $\varphi_2 : I_0 \to I$, $\varphi_3 : I_0 \to I$, $\varphi_4 : I_0 \to I$, and $\varphi_5 : I_0 \times I_0 \to I$, increasing and continuous functions with $\varphi(0) = \varphi_3(0) = 0$ so for all $x, y \in \mathbb{D}_0$

$$\begin{aligned}
&\|\lambda'(s_*)^{-1}(\lambda'(y) - \lambda'(x))\| \le \varphi(\|y - x\|),\\
&\|\lambda'(s_*)^{-1}\lambda'(x)\| \le \varphi_1(\|x - s_*\|),\\
&\|I - H(x)\| \le \varphi_2(\|x - s_*\|),\\
&\|\lambda'(s_*)^{-1}([x, y; \lambda] - \lambda'(x))\| \le \varphi_3(\|y - x\|),\\
&\|H^0(x)\| \le \varphi_4(\|x - s_*\|),
\end{aligned}$$

and

$$\|\lambda'(s_*)^{-1}[x, y; \lambda]\| \le \varphi_5(\|x - s_*\|, \|y - s_*\|),$$

(c_4) $S(s_*, R) \subseteq \mathbb{D}$, ρ_0, ρ_1 given, respectively by (6), (38) exist, (37) and (38) hold, and R is defined in (40).
(c_5) = (a_5).

Then, using the estimates

$$\begin{aligned} \|y_0 - s_*\| &= \|x_0 - s_* - \lambda'(x_0)^{-1}\lambda(x_0) + (I - H_0)\lambda'(x_0)^{-1}\lambda(x_0)\| \\ &\le \frac{\int_0^1 \varphi((1-\tau)\|x_0 - s_*\|)d\tau\|x_0 - s_*\|}{1 - \varphi_0(\|x_0 - s_*\|)} + \|I - H_0\|\|\lambda'(x_0)^{-1}\lambda'(s_*)\|\|\lambda'(s_*)^{-1}\lambda(x_0)\| \\ &\le \left[\frac{\int_0^1 \varphi((1-\tau)\|x_0 - s_*\|)d\tau + \varphi_2(\|x_0 - s_*\|)\int_0^1 \varphi_1(\tau\|x_0 - s_*\|)d\tau}{1 - \varphi_0(\|x_0 - s_*\|)}\right]\|x_0 - s_*\| \\ &\le \psi_1(\|x_0 - s_*\|)\|x_0 - s_*\| \le \|x_0 - s_*\| < R, \end{aligned} \tag{41}$$

and

$$\begin{aligned} \|x_1 - s_*\| &= \|z_0 - s_* - \lambda'(z_0)^{-1}\lambda(z_0) + \lambda'(z_0)^{-1}(\lambda'(x_0) - \lambda'(z_0))\lambda'(x_0)^{-1}\lambda(z_0)\| \\ &\quad + 2\lambda'(x_0)^{-1}([x_0, z_0; \lambda] - \lambda'(x_0))\lambda'(x_0)^{-1}\lambda(z_0) + H_j^0\lambda'(x_0)^{-1}[x_0, z_0; \lambda]\lambda'(x_0)^{-1}\lambda(z_0)\| \\ &\le \left[\frac{\int_0^1 \varphi((1-\tau)g(\|x_0 - s_*\|)\|x_0 - s_*\|)d\tau}{1 - \varphi_0(g(\|x_0 - s_*\|)\|x_0 - s_*\|)}\right. \\ &\quad + \frac{(\varphi_0(\|x_0 - s_*\|) + \varphi_0(g(\|x_0 - s_*\|)\|x_0 - s_*\|))\int_0^1 \varphi_1(\tau g(\|x_0 - s_*\|)\|x_0 - s_*\|)d\tau}{(1 - \varphi_0(g(\|x_0 - s_*\|)\|x_0 - s_*\|))(1 - \varphi_0(\|x_0 - s_*\|))} \\ &\quad + 2\frac{\varphi_3\left(\frac{(1+\psi_1(\|x_0 - s_*\|))\|x_0 - s_*\|}{2}\right)\int_0^1 \varphi_1(\tau g(\|x_0 - s_*\|)\|x_0 - s_*\|)d\tau}{(1 - \varphi_0(\|x_0 - s_*\|))^2} \\ &\quad \left.+ \frac{\varphi_4(\|x_0 - s_*\|)\varphi_5(\|x_0 - s_*\|, \|y_0 - s_*\|)\int_0^1 \varphi_1(\tau g(\|x_0 - s_*\|)\|x_0 - s_*\|)d\tau}{(1 - \varphi_0(\|x_0 - s_*\|))^2}\right]\|z_0 - s_*\| \\ &\le \psi_2(\|x_0 - s_*\|)\|x_0 - s_*\| \le \|x_0 - s_*\|. \end{aligned} \tag{42}$$

Here, recalling that $z_0 = \frac{x_0 + y_0}{2}$, we also used the estimates

$$\begin{aligned} \|z_0 - s_*\| &= \left\|\frac{x_0 + y_0}{2} - s_*\right\| \le \frac{1}{2}(\|x_0 - s_*\| + \|y_0 - s_*\|) \\ &\le \frac{1}{2}(1 + \psi_1(\|x_0 - s_*\|))\|x_0 - s_*\|, \end{aligned} \tag{43}$$

$$\begin{aligned} \alpha &= \lambda'(z_0)^{-1} - \lambda'(x_0)^{-1} = \lambda'(z_0)^{-1}\left[(\lambda'(x_0) - \lambda'(s_*)) + (\lambda'(s_*) - \lambda'(z_0))\right]\lambda'(x_0)^{-1}, \\ \beta &= (-2I + H_0\lambda'(x_0)^{-1}[x_0, z_0; \lambda])\lambda'(x_0)^{-1}, \end{aligned}$$

and

$$\begin{aligned} \gamma &= -2I + (2I + H_0^0)\lambda'(x_0)^{-1}[x_0, z_0; \lambda] \\ &= -2I + 2I\lambda'(x_0)^{-1}[x_0, z_0; \lambda] + 2H_0^0\lambda'(x_0)^{-1}[x_0, z_0; \lambda] \\ &= 2\lambda'(x_0)^{-1}([x_0, z_0; \lambda] - \lambda'(x_0)) + H_0^0\lambda'(x_0)^{-1}[x_0, z_0; \lambda] \end{aligned}$$

to obtain (41) and (42). □

Hence, we arrived at the next following Theorem.

Theorem 4. *Under the conditions* (C), *the conclusions of Theorem 1 hold for method* (5).

3. Numerical Applications

We test the theoretical results on many examples. In addition, we use five examples and out of them: The first one is a counter example where the earlier results are applicable; the next three are real life problems, e.g., a chemical engineering problem, an electron trajectory in the air gap among two parallel surfaces problem, and integral equation of Hammerstein problem, which are displayed in

Examples 1–5. The last one compares favorably (5) to the other three methods. Moreover, the solution to corresponding problem are also listed in the corresponding example which is correct up to 20 significant digits. However, the desired roots are available up to several number of significant digits (minimum one thousand), but due to the page restriction only 30 significant digits are displayed.

We compare the four methods namely (2)–(5), denoted by NM, HM, JM, and BM, respectively on the basis of radii of convergence ball and the approximated computational order of convergence $\rho = \frac{\log\left[\|x_{(j+1)}-x_{(j)}\|/\|x_{(j)}-x_{(j-1)}\|\right]}{\log\left[\|x_{(j)}-x_{(j-1)}\|/\|x_{(j-1)}-x_{(j-2)}\|\right]}, j = 2,3,4,\ldots$ (for the details please see Cordero and Torregrosa [5]) $(ACOC)$. We have included the radii of ball convergence in the following Tables 1–6 except, the Table 4 that belongs to the values of abscissas t_j and weights w_j. We use the *Mathematica* 9 programming package with multiple precision arithmetic for computing work.

We choose in all examples $H^0(x) = 0$ and $H(x) = 2I$, so $\varphi_2(t) = 1$ and $\varphi_4(t) = 0$. The divided difference is $[x, y; \lambda] = \int_0^1 \lambda'(y + \theta(x-y))d\theta$. In addition, we choose the following stopping criteria (i) $\|x_{j+1} - x_j\| < \epsilon$ and (ii) $\|\lambda(x_j)\| < \epsilon$, where $\epsilon = 10^{-250}$.

Example 1. *Set $\mathbb{X} = \mathbb{Y} = \mathbb{R}$. We suggest a function λ on $\mathbb{D} = [-\frac{1}{\pi}, \frac{2}{\pi}]$ as*

$$\lambda(x) = \begin{cases} 0, & x = 0 \\ x^5 \sin(1/x) + x^3 \log(\pi^2 x^2), & x \neq 0 \end{cases}.$$

But, $\lambda'''(x)$ is unbounded on Ω at $x = 0$. The solution of this problem is $s_* = \frac{1}{\pi}$. The results in Nedzhibov [1], Hueso et al. [2], Junjua et al. [3], and Behl et al. [4] cannot be utilized. In particular, conditions on the 5th derivative of λ or may be even higher are considered there to obtain the convergence of these methods. But, we need conditions on λ' according to our results. In additon, we can choose

$$H = \frac{80 + 16\pi + (\pi + 12\log 2)\pi^2}{2\pi + 1}, \quad \varphi_1(t) = 1 + Ht, \quad \varphi_0(t) = \varphi(t) = Ht,$$

$$\varphi_5(s,t) = \frac{1}{2}(\varphi_1(s) + \varphi_1(t)) \quad \text{and} \quad \varphi_3(t) = \frac{1}{2}\varphi_2(t).$$

The distinct radius of convergence, number of iterations n, and COC (ρ) are mentioned in Table 1.

Table 1. Comparison on the basis of different radius of convergence for Example 1.

Schemes	R_1	R_2	R	x_0	n	ρ
NM	0.011971	0.010253	0.010253	0.30831	4	4.0000
HM	0.011971	0.01329	0.011971	0.32321	4	4.0000
JM	0.011971	0.025483	0.011971	0.32521	4	4.0000
BM	0	0	0	-	-	-

Equation (39) is violated with these choices of φ_i. This is the reason that R is zero in the method BM. Therefore, our results hold only, if $x_0 = s_*$.

Example 2. *The function*

$$\lambda_2(x) = x^4 - 1.674 - 7.79075x^3 + 14.7445x^2 + 2.511x. \tag{44}$$

appears in the conversion to ammonia of hydrogen-nitrogen [6,7]. The function λ_2 has 4 zeros, but we choose $s_* \approx 3.9485424455620457727 + 0.3161235708970163733i$. Moreover, we have

$$\varphi_0(t) = \varphi(t) = 40.6469t, \quad \varphi_1(t) = 1 + 40.6469t, \quad \varphi_3(t) = \frac{1}{2}\varphi_2(t), \quad \text{and} \quad \varphi_5(s,t) = \frac{1}{2}(\varphi_1(s) + \varphi_1(t)).$$

The distinct radius of convergence, number of iterations n, and COC (ρ) are mentioned in Table 2.

Table 2. Comparison on the basis of different radius of convergence for Example 2.

Schemes	R_1	R_2	R	x_0	n	ρ
NM	0.0098841	0.0048774	0.0048774	$3.953 + 0.3197i$	4	4.0000
HM	0.0098841	0.016473	0.016473	$3.9524 + 0.32i$	4	4.0000
JM	0.0098841	0.0059094	0.0059094	$3.9436 + 0.3112i$	4	4.0000
BM	0	0	0	-	-	-

Equation (39) is violated with these choices of φ_i. This is the reason that R is zero in the method BM. Therefore, our results hold only, if $x_0 = s_*$.

Example 3. *An electron trajectory in the air gap among two parallel surfaces is formulated given as*

$$x(t) = x_0 + \left(v_0 + e\frac{E_0}{m\omega}\sin(\omega t_0 + \alpha)\right)(t - t_0) + e\frac{E_0}{m\omega^2}\big(\cos(\omega t + \alpha) + \sin(\omega + \alpha)\big), \tag{45}$$

where e, m, x_0, v_0, and $E_0 \sin(\omega t + \alpha)$ are the charge, the mass of the electron at rest, the position, velocity of the electron at time t_0, and the RF electric field among two surfaces, respectively. For particular values of these parameters, the following simpler expression is provided:

$$f_3(x) = x + \frac{\pi}{4} - \frac{1}{2}\cos(x). \tag{46}$$

The solution of function f_3 is $s_* \approx -0.30909327154179495274198680892 4$. Moreover, we have

$$\varphi(t) = \varphi_0(t) = 0.5869t,\ \ \varphi_1(t) = 1 + 0.5869t,\ \ \varphi_3(t) = \frac{1}{2}\varphi_2(t)\ \ \text{and}\ \ \varphi_5(s,t) = \frac{1}{2}(\varphi_1(s) + \varphi_1(t)).$$

The distinct radius of convergence, number of iterations n, and COC (ρ) are mentioned in Table 3.

Table 3. Comparison on the basis of different radius of convergence for Example 3.

Schemes	R_1	R_2	R	x_0	n	ρ
NM	0.678323	0.33473	0.33473	0.001	4	4.0000
HM	0.678323	1.13054	0.678323	-0.579	4	4.0000
JM	0.678323	0.40555	0.40555	0.091	5	4.0000
BM	0	7.60065×10^{-18}	0	-	-	-

Equation (39) is violated with these choices of φ_i. This is the reason that R is zero in the method BM. Therefore, our results hold only, if $x_0 = s_*$.

Example 4. *Considering mixed Hammerstein integral equation Ortega and Rheinbolt [8], as*

$$x(s) = 1 + \frac{1}{5}\int_0^1 U(s,t)x(t)^3 dt,\ \ x \in C[0,1],\ s,t \in [0,1], \tag{47}$$

where the kernel U is

$$U(s,t) = \begin{cases} s(1-t), s \le t, \\ (1-s)t, t \le s. \end{cases}$$

We phrase (47) by using the Gauss-Legendre quadrature formula with $\int_0^1 \phi(t)dt \simeq \sum_{k=1}^{10} w_k \phi(t_k)$, where t_k and w_k are the abscissas and weights respectively. Denoting the approximations of $x(t_i)$ with x_i $(i = 1,2,3,...,10)$, then we yield the following 8×8 system of nonlinear equations

$$5x_i - 5 - \sum_{k=1}^{10} a_{ik}x_k^3 = 0,\ i = 1,2,3...,10,$$

$$a_{ik} = \begin{cases} w_k t_k (1 - t_i), & k \le i, \\ w_k t_i (1 - t_k), & i < k. \end{cases}$$

The values of t_k and w_k can be easily obtained from Gauss-Legendre quadrature formula when $k = 8$ mentioned in Table 4.

Table 4. Values of abscissas t_j and weights w_j.

j	t_j	w_j
1	0.01304673574141413996101799	0.03333567215434406879678440
2	0.06746831665550774463395165	0.07472567457529029657288816
3	0.16029521585048779688283632	0.10954318125799102199776746
4	0.28330230293537640460036703	0.13463335965499817754561346
5	0.42556283050918439455758700	0.14776211235737643508694649
6	0.57443716949081560544241300	0.14776211235737643508694649
7	0.71669769706462359539963297	0.13463335965499817754561346
8	0.83970478414951220311716368	0.10954318125799102199776746
9	0.93253168334449225536604834	0.07472567457529029657288816
10	0.98695326425858586003898201	0.03333567215434406879678440

The required approximate root is $s_* \approx (1.001377, \ldots, 1.006756, \ldots, 1.014515, \ldots, 1.021982, \ldots, 1.026530, \ldots, 1.026530, \ldots, 1.021982, \ldots, 1.014515, \ldots, 1.006756, \ldots, 1.001377, \ldots)^T$. Moreover, we have

$$\varphi_0(t) = \varphi(t) = \frac{3}{20}t, \;\; \varphi_1(t) = 1 + \frac{3}{20}t, \; \varphi_3(t) = \frac{1}{2}\varphi_2(t) \;\; \text{and} \;\; \varphi_5(s,t) = \frac{1}{2}(\varphi_1(s) + \varphi_1(t)).$$

The distinct radius of convergence, number of iterations n, and COC (ρ) are mentioned in Table 5.

Table 5. Comparison on the basis of different radius of convergence for Example 4.

Schemes	R_1	R_2	R	x_0	n	ρ
NM	2.6667	1.3159	1.3159	(1,1,...,1)	4	4.0000
HM	2.6667	4.4444	2.6667	(1.9,1.9,...,1.9)	5	4.0000
JM	2.6667	1.5943	1.5943	(2.1,2.1,...,2.1)	5	4.0000
BM	0	0	0	-	-	-

Equation (39) is violated with these choices of φ_i. This is the reason that R is zero in the method BM. Therefore, our results hold only, if $x_0 = s_*$.

Example 5. *We consider a boundary value problem from [8], which is defined as follows:*

$$t'' = \frac{1}{2}t^3 + 3t' - \frac{3}{2-x} + \frac{1}{2}, \;\; t(0) = 0, \; t(1) = 1. \tag{48}$$

We assume the following partition on $[0,\ 1]$

$$x_0 = 0 < x_1 < x_2 < \cdots < x_j, \;\; \text{where} \;\; x_{j+1} = x_j + h, \; h = \frac{1}{j}.$$

We discretize this BVP (48) by

$$t'_i \approx \frac{t_{i+1} - t_{i-1}}{2h}, \; t''_i \approx \frac{t_{i-1} - 2t_i + t_{i+1}}{h^2}, \; i = 1,\ 2,\ \ldots,\ j-1.$$

Then, we obtain a $(k-1) \times (k-1)$ order nonlinear system, given by

$$t_{i+1} - 2t_i + t_{i-1} - \frac{h^2}{2}t_i^3 - \frac{3}{2-x_i}h^2 - 3\frac{t_{i+1} - t_{i-1}}{2}h - \frac{1}{h^2} = 0, \; i = 1, 2, \ldots, j-1,$$

where $t_0 = t(x_0) = 0,\ t_1 = t(x_1),\ \ldots,\ t_{j-1} = t(x_{j-1}),\ t_j = t(x_j) = 1$ *and initial approximation* $t_0^{(0)} = \left(\frac{1}{2}, \frac{1}{2}, \frac{1}{2}, \frac{1}{2}, \frac{1}{2}, \frac{1}{2}\right)^T$. *In particular, we choose* $k = 6$ *so that we can obtain a* 5×5 *nonlinear system. The required solution of this problem is*

$$\bar{x} \approx (0.09029825\ldots,\ 0.1987214\ldots,\ 0.3314239\ldots,\ 0.4977132\ldots,\ 0.7123306\ldots)^T.$$

The distinct radius of convergence, number of iterations n, and COC (ρ) are mentioned in Table 6.

Table 6. Convergence behavior of distinct fourth-order methods for Example 5.

Methods	j	$\|F(x^{(j)})\|$	$\|x^{(j+1)} - x^{(j)}\|$	ρ
	1	8.1 (−6)	2.0 (−4)	
MM	2	1.0 (−23)	3.1 (−23)	
	3	9.1 (−95)	2.4 (−94)	
	4	3.7 (−379)	9.0 (−379)	3.9996
	1	7.8 (−6)	1.9 (−5)	
HM	2	7.6 (−24)	2.4 (−23)	
	3	2.7 (−95)	7.2 (−95)	
	4	2.6 (−381)	6.3 (−381)	3.9997
	1	7.8 (−6)	1.9 (−5)	
JM	2	7.6 (−24)	2.4 (−23)	
	3	2.7 (−95)	7.2 (−95)	
	4	2.6 (−381)	6.3 (−381)	3.9997
	1	7.2 (−6)	1.7 (−5)	
BM	2	4.2 (−24)	1.3 (−23)	
	3	1.9 (−96)	5.2 (−96)	
	4	5.6 (−386)	1.4 (−385)	3.9997

4. Conclusions

The convergence order of iterative methods involves Taylor series, and the existence of high order derivatives. Consequently, upper error bounds on $\|x_j - s_*\|$ and uniqueness results are not reported with this technique. Hence, the applicability of these methods is limited to functions with high order derivatives. To address these problems, we present local convergence results based on the first derivative. Moreover, we compare methods (2)–(5). Notice that our convergence criteria are sufficient but not necessary. Therefore, if e.g., the radius of convergence for the method (5) is zero, that does not necessarily imply that the method does not converge for a particular numerical example. Our method can be adopted in order to expand the applicability of other methods in an analogous way.

Author Contributions: Both the authors have equal contribution for this paper.

References

1. Nedzhibov, G.H. A family of multi-point iterative methods for solving systems of nonlinear equations. *Comput. Appl. Math.* **2008**, *222*, 244–250. [CrossRef]
2. Hueso, J.L.; Martínez, E.; Teruel, C. Convergence, Efficiency and Dynamics of new fourth and sixth order families of iterative methods for nonlinear systems. *Comp. Appl. Math.* **2015**, *275*, 412–420. [CrossRef]
3. Junjua, M.; Akram, S.; Yasmin, N.; Zafar, F. A New Jarratt-Type Fourth-Order Method for Solving System of Nonlinear Equations and Applications. *Appl. Math.* **2015**, *2015*, 805278. [CrossRef]
4. Behl, R.; Cordero, A.; Torregrosa, J.R.; Alshomrani, A.S. New iterative methodsfor solving nonlinear problems with one and several unknowns. *Mathematics* **2018**, *6*, 296. [CrossRef]

5. Cordero, A.; Torregrosa, J.R. Variants of Newton's method using fifth-order quadrature formulas. *Appl. Math. Comput.* **2007**, *190*, 686–698. [CrossRef]
6. Balaji, G.V.; Seader, J.D. Application of interval Newton's method to chemical engineering problems. *Reliab. Comput.* **1995**, *1*, 215–223. [CrossRef]
7. Shacham, M. An improved memory method for the solution of a nonlinear equation. *Chem. Eng. Sci.* **1989**, *44*, 1495–1501. [CrossRef]
8. Ortega, J.M.; Rheinbolt, W.C. *Iterative Solutions of Nonlinears Equations in Several Variables*; Academic Press: Cambridge, MA, USA, 1970.

7

A New Class of Iterative Processes for Solving Nonlinear Systems by using One Divided Differences Operator

Alicia Cordero [1], Cristina Jordán [1], Esther Sanabria [2] and Juan R. Torregrosa [1,*]

1 Multidisciplinary Institute of Mathematics, Universitat Politènica de València, 46022 València, Spain
2 Department of Applied Mathematics, Universitat Politènica de València, 46022 València, Spain
* Correspondence: jrtorre@mat.upv.es

Abstract: In this manuscript, a new family of Jacobian-free iterative methods for solving nonlinear systems is presented. The fourth-order convergence for all the elements of the class is established, proving, in addition, that one element of this family has order five. The proposed methods have four steps and, in all of them, the same divided difference operator appears. Numerical problems, including systems of academic interest and the system resulting from the discretization of the boundary problem described by Fisher's equation, are shown to compare the performance of the proposed schemes with other known ones. The numerical tests are in concordance with the theoretical results.

Keywords: nonlinear systems; multipoint iterative methods; divided difference operator; order of convergence; Newton's method; computational efficiency index

---

1. Introduction

The design of iterative processes for solving scalar equations, $f(x) = 0$, or nonlinear systems, $F(x) = 0$, with n unknowns and equations, is an interesting challenge of numerical analysis. Many problems in Science and Engineering need the solution of a nonlinear equation or system in any step of the process. However, in general, both equations and nonlinear systems have no analytical solution, so we must resort to approximate the solution using iterative techniques. There are different ways to develop iterative schemes such as quadrature formulaes, Adomian polynomials, divided difference operator, weight function procedure, etc. have been used by many researchers for designing iterative schemes to solve nonlinear problems. For a good overview on the procedures and techniques as well as the different schemes developed in the last half century, one refers to some standard texts [1–5].

In this paper, , we want to design Jacobian-free iterative schemes for approximating the solution $\bar{x} = (x_1, x_2, \ldots, x_n)^T$ of a nonlinear system $F(x) = 0$, where $F : D \subseteq \mathbb{R}^n \rightarrow \mathbb{R}^n$ is a nonlinear multivariate function defined in a convex set D. The best known method for finding a solution $\bar{x} \in D$ is Newton's procedure,

$$x^{(k+1)} = x^{(k)} - [F'(x^{(k)})]^{-1}F(x^{(k)}),\ k = 0, 1, 2, \ldots,$$

$F'(x^{(k)})$ being the Jacobian of F evaluated in the kth iteration.

Based on Newton-type schemes and by using different techniques, several methods for approximating a solution of $F(x) = 0$ have been published recently. The main objective of all these processes is to speed the convergence or increase their computational efficiency. We are going to recall some of them that we will use in the last section for comparison purposes.

From a variant of Steffensen's method for systems introduced by Samanskii in [6] that replaces the Jacobian matrix $F'(x)$ by the divided difference operator defined as

$$[x, y; F](x-y) = F(x) - F(y),$$

being $x, y \in \mathbb{R}^n$, Wang and Fang in [7] designed a fourth-order scheme, denoted by WF4, whose iterative expression is

$$\begin{aligned} r^{(k)} &= x^{(k)} - [a^{(k)}, b^{(k)}; F]^{-1} F(x^{(k)}), \\ x^{(k+1)} &= r^{(k)} - \left(3I - 2[a^{(k)}, b^{(k)}; F]^{-1}[x^{(k)}, r^{(k)}; F]\right) [a^{(k)}, b^{(k)}; F]^{-1} F(r^{(k)}), \end{aligned} \tag{1}$$

where I is the identity matrix of size $n \times n$, $a^{(k)} = x^{(k)} + F(x^{(k)})$ and $b^{(k)} = x^{(k)} - F(x^{(k)})$. Let us observe that this method uses two functional evaluations and two divided difference operators per iteration. Let us remark that Samanskii in [6] defined also a third-order method with the same divided differences operator at the two steps.

Sharma and Arora in [8] added a new step in the previous method obtaining a sixth-order scheme, denoted by SA6, whose expression is

$$\begin{aligned} r^{(k)} &= x^{(k)} - [a^{(k)}, b^{(k)}; F]^{-1} F(x^{(k)}), \\ s^{(k)} &= r^{(k)} - \left(3I - 2[a^{(k)}, b^{(k)}; F]^{-1}[x^{(k)}, r^{(k)}; F]\right) [a^{(k)}, b^{(k)}; F]^{-1} F(r^{(k)}), \\ x^{(k+1)} &= s^{(k)} - \left(3I - 2[a^{(k)}, b^{(k)}; F]^{-1}[x^{(k)}, r^{(k)}; F]\right) [a^{(k)}, b^{(k)}; F]^{-1} F(s^{(k)}), \end{aligned} \tag{2}$$

where, as before, $a^{(k)} = x^{(k)} + F(x^{(k)})$ and $b^{(k)} = x^{(k)} - F(x^{(k)})$. In relation with WF4, a new functional evaluation, per iteration, is needed.

By replacing the third step of equation (2), Narang et al. in [9] proposed the following seventh-order scheme that uses two divided difference operators and three functional evaluations per iteration, which is denoted by NM7,

$$\begin{aligned} r^{(k)} &= x^{(k)} - Q^{-1} F(x^{(k)}), \\ s^{(k)} &= r^{(k)} - Q^{-1} F(r^{(k)}), \\ x^{(k+1)} &= s^{(k)} - \left(\frac{17}{4} I - Q^{-1} P \left(-\frac{27}{4} I + Q^{-1} P \left(\frac{19}{4} I - \frac{5}{4} Q^{-1} P\right)\right)\right) Q^{-1} F(s^{(k)}), \end{aligned} \tag{3}$$

where $Q = [a^{(k)}, b^{(k)}; F]$, being again $a^{(k)} = x^{(k)} + F(x^{(k)})$, $b^{(k)} = x^{(k)} - F(x^{(k)})$ and $P = [w^{(k)}, t^{(k)}; F]$, with $w^{(k)} = s^{(k)} + F(x^{(k)})$, $t^{(k)} = s^{(k)} - F(x^{(k)})$.

In a similar way, Wang et al. (see [10]) designed a scheme of order 7 that we denote by S7, modifying only the third step of expression (3). Its iterative expression is

$$\begin{aligned} r^{(k)} &= x^{(k)} - Q^{-1} F(x^{(k)}), \\ s^{(k)} &= r^{(k)} - \left(3I - 2Q^{-1}[r^{(k)}, x^{(k)}; F]\right) Q^{-1} F(r^{(k)}), \\ x^{(k+1)} &= s^{(k)} - \left(\frac{13}{4} I - Q^{-1} [s^{(k)}, r^{(k)}; F] \left(\frac{7}{2} I - \frac{5}{4} Q^{-1} [s^{(k)}, r^{(k)}; F]\right)\right) Q^{-1} F(s^{(k)}), \end{aligned} \tag{4}$$

where, as in the previous schemes, $Q = [a^{(k)}, b^{(k)}; F]$, with $a^{(k)} = x^{(k)} + F(x^{(k)})$ and $b^{(k)} = x^{(k)} - F(x^{(k)})$.

Different indices can be used to compare the efficiency of iterative processes. For example, in [11], Ostrowski introduced the efficiency index $EI = p^{1/d}$, where p is the convergence order and d is the quantity of functional evaluations at each iteration. Moreover, the matrix inversions appearing in the iterative expressions are in practice calculated by solving linear systems. Therefore, the amount of quotients/products, denoted by op, employed in each iteration play an important role. This is the reason why we presented in [12] the computational efficiency index, CEI, combining EI and the number of operations per iteration. This index is defined as $CEI = p^{1/(d+op)}$.

Our goal of this manuscript is to construct high-order Jacobian-free iterative schemes for solving nonlinear systems involving low computational cost on large systems.

We recall, in Section 2, some basic concepts that we will use in the rest of the manuscript. Section 3 is devoted to describe our proposed iterative methods for solving nonlinear systems and to analyze their convergence. The efficiency indices of our methods are studied in Section 4, as well as a comparative analysis with the schemes presented in the Introduction. Several numerical tests are shown in Section 5, for illustrating the performance of the new schemes. To get this aim, we use a discretized nonlinear one-dimensional heat conduction equation by means of approximations of the derivatives and also some systems of academic interest. We finish the manuscript with some conclusions.

2. Basic Concepts

If a sequence $\{x^{(k)}\}_{k\geq 0}$ in $\mathbb{R}^n$ converges to $\bar{x}$, it is said to be of order of convergence p, being $p \geq 1$, if $C > 0$ $(0 < C < 1$ for $p = 1)$ and k_0 exist satisfying

$$\|x^{(k+1)} - \bar{x}\| \leq C\|x^{(k)} - \bar{x}\|^p, \ \forall k \geq k_0,$$

or

$$\|e^{(k+1)}\| \leq C\|e^{(k)}\|^p, \ \forall k \geq k_0,$$

being $e^{(k)} = x^{(k)} - \bar{x}$.

Although this notation was presented by the authors in [12], we show it for the sake of completeness. Let $\Phi : D \subseteq \mathbb{R}^n \longrightarrow \mathbb{R}^n$ be sufficiently Fréchet differentiable in D. The qth derivative of Φ at $x \in \mathbb{R}^n$, $q \geq 1$, is the q-linear function $\Phi^{(q)}(x) : \mathbb{R}^n \times \cdots \times \mathbb{R}^n \longrightarrow \mathbb{R}^n$ such that $\Phi^{(q)}(x)(y_1, \ldots, y_q) \in \mathbb{R}^n$. Let us observe that

1. $\Phi^{(q)}(x)(y_1, \ldots, y_{q-1}, \cdot) \in \mathcal{L}(\mathbb{R}^n)$, where $\mathcal{L}(\mathbb{R}^n)$ denotes the set of linear mappings defined from $(\mathbb{R}^n)$ into $(\mathbb{R}^n)$.
2. $\Phi^{(q)}(x)(y_{\sigma(1)}, \ldots, y_{\sigma(q)}) = \Phi^{(q)}(x)(y_1, \ldots, y_q)$, for all permutation σ of $\{1, 2, \ldots, q\}$.

From the above properties, we can use the following notation (let us observe that y^p denotes $(y, \ldots, y)$ p times):

(a) $\Phi^{(q)}(x)(y_1, \ldots, y_q) = \Phi^{(q)}(x)y_1 \ldots y_q$,
(b) $\Phi^{(q)}(x)y^{q-1}\Phi^{(p)}y^p = \Phi^{(q)}(x)\Phi^{(p)}(x)y^{q+p-1}$.

Let us consider $\bar{x} + \varepsilon \in \mathbb{R}^n$ in a neighborhood of $\bar{x}$. By applying Taylor series and considering that $\Phi'(\bar{x})$ is nonsingular,

$$\Phi(\bar{x} + \varepsilon) = \Phi'(\bar{x}) \left[\varepsilon + \sum_{q=2}^{p-1} C_q \varepsilon^q\right] + \mathcal{O}(\varepsilon^p), \tag{5}$$

being $C_q = (1/q!)[\Phi'(\bar{x})]^{-1}\Phi^{(q)}(\bar{x})$, $q \geq 2$. Let us notice that $C_q\varepsilon^q \in \mathbb{R}^n$ as $\Phi^{(q)}(\bar{x}) \in \mathcal{L}(\mathbb{R}^n \times \cdots \times \mathbb{R}^n, \mathbb{R}^n)$ and $[\Phi'(\bar{x})]^{-1} \in \mathcal{L}(\mathbb{R}^n)$.

Moreover, we express Φ' as

$$\Phi'(\bar{x} + \varepsilon) = \Phi'(\bar{x}) \left[I + \sum_{q=2}^{p-1} qC_q\varepsilon^{q-1}\right] + \mathcal{O}(\varepsilon^{p-1}), \tag{6}$$

the identity matrix being denoted by I. Then, $qC_q\varepsilon^{q-1} \in \mathcal{L}(\mathbb{R}^n)$. From expression (6), we get

$$[\Phi'(\bar{x} + \varepsilon)]^{-1} = \left[I + Y_2\varepsilon + Y_3\varepsilon^2 + Y_4\varepsilon^4 + \cdots\right] [\Phi'(\bar{x})]^{-1} + \mathcal{O}(\varepsilon^{p-1}), \tag{7}$$

where

$$\begin{aligned} Y_2 &= -2C_2, \\ Y_3 &= 4C_2^2 - 3C_3, \\ Y_4 &= -8C_2^3 + 6C_2C_3 + 6C_3C_2 - 4C_4. \\ &\vdots \end{aligned}$$

The equation

$$e^{(k+1)} = Ke^{(k)^p} + \mathcal{O}(e^{(k)^{p+1}}),$$

where K is a p-linear operator $K \in \mathcal{L}(\mathbb{R}^n \times \cdots \times \mathbb{R}^n, \mathbb{R}^n)$, known as error equation and p is the order of convergence.In addition, we denote $e^{(k)^p}$ by $(e^{(k)}, e^{(k)}, \cdots, e^{(k)})$.

Divided difference operator of function Φ (see, for example, [2]) is defined as a mapping $[\cdot,\cdot;\Phi]$: $D \times D \subseteq \mathbb{R}^n \times \mathbb{R}^n \to \mathcal{L}(\mathbb{R}^n)$ satisfying

$$[x,y;\Phi](x-y) = \Phi(x) - \Phi(y), \quad \text{for all } x,y \in D.$$

In addition, by using the formula of Gennochi-Hermite [13] and Taylor series expansions around x, the divided difference operator is defined for all x, $x+h \in \mathbb{R}^n$ as follows:

$$[x+\varepsilon, x;\Phi] = \int_0^1 \Phi'(x+t\varepsilon)dt = \Phi'(x) + \frac{1}{2}\Phi''(x)\varepsilon + \frac{1}{6}\Phi'''(x)\varepsilon^2 + \frac{1}{24}\Phi^{(iv)}(x)\varepsilon^3 + \mathcal{O}(\varepsilon^4). \tag{8}$$

Being $a^{(k)} = x^{(k)} + \Phi(x^{(k)})$ and $b^{(k)} = x^{(k)} - \Phi(x^{(k)})$, the divided difference operator for points $a^{(k)}$ and $b^{(k)}$ is

$$\begin{aligned} [a^{(k)}, b^{(k)};\Phi] &= \Phi'(b^{(k)}) + \frac{1}{2}\Phi''(b^{(k)})(a^{(k)} - b^{(k)}) + \frac{1}{6}\Phi'''(b^{(k)})(a^{(k)} - b^{(k)})^2 + \mathcal{O}((a^{(k)} - b^{(k)})^3) \\ &= \Phi'(\bar{x})[I + A_1 e^{(k)} + A_2 e^{(k)^2} + A_3 e^{(k)^3}] + \mathcal{O}(e^{(k)^4}), \end{aligned} \tag{9}$$

where

$$\begin{aligned} A_1 &= 2C_2, \\ A_2 &= C_3(3+\Phi'(\bar{x})^2), \\ A_3 &= 4C_4(1+\Phi'(\bar{x})^2 + C_3\Phi'(\bar{x})^2C_2 + C_3\Phi'(\bar{x})C_2\Phi'(\bar{x}) \end{aligned}$$

are obtained by replacing the Taylor expansion of the different terms that appear in development (9) and doing algebraic manipulations.

For computational purposes, the following expression (see [2]) is used

$$[y,x;F]_{i,j} = \frac{F_i(y_1,\ldots,y_{j-1},y_j,x_{j+1},\ldots,x_n) - F_i(y_1,\ldots,y_{j-1},x_j,x_{j+1},\ldots,x_n)}{y_j - x_j},$$

where $x = (x_1,\ldots,x_{kj-1},x_j,x_{j+1},\ldots,x_n)$ and $y = (y_1,\ldots,y_{j-1},y_j,y_{j+1},\ldots,y_n)$ and $1 \le i,j \le n$.

3. Proposed Methods and Their Convergence

From a Samanskii-type method and by using the composition procedure "frozening" the divided difference operator (we hold the same divided difference operator in all the steps of the method), we propose the following four-steps iterative class with the aim of reaching order five:

$$\begin{aligned} y^{(k)} &= x^{(k)} - \mu[a^{(k)}, b^{(k)}; F]^{-1}F(x^{(k)}), \\ z^{(k)} &= y^{(k)} - \alpha\,[a^{(k)}, b^{(k)}; F]^{-1}F(y^{(k)}), \\ t^{(k)} &= z^{(k)} - \beta\,[a^{(k)}, b^{(k)}; F]^{-1}F(z^{(k)}), \\ x^{(k+1)} &= t^{(k)} - \gamma\,[a^{(k)}, b^{(k)}; F]^{-1}F(t^{(k)}), \end{aligned} \tag{10}$$

where μ, α, β and γ are real parameters, $a^{(k)} = x^{(k)} + F(x^{(k)})$ and $b^{(k)} = x^{(k)} - F(x^{(k)})$.

It is possible to prove that, under some assumptions, we can reach order five. We have used different combinations of these steps trying to preserve order 5 and reducing the computational cost. The best result we have been able to achieve is the following:

$$\begin{aligned} y^{(k)} &= x^{(k)} - [a^{(k)}, b^{(k)}; F]^{-1} F(x^{(k)}), \\ z^{(k)} &= y^{(k)} - \alpha\, [a^{(k)}, b^{(k)}; F]^{-1} F(y^{(k)}), \\ t^{(k)} &= z^{(k)} - \beta\, [a^{(k)}, b^{(k)}; F]^{-1} F(y^{(k)}), \\ x^{(k+1)} &= z^{(k)} - \gamma\, [a^{(k)}, b^{(k)}; F]^{-1} F(t^{(k)}), \end{aligned} \tag{11}$$

where α, β and γ are real parameters, $a^{(k)} = x^{(k)} + F(x^{(k)})$ and $b^{(k)} = x^{(k)} - F(x^{(k)})$. The convergence of class (11) is presented in the following result.

Theorem 1. *Let us assume $F : D \subseteq \mathbb{R}^n \longrightarrow \mathbb{R}^n$ being a differentiable enough operator at each point of the open neighborhood D of the solution $\bar{x}$ of the system $F(x) = 0$. Let us suppose that $F'(x)$ is continuous and nonsingular in $\bar{x}$ and the initial estimation $x^{(0)}$ is near enough to $\bar{x}$. Therefore, sequence $\{x^{(k)}\}_{k\geq 0}$ calculated from expression (11) converges to $\bar{x}$ with order 4 if $\alpha = 2 - \gamma$, $\beta = \dfrac{(\gamma-1)^2}{\gamma}$ and for all $\gamma \in \mathbb{R}$, the error equation being*

$$e^{(k+1)} = \Big(\frac{5\gamma - 1}{\gamma}\, C_2^3\Big) {e^{(k)}}^4 + \mathcal{O}({e^{(k)}}^5).$$

In addition, if $\gamma = \frac{1}{5}$, the order of convergence is five and the error equation is

$$e^{(k+1)} = \Big(14C_2^4 - 2C_2C_3C_2 + 6C_3C_2^2 - 2C_2C_3F'(\bar{x})^2C_2 + 2C_3F'(\bar{x})^2C_2^2\Big) {e^{(k)}}^5 + \mathcal{O}({e^{(k)}}^6),$$

where $C_j = \dfrac{1}{j!}[F'(\bar{x})]^{-1} F^{(j)}(\bar{x})$, $j = 2, 3, \ldots$

Proof. By using the Taylor expansion of $F(x^{(k)})$ and its derivatives around $\bar{x}$:

$$\begin{aligned} F(x^{(k)}) &= F'(\bar{x})\Big[e^{(k)} + C_2{e^{(k)}}^2 + C_3{e^{(k)}}^3 + C_4{e^{(k)}}^4 + C_5{e^{(k)}}^5\Big] + \mathcal{O}({e^{(k)}}^6), \\ F'(x^{(k)}) &= F'(\bar{x})\Big[I + 2C_2e^{(k)} + 3C_3{e^{(k)}}^2 + 4C_4{e^{(k)}}^3 + 5C_5{e^{(k)}}^4\Big] + \mathcal{O}({e^{(k)}}^5), \\ F''(x^{(k)}) &= F'(\bar{x})\Big[2C_2 + 6C_3e^{(k)} + 12C_4{e^{(k)}}^2 + 20C_5{e^{(k)}}^3\Big] + \mathcal{O}({e^{(k)}}^4), \\ F'''(x^{(k)}) &= F'(\bar{x})\Big[6C_3 + 24C_4e^{(k)} + 60C_5{e^{(k)}}^2\Big] + \mathcal{O}({e^{(k)}}^3), \\ F^{(iv)}(x^{(k)}) &= F'(\bar{x})\Big[24C_4 + 120C_5e^{(k)}\Big] + \mathcal{O}({e^{(k)}}^2), \\ F^{(v)}(x^{(k)}) &= F'(\bar{x})\Big[120C_5\Big] + \mathcal{O}(e^{(k)}). \end{aligned}$$

From the above expression, by replacing in the first order divided difference operator $[a^{(k)}, b^{(k)}; F]$, the values $a^{(k)} = x^{(k)} + F(x^{(k)})$, $b^{(k)} = x^{(k)} - F(x^{(k)})$, we obtain:

$$\begin{aligned} \Big[a^{(k)}, b^{(k)}; F\Big] = F'(\overline{x})\Big[&I + 2C_2e^{(k)} + \Big(3C_3 + C_3F'(\overline{x})^2\Big){e^{(k)}}^2 \\ &+ \Big(4C_4 + 4C_4F'(\overline{x})^2 + C_3F'(\overline{x})^2C_2 + C_3F'(\overline{x})C_2F'(\overline{x})\Big){e^{(k)}}^3 \\ &+ \Big(5C_5 + C_5F'(\overline{x})^4 + 10C_5F'(\overline{x})^2 + 4C_4F'(\overline{x})^2C_2 + 4C_4F'(\overline{x})C_2F'(\overline{x}) \\ &+ C_3F'(\overline{x})^2C_3 + C_3F'(\overline{x})C_2F'(\overline{x})C_2 + C_3F'(\overline{x})C_3F'(\overline{x})\Big){e^{(k)}}^4\Big] + \mathcal{O}({e^{(k)}}^5). \end{aligned}$$

From the above expression, we have

$$\Big[a^{(k)}, b^{(k)}; F\Big]^{-1} = \Big[I + X_2e^{(k)} + X_3{e^{(k)}}^2\Big][F'(\overline{x})]^{-1} + \mathcal{O}({e^{(k)}}^3),$$

where

$$X_2 = -2C_2,$$
$$X_3 = -3C_3 + 4C_2^2 - C_3F'(\overline{x})^2,$$
$$X_4 = -4C_4 + 6C_2C_3 + 6C_3C_2 - 8C_2^3 + C_3F'(\overline{x})^2C_2 - C_3F'(\overline{x})C_2F'(\overline{x}) + 2C_2C_3F'(\overline{x})^2 - 4C_4F'(\overline{x})^2.$$

Then,

$$\begin{aligned}\left[a^{(k)},b^{(k)};F\right]^{-1}F(x^{(k)}) &= e^{(k)} - C_2{e^{(k)}}^2 + \left(-2C_3 + 2C_2^2 - C_3F'(\overline{x})^2\right){e^{(k)}}^3\\ &+ \left(-3C_4 + 4C_2C_3 + 3C_3C_2 - 4C_2^3 - C_3F'(\overline{x})C_2F'(\overline{x}) + 2C_2C_3F'(\overline{x})^2 - 4C_4F'(\overline{x})^2\right){e^{(k)}}^4 + \mathcal{O}({e^{(k)}}^5).\end{aligned}$$

Thus,

$$\begin{aligned}y^{(k)} - \bar{x} &= C_2{e^{(k)}}^2 + \left(2C_3 - 2C_2^2 + C_3F'(\overline{x})^2\right){e^{(k)}}^3\\ &+ \left(3C_4 - 4C_2C_3 - 3C_3C_2 + 4C_2^3 + C_3F'(\overline{x})C_2F'(\overline{x}) - 2C_2C_3F'(\overline{x})^2 + 4C_4F'(\overline{x})^2\right){e^{(k)}}^4 + \mathcal{O}({e^{(k)}}^5),\\ (y^{(k)} - \bar{x})^2 &= C_2^2{e^{(k)}}^4 + \mathcal{O}({e^{(k)}}^5),\end{aligned}$$

and

$$\begin{aligned}F(y^{(k)}) &= F'(\bar{x})\left[(y^{(k)} - \bar{x}) + C_2(y^{(k)} - \bar{x})^2\right] + \mathcal{O}((y^{(k)} - \bar{x})^3)\\ &= F'(\bar{x})\Big[C_2{e^{(k)}}^2 + \left(2C_3 - 2C_2^2 + C_3F'(\overline{x})^2\right){e^{(k)}}^3\\ &+\left(3C_4 - 4C_2C_3 - 3C_3C_2 + 5C_2^3 + C_3F'(\overline{x})C_2F'(\overline{x}) - 2C_2C_3F'(\overline{x})^2 + 4C_4F'(\overline{x})^2\right){e^{(k)}}^4\Big] + \mathcal{O}({e^{(k)}}^5).\end{aligned}$$

From the values of $z^{(k)}$ and $t^{(k)}$ in expression (11), we have

$$t^{(k)} = y^{(k)} - (\alpha+\beta)\,[a^{(k)},b^{(k)};F]^{-1}F(y^{(k)}).$$

Then,

$$\begin{aligned}\left[a^{(k)},b^{(k)};F\right]^{-1}F(y^{(k)}) &= C_2{e^{(k)}}^2 + \left(2C_3 - 4C_2^2 + C_3F'(\overline{x})^2\right){e^{(k)}}^3\\ &+\Big(3C_4 - 8C_2C_3 - 6C_3C_2 + 13C_2^3\\ &+C_3F'(\overline{x})C_2F'(\overline{x}) - 4C_2C_3F'(\overline{x})^2 + 4C_4F'(\overline{x})^2 - C_3F'(\overline{x})^2C_2\Big){e^{(k)}}^4 + \mathcal{O}({e^{(k)}}^5).\end{aligned}$$

Similarly, we obtain

$$\begin{aligned}t^{(k)} - \bar{x} &= \left(1-(\alpha+\beta)\right)C_2{e^{(k)}}^2 + \left(\left(1-(\alpha+\beta)\right)\left(2C_3 - 2C_2^2 + C_3F'(\overline{x})^2\right) + 2(\alpha+\beta)C_2^2\right){e^{(k)}}^3\\ &+\Big(\left(1-(\alpha+\beta)\right)\left(3C_4 - 4C_2C_3 - 3C_3C_2 + 4C_2^3 + C_3F'(\overline{x})C_2F'(\overline{x}) - 2C_2C_3F'(\overline{x})^2 + 4C_4F'(\overline{x})^2\right)\\ &-(\alpha+\beta)\left(-4C_2C_3 - 3C_3C_2 + 9C_2^3 - 2C_2C_3F'(\overline{x})^2 - C_3F'(\overline{x})^2C_2\right){e^{(k)}}^4 + \mathcal{O}({e^{(k)}}^5),\\ (t^{(k)} - \bar{x})^2 &= \left(1-(\alpha+\beta)\right)^2C_2^2{e^{(k)}}^4 + \mathcal{O}({e^{(k)}}^5)\end{aligned}$$

and

$$\begin{aligned}F(t^{(k)}) &= F'(\bar{x})\Big[\left(1-(\alpha+\beta)\right)C_2{e^{(k)}}^2 + \left(\left(1-(\alpha+\beta)\right)\left(2C_3 - 2C_2^2 + C_3F'(\overline{x})^2\right) + 2(\alpha+\beta)C_2^2\right){e^{(k)}}^3\\ &+\Big(\left(1-(\alpha+\beta)\right)\left(3C_4 - 4C_2C_3 - 3C_3C_2 + 4C_2^3 + C_3F'(\overline{x})C_2F'(\overline{x}) - 2C_2C_3F'(\overline{x})^2 + 4C_4F'(\overline{x})^2\right)\\ &-(\alpha+\beta)\left(-4C_2C_3 - 3C_3C_2 + 9C_2^3 - 2C_2C_3F'(\overline{x})^2 - C_3F'(\overline{x})^2C_2\right) + \left(1-(\alpha+\beta)\right)^2C_2^3\Big){e^{(k)}}^4\Big] + \mathcal{O}({e^{(k)}}^5).\end{aligned}$$

Thus,

$$\begin{aligned}\left[a^{(k)}, b^{(k)}; F\right]^{-1} F(t^{(k)}) &= \Big(1-(\alpha+\beta)\Big)C_2 {e^{(k)}}^2 \\ &+\Big(\Big(1-(\alpha+\beta)\Big)\Big(2C_3-4C_2^2+C_3F'(\overline{x})^2\Big)+2(\alpha+\beta)C_2^2\Big){e^{(k)}}^3 \\ &+\Big(\Big(1-(\alpha+\beta)\Big)\Big(3C_4-8C_2C_3-6C_3C_2+12C_2^3+C_3F'(\overline{x})C_2F'(\overline{x})-4C_2C_3F'(\overline{x})^2 \\ &-C_3F'(\overline{x})^2C_2+4C_4F'(\overline{x})^2\Big)+\Big(1-(\alpha+\beta)\Big)^2C_2^3 \\ &-(\alpha+\beta)\Big(-4C_2C_3-3C_3C_2+13C_2^3-C_3F'(\overline{x})^2C_2-2C_2C_3F'(\overline{x})^2\Big)\Big){e^{(k)}}^4+\mathcal{O}({e^{(k)}}^5).\end{aligned}$$

Therefore, we obtain

$$\begin{aligned}e^{(k+1)} &= e^{(k)}-\left[a^{(k)}, b^{(k)}; F\right]^{-1}F(x^{(k)})-\alpha\left[a^{(k)}, b^{(k)}; F\right]^{-1}F(y^{(k)})-\gamma\left[a^{(k)}, b^{(k)}; F\right]^{-1}F(t^{(k)}) \\ &= \Big(1-\alpha-\gamma\Big(1-(\alpha+\beta)\Big)\Big)C_2{e^{(k)}}^2+\Big[\Big(1-\alpha-\gamma\Big(1-(\alpha+\beta)\Big)\Big)\Big(2C_3-2C_2^2+C_3F'(\overline{x})^2\Big) \\ &-2\Big(-\alpha+\gamma(\alpha+\beta)-\gamma\Big(1-(\alpha+\beta)\Big)\Big)C_2^2\Big]{e^{(k)}}^3 \\ &+\Big[\Big(1-\alpha-\gamma\Big(1-(\alpha+\beta)\Big)\Big)\Big(3C_4-4C_2C_3-3C_3C_2+4C_2^3+C_3F'(\overline{x})C_2F'(\overline{x})-2C_2C_3F'(\overline{x})^2+4C_4F'(\overline{x})^2\Big) \\ &+\Big(-\alpha+\gamma(\alpha+\beta)-\gamma\Big(1-(\alpha+\beta)\Big)\Big)\Big(-4C_2C_3-3C_3C_2+8C_2^3-2C_2C_3F'(\overline{x})^2-C_3F'(\overline{x})^2C_2\Big) \\ &\Big(-\alpha+5\gamma(\alpha+\beta)-\gamma\Big(1-(\alpha+\beta)\Big)^2\Big)C_2^3\Big]{e^{(k)}}^4+\mathcal{O}({e^{(k)}}^5).\end{aligned}$$

Thus, by requiring that the coefficients of ${e^{(k)}}^2$ and ${e^{(k)}}^3$ are null, we get $\alpha = 2-\gamma$ and $\beta = \dfrac{(\gamma-1)^2}{\gamma}$ for all $\gamma \in \mathbb{R}$, 4 being its order,

$$e^{(k+1)} = \Big(\frac{5\gamma-1}{\gamma}C_2^3\Big){e^{(k)}}^4+\mathcal{O}({e^{(k)}}^5).$$

By adding the coefficient of ${e^{(k)}}^4$ to the above system, we get $\gamma = \frac{1}{5}$, 5 being the order of convergence with error equation in this case:

$$e^{(k+1)} = \Big(14C_2^4-2C_2C_3C_2+6C_3C_2^2-2C_2C_3F'(\overline{x})^2C_2+2C_3F'(\overline{x})^2C_2^2\Big){e^{(k)}}^5+\mathcal{O}({e^{(k)}}^6).$$

□

4. Efficiency Indices

As we have mentioned in the Introduction, we use indices $EI = p^{1/d}$ and CEI to compare the different iterative methods.

To evaluate function F, n scalar functions are calculated and $n(n-1)$ for the first order divided difference $[\cdot,\cdot;F]$. In addition, to calculate an inverse linear operator, an $n \times n$ linear system must be solved; then, we have to do $\frac{1}{3}n^3+n^2-\frac{1}{3}n$ quotients/products for getting LU decomposition and solving the corresponding triangular linear systems. Moreover, for solving m linear systems with the same matrix of coefficients, we need to do $\frac{1}{3}n^3+mn^2-\frac{1}{3}n$ products-quotients. In addition, we need n^2 products for each matrix-vector multiplication and n^2 quotients for evaluating a divided difference operator.

According to the last considerations, we calculate the efficiency indices EI of methods CJST5, NM7, S7, SA6 and WF4. In case CJST5, for each iteration, we evaluate F three times and once $[\cdot,\cdot;F]$, so n^2+2n functional evaluations are needed. Therefore, $EI_{CJST5} = 5^{\frac{1}{n^2+2n}}$. The indices obtained for the mentioned methods are also calculated and shown in Table 1.

Table 1. Efficiency indices for different methods.

Method	Order	NFE	*EI*
NM7	7	$2n^2+n$	$7^{\frac{1}{2n^2+n}}$
S7	7	$3n^2$	$7^{\frac{1}{3n^2}}$
SA6	6	$2n^2+n$	$6^{\frac{1}{2n^2+n}}$
WF4	4	$2n^2$	$4^{\frac{1}{2n^2}}$
CJST5	5	n^2+2n	$5^{\frac{1}{n^2+2n}}$

In Table 2, we present the indices CEI of schemes NM7, SA6, S7, WF4 and CJST5. In it, the amount of functional evaluations is denoted by NFE, the number of linear systems with the same $[\cdot,\cdot;F]$ as the matrix of coefficients is $NLS1$ and $M \times V$ represents the quantity of products matrix-vector. Then, in case CJST5, for each iteration, n^2+2n functional evaluations are needed, since we evaluate three times the function F and one divided difference of first order $[a^{(k)}, b^{(k)}, F]$. In addition, we must solve three linear systems with $[a^{(k)}, b^{(k)}, F]$ as coefficients matrix (that is $\frac{1}{3}n^3+3n^2-\frac{1}{3}n$). Thus, the value of CEI for CJST5 is

$$CEI_{CJST5} = 5^{\frac{1}{\frac{1}{3}n^3+5n^2+\frac{5}{3}n}}.$$

Analogously, we obtain the indices CEI of the other methods. In Figure 1, we observe the computational efficiency index of the different methods of size 5 to 80. The best index corresponds to our proposed scheme.

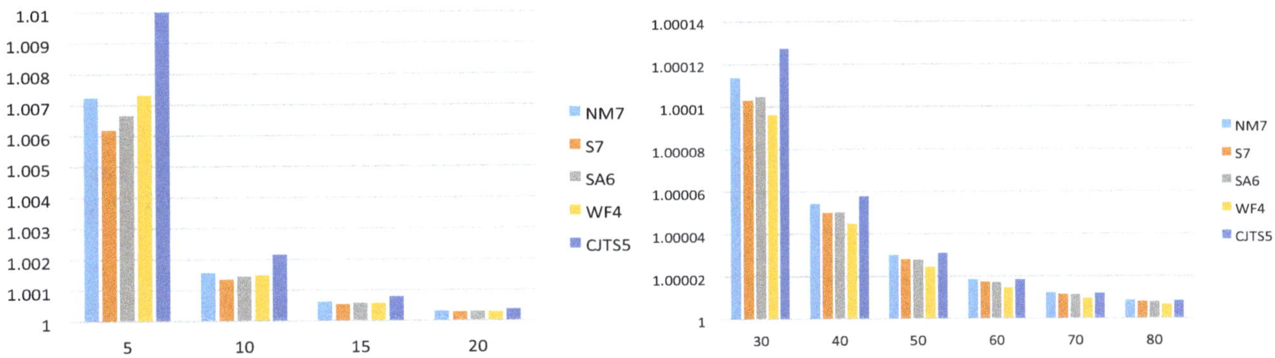

(**a**) Size from 5 to 20 (**b**) Size from 30 to 80

Figure 1. CEI for several sizes of the system.

Table 2. Computational cost of the procedures.

Method	Order	*NFE*	*NLS1*	$M \times V$	*CEI*
NM7	7	$2n^2+3n$	4	3	$7^{\frac{1}{\frac{1}{3}n^3+11n^2+\frac{2}{3}n}}$
S7	7	$3n^2$	6	3	$7^{\frac{1}{\frac{1}{3}n^3+14n^2-\frac{1}{3}n}}$
SA6	6	$2n^2+n$	4	2	$6^{\frac{1}{\frac{1}{3}n^3+11n^2+\frac{2}{3}n}}$
WF4	4	$2n^2$	3	1	$4^{\frac{1}{\frac{1}{3}n^3+8n^2-\frac{1}{3}n}}$
CJST5	5	n^2+2n	3	0	$5^{\frac{1}{\frac{1}{3}n^3+5n^2+\frac{5}{3}n}}$

5. Numerical Examples

We begin this section checking the performance of the new method on the resulting system obtained by the discretization of Fisher's partial differential equation. Thereafter, we compare its behavior with that of other known methods on some academic problems. For the computations, we have used *Matlab* R2015a (Natick, Massachusetts, USA) with variable precision arithmetic, with 1000 digits of mantissa. The characteristics of the computer are, regarding the processor, Intel(R) Core(TM) i7-7700 CPU @ 3.6 GHz, 3.601 Mhz, four processors and RAM 16 GB.

We use the estimation of the theoretical order of convergence p, called Computational Order of Convergence (COC), introduced by Jay [14] with the following expression:

$$p \approx COC = \frac{\ln\left(\|F(x^{(k+1)})\|_2/\|F(x^{(k)})\|_2\right)}{\ln\left(\|F(x^{(k)})\|_2/\|F(x^{(k-1)})\|_2\right)}, \quad k = 1, 2, \ldots$$

and the Approximated Computational Order of Convergence (ACOC), defined by Cordero and Torregrosa in [15]

$$p \approx ACOC = \frac{\ln\left(\|x^{(k+1)} - x^{(k)}\|_2/\|x^{(k)} - x^{(k-1)}\|_2\right)}{\ln\left(\|x^{(k)} - x^{(k-1)}\|_2/\|x^{(k-1)} - x^{(k-2)}\|_2\right)}.$$

Example 1. *Fisher's equation,*

$$u_t = Du_{xx} + ru\left(1 - \frac{u}{p}\right), \quad x \in [a,b] \; t \geq 0, \tag{12}$$

was proposed in [16] by Fisher to model the diffusion process in population dynamics. In it, $D > 0$ is the diffusion constant, r is the level of growth of the species and p is the carrying capacity. Lately, this formulation has proven to be fruitful for many other problems as wave genetics, economy or propagation.

Now, we study a particular case of this equation, when $r = p = 1$ and the spatial interval is $[0, 1]$, $u(x, 0) = sech^2(\pi x)$ and null boundary conditions.

We transform Example 1 in a set of nonlinear systems by applying an implicit method of finite differences, providing the estimated solution in the instant t_k from the estimated one in t_{k-1}. The spacial step $h = 1/n_x$ is selected and the temporal step is $k = T_{max}/n_t$, n_x and n_t being the quantity of subintervals in x and t, respectively, and T_{max} is the final instant. Therefore, a grid of domain $[0, 1] \times [0, T_{max}]$ with points (x_i, t_j), is selected:

$$x_i = 0 + ih, \; i = 0, 1, \ldots, n_x, \qquad t_j = 0 + jk, \; j = 0, 1, \ldots, n_t.$$

Our purpose is to estimate the solution of problem (12) at these points, by solving many nonlinear systems, as much as the number of temporal nodes t_j. For it, we use the following finite differences of order $\mathcal{O}(k + h^2)$:

$$u_t(x,t) \approx \frac{u(x,t) - u(x,t-k)}{k},$$
$$u_{xx}(x,t) \approx \frac{u(x+h,t) - 2u(x,t) + u(x-h,t)}{h^2}.$$

By denoting the approximation of the solution at (x_i, t_j) as $u_{i,j}$, and, by replacing it in Example 1, we get the system

$$ku_{i+1,j} + (kh^2 - 2k - h^2)u_{i,j} - kh^2u_{i,j}^2 + ku_{i-1,j} = -h^2u_{i,j-1},$$

for $i = 1, 2, \ldots, n_x - 1$ and $j = 1, 2, \ldots, n_t$. The unknowns of this system are $u_{1,j}, u_{2,j}, \ldots, u_{n_x-1,j}$, that is, the approximations of the solution in each spatial node for the fixed instant t_j. Let us remark that, for solving this system, the knowledge of solution in t_{j-1} is required.

Let us observe (Table 3) that the results improve when the temporal step is smaller. In this case, the COC is not a good estimation of the theoretical error. In Figure 2, we show the approximated solution of the problem when $T_{max} = 10$, by taking $n_t = 50$, $n_x = 10$ and using method CJST5.

Table 3. Fisher results by CJST5 and different T_{max}.

T_{max}	n_x	n_t	$\|F(x^1)\|$	$\|F(x^2)\|$	$\|F(x^3)\|$	COC
0.1	10	20	8.033×10^{-9}	2.356×10^{-35}	3.243×10^{-68}	1.2385
0.1	10	200	8.679×10^{-13}	3.203×10^{-44}	7.623×10^{-77}	1.0379
1	10	20	4.158×10^{-5}	3.4×10^{-25}	1.679×10^{-56}	1.5585
1	10	200	8.033×10^{-9}	2.356×10^{-35}	3.243×10^{-68}	1.2385
10	10	20	nc			
10	10	50	0.01945	2.757×10^{-11}	1.953×10^{-38}	3.0683

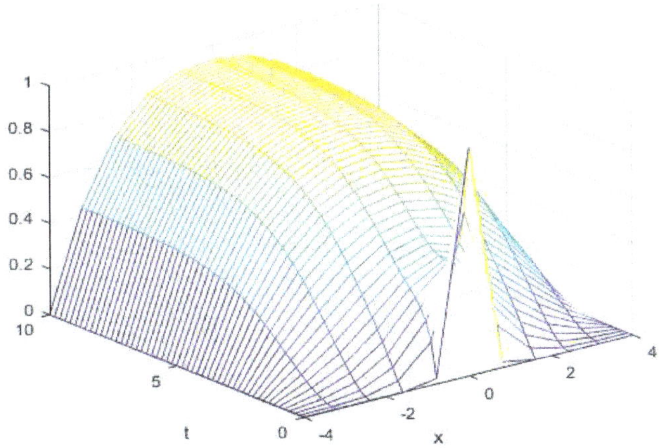

Figure 2. Approximated solution of Example 1.

In the rest of examples, we are going to compare the performance of the proposed method with the schemes presented in the Introduction as well as with the Newton-type method replacing the Jacobian matrix by the divided difference operator, that is, the Samanskii's scheme (see [6]).

Example 2. *Let us define the nonlinear system*

$$\begin{cases} \cos(x_2) - \sin(x_1) = 0, \\ x_3^{x_1} - 1/x_2 = 0, \\ e^{x_1} - x_3^2 = 0. \end{cases}$$

We use, in this example, the starting estimation $x^{(0)} = (1.25, 1.25, 1.25)^T$, the solution being $\bar{x} \approx (0.9096, 0.6612, 1.576)^T$. Table 4 shows the residuals $\|x^{(k)} - x^{(k-1)}\|$ and $\|F(x^{(k)})\|$ for $k = 1, 2, 3$ as well as ACOC and COC. We observe that the COC index is better than the corresponding ACOC of the other methods. In addition, the value $\|x^3 - x^2\|$ is better or similar to that of S7 and NM7 methods, both of them of order 7.

Table 4. Numerical results for Example 2.

	Samanskii	CJST5	WF4	SA6	S7	NM7
$\|x^{(1)} - x^{(0)}\|$	1.415	0.8848	0.8539	0.8934	0.9148	0.9355
$\|x^{(2)} - x^{(1)}\|$	0.5427	0.247	0.2039	0.2689	0.2942	0.3098
$\|x^{(3)} - x^{(2)}\|$	0.1738	0.01159	0.006249	0.005301	0.01069	0.02667
ACOC	1.1875	2.3976	2.433	3.2705	2.9221	4.25
$\|F(x^{(1)})\|$	0.1954	0.1282	0.1038	0.1385	0.1817	0.2098
$\|F(x^{(2)})\|$	0.02369	0.009956	0.004815	0.003584	0.006462	0.01669
$\|F(x^{(3)})\|$	0.00269	2.805×10^{-8}	7.663×10^{-8}	4.009×10^{-11}	8.074×10^{-12}	1.333×10^{-9}
COC	1.0313	5.0015	3.5983	5.0104	6.1445	6.4561

Example 3. *We consider now*

$$\sum_{j=1}^{n} x_j - x_i - e^{-x_i} x_i = 0, \quad i = 1, 2, \ldots, n.$$

The numerical results are displayed in Table 5. The initial estimation is $x^{(0)} = (0.25, 0.25, \ldots, 0.25)^T$ and the size of the system is $n = 10$, the solution being $\bar{x} = (0, 0, \ldots, 0)^T$. We show the same information as in the previous example.

Example 4. *The third example is given by the system:*

$$x_i + 1 - 2\log\left(1 + \sum_{j=1}^{n} x_j - x_i\right) = 0, \quad i = 1, 2, \ldots, n.$$

Its solution is $\bar{x} \approx (9.376, 9.376, \ldots, 9.376)^T$. By using the starting guess $x^{(0)} = (1, 1, \ldots, 1)^T$ with $n = 10$, we obtain the results appearing in Table 6.

Table 5. Numerical results for Example 3.

	Samanskii	**CJST5**	**WF4**	**SA6**	**S7**	**NM7**
$\|\|x^{(1)} - x^{(0)}\|\|$	1.036	0.8249	0.9116	0.8499	0.7847	0.7897
$\|\|x^{(2)} - x^{(1)}\|\|$	0.2552	0.03432	0.121	0.05932	0.00583	0.0008995
$\|\|x^{(3)} - x^{(2)}\|\|$	0,009667	1.487×10^{-11}	9.264×10^{-6}	5.367×10^{-10}	2.529×10^{-21}	1.048×10^{-28}
ACOC	2.3361	6.7807	4.6937	6.9572	8.6247	4.25
$\|\|F(x^{(1)})\|\|$	1.944	0.2742	0.9634	0.4735	0.04665	0.007196
$\|\|F(x^{(2)})\|\|$	0.0773	1.19×10^{-10}	7.411×10^{-5}	4.293×10^{-9}	2.023×10^{-20}	8.381×10^{-28}
$\|\|F(x^{(3)})\|\|$	2.65×10^{-5}	2.839×10^{-59}	9.644×10^{-22}	1.3×10^{-57}	9.093×10^{-149}	2.621×10^{-202}
COC	2.4743	5.1932	4.1045	6.0328	6.9895	6.9987

Table 6. Numerical results for Example 4.

	Samanskii	**CJST5**	**WF4**	**SA6**	**S7**	**NM7**
$\|\|x^{(1)} - x^{(0)}\|\|$	6.013	40.68	67.88	73.31	34.3	39.23
$\|\|x^{(2)} - x^{(1)}\|\|$	12.15	13.83	36.07	56.74	17.15	7.369
$\|\|x^{(3)} - x^{(2)}\|\|$	10.11	0.0002872	0.06643	0.02688	0.0001485	5.422×10^{-8}
ACOC	-	9.9941	9.9587	29.874	16.819	4.25
$\|\|F(x^{(1)})\|\|$	10.84	11.03	30.43	48.74	13.42	5.842
$\|\|F(x^{(2)})\|\|$	6.084	0.0002263	0.05234	0.02117	0.000117	4.272×10^{-8}
$\|\|F(x^{(3)})\|\|$	1.162	9.286×10^{-28}	6.508×10^{-12}	1.807×10^{-20}	3.105×10^{-40}	7.552×10^{-65}
COC	2.8651	4.9888	3.583	5.3744	7.0313	6.9755

The different methods give us the expected results, according to their order of convergence.

Example 5. *Finally, the last example that we consider is:*

$$\arctan(x_i) + 1 - 2\left(\sum_{j=1}^{n} x_j^2 - x_i^2\right) = 0.$$

The solution of it is $\bar{x} \approx (0.1758, 0.1758, \ldots, 0.1758)^T$. By using the initial estimation $x^{(0)} = (0.5, 0.5, \ldots, 0.5)^T$ with $n = 20$, we obtain the numerical results displayed in Table 7.

Table 7. Numerical results for Example 5.

	Samanskii	**CJST5**	**WF4**	**SA6**	**S7**	**NM7**
$\|\|x^{(1)} - x^{(0)}\|\|$	0.9503	1.323	1.272	1.368	1.394	1.393
$\|\|x^{(2)} - x^{(1)}\|\|$	0.3912	0.1266	0.177	0.0821	0.05639	0.05732
$\|\|x^{(3)} - x^{(2)}\|\|$	0.1013	4.988×10^{-5}	0.0007407	6.903×10^{-7}	7.214×10^{-9}	6.655×10^{-9}
ACOC	1.5229	3.3404	2.776	4.1543	4.9485	4.25
$\|\|F(x^{(1)})\|\|$	8.324	1.706	2.471	1.075	0.7257	0.7381
$\|\|F(x^{(2)})\|\|$	1.445	0.0006179	0.009181	8.552×10^{-6}	8.937×10^{-8}	8.245×10^{-8}
$\|\|F(x^{(3)})\|\|$	0.0902	1.206×10^{-20}	5.635×10^{-12}	5.437×10^{-36}	8.115×10^{-56}	3.521×10^{-56}
COC	1.5839	4.8559	3.791	5.9219	6.953	6.9577

It is observed in Tables 5–7 that, for the proposed academic problems, the introduced method (CJST5) shows a good performance comparable with higher-order methods. Of course, the worst results are those obtained by Samanskii's method, but it has been included because it is the Jacobian-free version of Newton's scheme and it is also the first step of our proposed scheme. Let us also remark that, when only three iterations are calculated, the index COC gives more reliable information than the ACOC one in all of the examples.

6. Conclusions

In this paper, we design a family of iterative methods for solving nonlinear systems with fourth-order convergence. This family does not use Jacobian matrices and one of its elements has order five. The relationship between the proposed method and other known ones in terms of efficiency index and computational efficiency index allows us to see that our method is more efficient than the other ones. In addition, its error bounds are smaller with the same number of iterations in some cases. Thus, our proposal is competitive mostly for big size systems.

Author Contributions: The individual contributions of the authors are as follows: conceptualization, J.R.T.; writing—original draft preparation, C.J. and E.S.; validation, A.C. and J.R.T.; formal analysis, A.C.; numerical experiments, C.J. and E.S.

Acknowledgments: The authors would like to thank the anonymous reviewers for their useful comments and suggestions that have improved the final version of this manuscript.

References

1. Traub, J.F. *Iterative Methods for the Solution of Equations*; Prentice-Hall: Englewood Cliffs, NJ, USA, 1964.
2. Ortega, J.M.; Rheinbolt, W.C. *Iterative Solutions of Nonlinears Equations in Several Variables*; Academic Press: New York, NY, USA, 1970.
3. Kelley, C.T. *Iterative Methods for Linear and Nonlinear Equations*; SIAM: Philadelphia, PA, USA, 1995.
4. Petković, M.S.; Neta, B.; Petković, L.D.; Džunić, J. *Multipoint Methods for Solving Nonlinear Equations*; Academic Press: New York, NY, USA, 2013.
5. Amat, S.; Busquier, S. *Advances in Iterative Methods for Nonlinear Equations*; SEMA SIMAI Springer Series; Springer International Publishing: Cham, Switzerland, 2016; Volume 10.
6. Samanskii, V. On a modification of the Newton method. *Ukrain. Mat.* **1967**, *19*, 133–138.
7. Wang, X.; Fang, X. Two Efficient Derivative-Free Iterative Method for Solving Nonlinear Systems. *Algorithms* **2016**, *9*, 14. [CrossRef]
8. Sharma, J.R.; Arora, H. Efficient derivative-free numerical methods for solving systems of nonlinear equations. *Comp. Appl. Math.* **2016**, *35*, 269–284. [CrossRef]
9. Narang, M.; Bathia S.; Kanwar, V. New efficient derivative free family of seventh-order methods for solving systems of nonlinear equations. *Numer. Algorithms* **2017**, *76*, 283–307. [CrossRef]
10. Wang, X.; Zhang, T.; Qian, W.; Teng, M. Seventh-order derivative-free iterative method for solving nonlinear systems. *Numer. Algorithms* **2015**, *70*, 545–558. [CrossRef]
11. Ostrowski, A.M. *Solution of Equations and Systems of Equations*; Prentice-Hall: Englewood Cliffs, NJ, USA, 1964.
12. Cordero, A.; Hueso, J.L.; Martínez, E.; Torregrosa, J.R. A modified Newton-Jarratt's composition. *Numer. Algorithms* **2010**, *55*, 87–99. [CrossRef]
13. Hermite, C. Sur la formule d'interpolation de Lagrange. *Reine Angew. Math.* **1878**, *84*, 70–79. [CrossRef]
14. Jay, L.O. A note of Q-order of convergence. *BIT Numer. Math.* **2001**, *41*, 422–429. [CrossRef]
15. Cordero, A; Torregrosa, J.R. Variants of Newton's method using fifth-order quadrature formulas. *Appl. Math. Comput.* **2007**, *190*, 686–698. [CrossRef]
16. Fisher, R.A. The wave of advance of advantageous genes. *Ann. Eugen.* **1937**, *7*, 353–369. [CrossRef]

8

Optimal Fourth, Eighth and Sixteenth Order Methods by using Divided Difference Techniques and their Basins of Attraction and its Application

Yanlin Tao [1,†] and Kalyanasundaram Madhu [2,*,†]

1 School of Computer Science and Engineering, Qujing Normal University, Qujing 655011, China; taoyanl@126.com
2 Department of Mathematics, Saveetha Engineering College, Chennai 602105, India
* Correspondence: kalyan742@pec.edu.
† These authors contributed equally to this work.

Abstract: The principal objective of this work is to propose a fourth, eighth and sixteenth order scheme for solving a nonlinear equation. In terms of computational cost, per iteration, the fourth order method uses two evaluations of the function and one evaluation of the first derivative; the eighth order method uses three evaluations of the function and one evaluation of the first derivative; and sixteenth order method uses four evaluations of the function and one evaluation of the first derivative. So these all the methods have satisfied the Kung-Traub optimality conjecture. In addition, the theoretical convergence properties of our schemes are fully explored with the help of the main theorem that demonstrates the convergence order. The performance and effectiveness of our optimal iteration functions are compared with the existing competitors on some standard academic problems. The conjugacy maps of the presented method and other existing eighth order methods are discussed, and their basins of attraction are also given to demonstrate their dynamical behavior in the complex plane. We apply the new scheme to find the optimal launch angle in a projectile motion problem and Planck's radiation law problem as an application.

Keywords: non-linear equation; basins of attraction; optimal order; higher order method; computational order of convergence

MSC: 65H05, 65D05, 41A25

---

1. Introduction

One of the most frequent problems in engineering, scientific computing and applied mathematics, in general, is the problem of solving a nonlinear equation $f(x) = 0$. In most of the cases, whenever real problems are faced, such as weather forecasting, accurate positioning of satellite systems in the desired orbit, measurement of earthquake magnitudes and other high-level engineering problems, only approximate solutions may get resolved. However, only in rare cases, it is possible to solve the governing equations exactly. The most familiar method of solving non linear equation is Newton's iteration method. The local order of convergence of Newton's method is two and it is an optimal method with two function evaluations per iterative step.

In the past decade, several higher order iterative methods have been developed and analyzed for solving nonlinear equations that improve classical methods such as Newton's method, Chebyshev method, Halley's iteration method, etc. As the order of convergence increases, so does the number of function evaluations per step. Hence, a new index to determine the efficiency called the efficiency index is introduced in [1] to measure the balance between these quantities. Kung-Traub [2] conjectured that the order of convergence of any multi-point without memory method with d function evaluations cannot exceed the bound 2^{d-1}, the optimal order. Thus the optimal order for three evaluations per iteration would be four, four evaluations per iteration would be eight, and so on. Recently, some fourth and eighth order optimal iterative methods have been developed (see [3–14] and references therein). A more extensive list of references as well as a survey on the progress made in the class of multi-point methods is found in the recent book by Petkovic et al. [11].

This paper is organized as follows. An optimal fourth, eighth and sixteenth order methods are developed by using divided difference techniques in Section 2. In Section 3, convergence order is analyzed. In Section 4, tested numerical examples to compare the proposed methods with other known optimal methods. The problem of Projectile motion is discussed in Section 5 where the presented methods are applied on this problem with some existing ones. In Section 6, we obtain the conjugacy maps of these methods to make a comparison from dynamical point of view. In Section 7, the proposed methods are studied in the complex plane using basins of attraction. Section 8 gives concluding remarks.

2. Design of an Optimal Fourth, Eighth and Sixteenth Order Methods

Definition 1 ([15]). *If the sequence $\{x_n\}$ tends to a limit x^* in such a way that*

$$\lim_{n\to\infty} \frac{x_{n+1} - x^*}{(x_n - x^*)^p} = C$$

for $p \geq 1$, then the order of convergence of the sequence is said to be p, and C is known as the asymptotic error constant. If $p = 1$, $p = 2$ or $p = 3$, the convergence is said to be linear, quadratic or cubic, respectively.
Let $e_n = x_n - x^*$, then the relation

$$e_{n+1} = C\, e_n^p + O\left(e_n^{p+1}\right) = O\left(e_n^p\right). \tag{1}$$

is called the error equation. The value of p is called the order of convergence of the method.

Definition 2 ([1]). *The Efficiency Index is given by*

$$EI = p^{\frac{1}{d}}, \tag{2}$$

where d is the total number of new function evaluations (the values of f and its derivatives) per iteration.

Let $x_{n+1} = \psi(x_n)$ define an Iterative Function (IF). Let x_{n+1} be determined by new information at $x_n, \phi_1(x_n), ..., \phi_{\mathrm{i}}(x_n), \mathrm{i} \geq 1$. No old information is reused. Thus,

$$x_{n+1} = \psi(x_n, \phi_1(x_n), ..., \phi_{\mathrm{i}}(x_n)). \tag{3}$$

Then ψ is called a multipoint IF without memory.
The Newton (also called Newton-Raphson) IF ($2^{nd}NR$) is given by

$$\psi_{2^{nd}NR}(x) = x - \frac{f(x)}{f'(x)}. \tag{4}$$

The $2^{nd}NR$ IF is one-point IF with two function evaluations and it satisfies the Kung-Traub conjecture with $d = 2$. Further, $EI_{2^{nd}NR} = 1.414$.

2.1. An Optimal Fourth Order Method

We attempt to get a new optimal fourth order IF as follows, let us consider two step Newton's method

$$\psi_{4^{th}NR}(x) = \psi_{2^{nd}NR}(x) - \frac{f(\psi_{2^{nd}NR}(x))}{f'(\psi_{2^{nd}NR}(x))}. \tag{5}$$

The above one is having fourth order convergence with four function evaluations. But, this is not an optimal method. To get an optimal, need to reduce a function and preserve the same convergence order, and so we estimate $f'(\psi_{2^{nd}NR}(x))$ by the following polynomial

$$q(t) = a_0 + a_1(t-x) + a_2(t-x)^2, \tag{6}$$

which satisfies

$$q(x) = f(x),\ q'(x) = f'(x),\ q(\psi_{2^{nd}NR}(x)) = f(\psi_{2^{nd}NR}(x)).$$

On implementing the above conditions on Equation (6), we obtain three unknowns a_0, a_1 and a_2. Let us define the divided differences

$$f[y,x] = \frac{f(y)-f(x)}{y-x},\ f[y,x,x] = \frac{f[y,x]-f'(x)}{y-x}.$$

From conditions, we get $a_0 = f(x)$, $a_1 = f'(x)$ and $a_2 = f[\psi_{2^{nd}NR}(x), x, x]$, respectively, by using divided difference techniques. Now, we have the estimation

$$f'(\psi_{2^{nd}NR}(x)) \approx q'(\psi_{2^{nd}NR}(x)) = a_1 + 2a_2(\psi_{2^{th}NR}(x) - x).$$

Finally, we propose a new optimal fourth order method as

$$\psi_{4^{th}YM}(x) = \psi_{2^{nd}NR}(x) - \frac{f(\psi_{2^{nd}NR}(x))}{f'(x) + 2f[\psi_{2^{nd}NR}(x), x, x](\psi_{2^{th}NR}(x) - x)}. \tag{7}$$

The efficiency of the method (7) is $EI_{4thYM} = 1.587$.

2.2. An Optimal Eighth Order Method

Next, we attempt to get a new optimal eighth order IF as following way

$$\psi_{8^{th}YM}(x) = \psi_{4^{th}YM}(x) - \frac{f(\psi_{4^{th}YM}(x))}{f'(\psi_{4^{th}YM}(x))}.$$

The above one is having eighth order convergence with five function evaluations. But, this is not an optimal method. To get an optimal, need to reduce a function and preserve the same convergence order, and so we estimate $f'(\psi_{4^{th}YM}(x))$ by the following polynomial

$$q(t) = b_0 + b_1(t-x) + b_2(t-x)^2 + b_3(t-x)^3, \tag{8}$$

which satisfies

$$q(x) = f(x),\ q'(x) = f'(x),\ q(\psi_{2^{nd}NR}(x)) = f(\psi_{2^{nd}NR}(x)),\ q(\psi_{4^{th}YM}(x)) = f(\psi_{4^{th}YM}(x)).$$

On implementing the above conditions on (8), we obtain four linear equations with four unknowns b_0, b_1, b_2 and b_3. From conditions, we get $b_0 = f(x)$ and $b_1 = f'(x)$. To find b_2 and b_3, we solve the following equations:

$$\begin{aligned} f(\psi_{2^{nd}NR}(x)) &= f(x) + f'(x)(\psi_{2^{nd}NR}(x) - x) + b_2(\psi_{2^{nd}NR}(x) - x)^2 + b_3(\psi_{2^{nd}NR}(x) - x)^3 \\ f(\psi_{4^{th}YM}(x)) &= f(x) + f'(x)(\psi_{4^{th}YM}(x) - x) + b_2(\psi_{4^{th}YM}(x) - x)^2 + b_3(\psi_{4^{th}YM}(x) - x)^3. \end{aligned}$$

Thus by applying divided differences, the above equations simplifies to

$$\begin{aligned} b_2 + b_3(\psi_{2^{nd}NR}(x) - x) &= f[\psi_{2^{nd}NR}(x), x, x] &(9)\\ b_2 + b_3(\psi_{4^{th}YM}(x) - x) &= f[\psi_{4^{th}YM}(x), x, x] &(10)\end{aligned}$$

Solving Equations (9) and (14), we have

$$\begin{aligned} b_2 &= \frac{f[\psi_{2^{nd}NR}(x), x, x](\psi_{4^{th}PM}(x) - x) - f[\psi_{4^{th}YM}(x), x, x](\psi_{2^{nd}NR}(x) - x)}{\psi_{4^{th}YM}(x) - \psi_{2^{nd}NR}(x)},\\ b_3 &= \frac{f[\psi_{4^{th}YM}(x), x, x] - f[\psi_{2^{nd}NR}(x), x, x]}{\psi_{4^{th}YM}(x) - \psi_{2^{nd}NR}(x)}. \end{aligned} \tag{11}$$

Further, using Equation (11), we have the estimation

$$f'(\psi_{4^{th}YM}(x)) \approx q'(\psi_{4^{th}YM}(x)) = b_1 + 2b_2(\psi_{4^{th}YM}(x) - x) + 3b_3(\psi_{4^{th}YM}(x) - x)^2.$$

Finally, we propose a new optimal eighth order method as

$$\psi_{8^{th}YM}(x) = \psi_{4^{th}YM}(x) - \frac{f(\psi_{4^{th}YM}(x))}{f'(x) + 2b_2(\psi_{4^{th}YM}(x) - x) + 3b_3(\psi_{4^{th}YM}(x) - x)^2}. \tag{12}$$

The efficiency of the method (12) is $EI_{8thYM} = 1.682$. Remark that the method is seems a particular case of the method of Khan et al. [16], they used weight function to develop their methods. Whereas we used finite difference scheme to develop proposed methods. We can say the methods $4^{th}YM$ and $8^{th}YM$ are reconstructed of Khan et al. [16] methods.

2.3. An Optimal Sixteenth Order Method

Next, we attempt to get a new optimal sixteenth order IF as following way

$$\psi_{16^{th}YM}(x) = \psi_{8^{th}YM}(x) - \frac{f(\psi_{8^{th}YM}(x))}{f'(\psi_{8^{th}YM}(x))}.$$

The above one is having eighth order convergence with five function evaluations. However, this is not an optimal method. To get an optimal, need to reduce a function and preserve the same convergence order, and so we estimate $f'(\psi_{8^{th}YM}(x))$ by the following polynomial

$$q(t) = c_0 + c_1(t - x) + c_2(t - x)^2 + c_3(t - x)^3 + c_4(t - x)^4, \tag{13}$$

which satisfies

$$q(x) = f(x),\ q'(x) = f'(x),\ q(\psi_{2^{nd}NR}(x)) = f(\psi_{2^{nd}NR}(x)),$$
$$q(\psi_{4^{th}YM}(x)) = f(\psi_{4^{th}YM}(x)), q(\psi_{8^{th}YM}(x)) = f(\psi_{8^{th}YM}(x)).$$

On implementing the above conditions on (13), we obtain four linear equations with four unknowns c_0, c_1, c_2 and c_3. From conditions, we get $c_0 = f(x)$ and $c_1 = f'(x)$. To find c_2, c_3 and c_4, we solve the following equations:

$$\begin{aligned} f(\psi_{2^{nd}NR}(x)) &= f(x) + f'(x)(\psi_{2^{nd}NR}(x) - x) + c_2(\psi_{2^{nd}NR}(x) - x)^2 + c_3(\psi_{2^{nd}NR}(x) - x)^3 + c_4(\psi_{2^{nd}NR}(x) - x)^4\\ f(\psi_{4^{th}YM}(x)) &= f(x) + f'(x)(\psi_{4^{th}YM}(x) - x) + c_2(\psi_{4^{th}YM}(x) - x)^2 + c_3(\psi_{4^{th}YM}(x) - x)^3 + c_4(\psi_{4^{th}YM}(x) - x)^4\\ f(\psi_{8^{th}YM}(x)) &= f(x) + f'(x)(\psi_{8^{th}YM}(x) - x) + c_2(\psi_{8^{th}YM}(x) - x)^2 + c_3(\psi_{8^{th}YM}(x) - x)^3 + c_4(\psi_{8^{th}YM}(x) - x)^4. \end{aligned}$$

Thus by applying divided differences, the above equations simplifies to

$$\begin{aligned} c_2 + c_3(\psi_{2^{nd}NR}(x) - x) + c_4(\psi_{2^{nd}NR}(x) - x)^2 &= f[\psi_{2^{nd}NR}(x), x, x] \\ c_2 + c_3(\psi_{4^{th}YM}(x) - x) + c_4(\psi_{4^{th}YM}(x) - x)^2 &= f[\psi_{4^{th}YM}(x), x, x] \\ c_2 + c_3(\psi_{8^{th}YM}(x) - x) + c_4(\psi_{8^{th}YM}(x) - x)^2 &= f[\psi_{8^{th}YM}(x), x, x] \end{aligned} \tag{14}$$

Solving Equation (14), we have

$$\begin{aligned} c_2 &= \frac{\Big(f[\psi_{2^{nd}NR}(x), x, x]\big(-S_2^2S_3 + S_2S_3^2\big) + f[\psi_{4^{th}YM}(x), x, x]\big(S_1^2S_3 - S_1S_3^2\big) + f[\psi_{8^{th}YM}(x), x, x]\big(-S_1^2S_2 + S_1S_2^2\big)\Big)}{-S_1^2S_2 + S_1S_2^2 + S_1^2S_3 - S_2^2S_3 - S_1S_3^2 + S_2S_3^2}, \\ c_3 &= \frac{\Big(f[\psi_{2^{nd}NR}(x), x, x]\big(S_2^2 - S_3^2\big) + f[\psi_{4^{th}YM}(x), x, x]\big(-S_1^2 + S_3^2\big) + f[\psi_{8^{th}YM}(x), x, x]\big(S_1^2 - S_2^2\big)\Big)}{-S_1^2S_2 + S_1S_2^2 + S_1^2S_3 - S_2^2S_3 - S_1S_3^2 + S_2S_3^2}, \\ c_4 &= \frac{\Big(f[\psi_{2^{nd}NR}(x), x, x]\big(-S_2 + S_3\big) + f[\psi_{4^{th}YM}(x), x, x]\big(S_1 - S_3\big) + f[\psi_{8^{th}YM}(x), x, x]\big(-S_1 + S_2\big)\Big)}{-S_1^2S_2 + S_1S_2^2 + S_1^2S_3 - S_2^2S_3 - S_1S_3^2 + S_2S_3^2}, \\ S_1 &= \psi_{2^{nd}NR}(x) - x,\ S_2 = \psi_{4^{th}YM}(x) - x,\ S_3 = \psi_{8^{th}YM}(x) - x. \end{aligned} \tag{15}$$

Further, using Equation (15), we have the estimation

$$f'(\psi_{8^{th}YM}(x)) \approx q'(\psi_{8^{th}YM}(x)) = c_1 + 2c_2(\psi_{8^{th}YM}(x) - x) + 3c_3(\psi_{8^{th}YM}(x) - x)^2 + 4c_4(\psi_{8^{th}YM}(x) - x)^3.$$

Finally, we propose a new optimal sixteenth order method as

$$\psi_{16^{th}YM}(x) = \psi_{8^{th}YM}(x) - \frac{f(\psi_{8^{th}YM}(x))}{f'(x) + 2c_2(\psi_{8^{th}YM}(x) - x) + 3c_3(\psi_{8^{th}YM}(x) - x)^2 + 4c_4(\psi_{8^{th}YM}(x) - x)^3}. \tag{16}$$

The efficiency of the method (16) is $EI_{16thYM} = 1.741$.

3. Convergence Analysis

In this section, we prove the convergence analysis of proposed IFs with help of Mathematica software.

Theorem 1. *Let $f : D \subset \mathbb{R} \to \mathbb{R}$ be a sufficiently smooth function having continuous derivatives. If $f(x)$ has a simple root x^* in the open interval D and x_0 chosen in sufficiently small neighborhood of x^*, then the method $4^{th}YM$ IFs (7) is of local fourth order convergence, and the $8^{th}YM$ IFs (12) is of local eighth order convergence.*

Proof. Let $e = x - x^*$ and $c[j] = \dfrac{f^{(j)}(x^*)}{j!f'(x^*)}$, $j = 2, 3, 4, \ldots$. Expanding $f(x)$ and $f'(x)$ about x^* by Taylor's method, we have

$$f(x) = f'(x^*)\Big(e + e^2c[2] + e^3c[3] + e^4c[4] + e^5c[5] + e^6c[6] + e^7c[7] + e^8c[8] + \ldots\Big) \tag{17}$$

and

$$f'(x) = f'(x^*)\Big(1 + 2e\,c[2] + 3e^2c[3] + 4e^3c[4] + 5e^4c[5] + 6e^5c[6] + 7e^6c[7] + 8e^7c[8] + 9e^8c[9] + \ldots\Big) \tag{18}$$

Thus,

$$\psi_{2^{nd}NR}(x) = x^* + c[2]e^2 + \Big(-2c[2]^2 + 2c[3]\Big)e^3 + \Big(4c[2]^3 - 7c[2]c[3] + 3c[4]\Big)e^4 + \Big(-8c[2]^4 + 20c[2]^2c[3] - 6c[3]^2 - 10c[2]c[4] + 4c[5]\Big)e^5 + \Big(16c[2]^5 - 52c[2]^3c[3] + 28c[2]^2c[4] - 17c[3]c[4] + c[2](33c[3]^2 - 13c[5]) + 5c[6]\Big)e^6 - 2\Big(16c[2]^6 - 64c[2]^4c[3] - 9c[3]^3 + 36c[2]^3c[4] + 6c[4]^2 + 9c[2]^2(7c[3]^2 - 2c[5]) + 11c[3]c[5] + c[2](-46c[3]c[4] + 8c[6]) - 3c[7]\Big)e^7 + \Big(64c[2]^7 - 304c[2]^5c[3] + 176c[2]^4c[4] + 75c[3]^2c[4] + c[2]^3(408c[3]^2 - 92c[5]) - 31c[4]c[5] - 27c[3]c[6] + c[2]^2(-348c[3]c[4] + 44c[6]) + c[2](-135c[3]^3 + 64c[4]^2 + 118c[3]c[5] - 19c[7]) + 7c[8]\Big)e^8 + \dots. \tag{19}$$

Expanding $f(\psi_{2^{nd}NR}(x))$ about x^* by Taylor's method, we have

$$f(\psi_{2^{nd}NR}(x)) = f'(x^*)\Big(c[2]e^2 + \Big(-2c[2]^2 + 2c[3]\Big)e^3 + \Big(5c[2]^3 - 7c[2]c[3] + 3c[4]\Big)e^4 - 2\Big(6c[2]^4 - 12c[2]^2c[3] + 3c[3]^2 + 5c[2]c[4] - 2c[5]\Big)e^5 + \Big(28c[2]^5 - 73c[2]^3c[3] + 34c[2]^2c[4] - 17c[3]c[4] + c[2](37c[3]^2 - 13c[5]) + 5c[6]\Big)e^6 - 2\Big(32c[2]^6 - 103c[2]^4c[3] - 9c[3]^3 + 52c[2]^3c[4] + 6c[4]^2 + c[2]^2(80c[3]^2 - 22c[5]) + 11c[3]c[5] + c[2](-52c[3]c[4] + 8c[6]) - 3c[7]\Big)e^7 + \Big(144c[2]^7 - 552c[2]^5c[3] + 297c[2]^4c[4] + 75c[3]^2c[4] + 2c[2]^3(291c[3]^2 - 67c[5]) - 31c[4]c[5] - 27c[3]c[6] + c[2]^2(-455c[3]c[4] + 54c[6]) + c[2](-147c[3]^3 + 73c[4]^2 + 134c[3]c[5] - 19c[7]) + 7c[8]\Big)e^8 + \dots\Big) \tag{20}$$

Using Equations (17)–(20) in divided difference techniques. We have

$$f[\psi_{2^{nd}NR}(x), x, x] = f'(x^*)\Big(c[2] + 2c[3]e + \Big(c[2]c[3] + 3c[4]\Big)e^2 + 2\Big(-c[2]^2c[3] + c[3]^2 + c[2]c[4] + 2c[5]\Big)e^3 + \Big(4c[2]^3c[3] - 3c[2]^2c[4] + 7c[3]c[4] + c[2](-7c[3]^2 + 3c[5]) + 5c[6]\Big)e^4 + \Big(-8c[2]^4c[3] - 6c[3]^3 + 4c[2]^3c[4] + 4c[2]^2(5c[3]^2 - c[5]) + 10c[3]c[5] + 4c[2](-5c[3]c[4] + c[6]) + 6(c[4]^2 + c[7])\Big)e^5 + \Big(16c[2]^5c[3] - 4c[2]^4c[4] - 25c[3]^2c[4] + 17c[4]c[5] + c[2]^3(-52c[3]^2 + 5c[5]) + c[2]^2(46c[3]c[4] - 5c[6]) + 13c[3]c[6] + c[2](33c[3]^3 - 14c[4]^2 - 26c[3]c[5] + 5c[7]) + 7c[8]\Big)e^6 + \dots\Big) \tag{21}$$

Substituting Equations (18)–(21) into Equation (7), we obtain, after simplifications,

$$\psi_{4^{th}YM}(x) = x^* + \Big(c[2]^3 - c[2]c[3]\Big)e^4 - 2\Big(2c[2]^4 - 4c[2]^2c[3] + c[3]^2 + c[2]c[4]\Big)e^5 + \Big(10c[2]^5 - 30c[2]^3c[3] + 12c[2]^2c[4] - 7c[3]c[4] + 3c[2](6c[3]^2 - c[5])\Big)e^6 - 2\Big(10c[2]^6 - 40c[2]^4c[3] - 6c[3]^3 + 20c[2]^3c[4] + 3c[4]^2 + 8c[2]^2(5c[3]^2 - c[5]) + 5c[3]c[5] + c[2](-26c[3]c[4] + 2c[6])\Big)e^7 + \Big(36c[2]^7 - 178c[2]^5c[3] + 101c[2]^4c[4] + 50c[3]^2c[4] + 3c[2]^3(84c[3]^2 - 17c[5]) - 17c[4]c[5] - 13c[3]c[6] + c[2]^2(-209c[3]c[4] + 20c[6]) + c[2](-91c[3]^3 + 37c[4]^2 + 68c[3]c[5] - 5c[7])\Big)e^8 + \dots. \tag{22}$$

Expanding $f(\psi_{4^{th}YM}(x))$ about x^* by Taylor's method, we have

$$\begin{aligned} f(\psi_{4^{th}YM}(x)) = f'(x^*)\Big(&\Big(c[2]^3 - c[2]c[3]\Big)e^4 - 2\Big(2c[2]^4 - 4c[2]^2c[3] + c[3]^2 + c[2]c[4]\Big)e^5 + \Big(10c[2]^5 \\ &- 30c[2]^3c[3] + 12c[2]^2c[4] - 7c[3]c[4] + 3c[2](6c[3]^2 - c[5])\Big)e^6 - 2\Big(10c[2]^6 - 40c[2]^4c[3] \\ &- 6c[3]^3 + 20c[2]^3c[4] + 3c[4]^2 + 8c[2]^2(5c[3]^2 - c[5]) + 5c[3]c[5] + c[2](-26c[3]c[4] + 2c[6])\Big)e^7 \\ &+ \Big(37c[2]^7 - 180c[2]^5c[3] + 101c[2]^4c[4] + 50c[3]^2c[4] + c[2]^3(253c[3]^2 - 51c[5]) - 17c[4]c[5] \\ &- 13c[3]c[6] + c[2]^2(-209c[3]c[4] + 20c[6]) + c[2](-91c[3]^3 + 37c[4]^2 + 68c[3]c[5] - 5c[7])\Big)e^8 + \dots\Big) \end{aligned} \tag{23}$$

Now,

$$\begin{aligned} f[\psi_{4^{th}YM}(x), x, x] = f'(x^*)\Big(&c[2] + 2c[3]e + 3c[4]e^2 + 4c[5]e^3 + \Big(c[2]^3c[3] - c[2]c[3]^2 + 5c[6]\Big)e^4 \\ &+ \Big(-4c[2]^4c[3] + 8c[2]^2c[3]^2 - 2c[3]^3 + 2c[2]^3c[4] - 4c[2]c[3]c[4] + 6c[7]\Big)e^5 \\ &+ \Big(10c[2]^5c[3] - 8c[2]^4c[4] + 28c[2]^2c[3]c[4] - 11c[3]^2c[4] + c[2]^3(-30c[3]^2 + 3c[5]) + \\ &2c[2](9c[3]^3 - 2c[4]^2 - 3c[3]c[5]) + 7c[8]\Big)e^6 + \dots\Big) \end{aligned} \tag{24}$$

Substituting Equations (19)–(21), (23) and (24) into Equation (12), we obtain, after simplifications,

$$\psi_{8^{th}YM}(x) - x^* = c[2]^2\Big(c[2]^2 - c[3]\Big)\Big(c[2]^3 - c[2]c[3] + c[4]\Big)e^8 + O(e^9) \tag{25}$$

Hence from Equations (22) and (25), we concluded that the convergence order of the $4^{th}YM$ and $8^{th}YM$ are four and eight respectively. □

The following theorem is given without proof, which can be worked out with the help of Mathematica.

Theorem 2. *Let $f : D \subset \mathbb{R} \to \mathbb{R}$ be a sufficiently smooth function having continuous derivatives. If $f(x)$ has a simple root x^* in the open interval D and x_0 chosen in sufficiently small neighborhood of x^*, then the method* (16) *is of local sixteenth order convergence and and it satisfies the error equation*

$$\psi_{16^{th}YM}(x) - x^* = \Big((c[2]^4)((c[2]^2 - c[3])^2)(c[2]^3 - c[2]c[3] + c[4])(c[2]^4 - c[2]^2c[3] + c[2]c[4] - c[5])\Big)e^{16} + O(e^{17}).$$

4. Numerical Examples

In this section, numerical results on some test functions are compared for the new methods $4^{th}YM$, $8^{th}YM$ and $16^{th}YM$ with some existing eighth order methods and Newton's method. Numerical computations have been carried out in the MATLAB software with 500 significant digits. We have used the stopping criteria for the iterative process satisfying $error = |x_N - x_{N-1}| < \epsilon$, where $\epsilon = 10^{-50}$ and N is the number of iterations required for convergence. The computational order of convergence is given by ([17])

$$\rho = \frac{\ln|(x_N - x_{N-1})/(x_{N-1} - x_{N-2})|}{\ln|(x_{N-1} - x_{N-2})/(x_{N-2} - x_{N-3})|}.$$

We consider the following iterative methods for solving nonlinear equations for the purpose of comparison: $\psi_{4^{th}SB}$, a method proposed by Sharma et al. [18]:

$$y = x - \frac{2f(x)}{3f'(x)}, \psi_{4^{th}SB}(x) = x - \Big(-\frac{1}{2} + \frac{9}{8}\frac{f'(x)}{f'(y)} + \frac{3}{8}\frac{f'(y)}{f'(x)}\Big)\frac{f(x)}{f'(x)}. \tag{26}$$

$\psi_{4^{th}CLND}$, a method proposed by Chun et al. [19]:

$$y = x - \frac{2f(x)}{3f'(x)}, \psi_{4^{th}CLND}(x) = x - \frac{16f(x)f'(x)}{-5f'(x)^2 + 30f'(x)f'(y) - 9f'(y)^2}. \tag{27}$$

$\psi_{4^{th}SJ}$, a method proposed by Singh et al. [20]:

$$y = x - \frac{2}{3}\frac{f(x)}{f'(x)}, \psi_{4^{th}SJ}(x) = x - \left(\frac{17}{8} - \frac{9}{4}\frac{f'(y)}{f'(x)} + \frac{9}{8}\left(\frac{f'(y)}{f'(x)}\right)^2\right)\left(\frac{7}{4} - \frac{3}{4}\frac{f'(y)}{f'(x_n)}\right)\frac{f(x)}{f'(x)}. \tag{28}$$

$\psi_{8^{th}KT}$, a method proposed by Kung-Traub [2]:

$$\begin{aligned} y &= x - \frac{f(x)}{f'(x)}, z = y - \frac{f(y) * f(x)}{(f(x) - f(y))^2}\frac{f(x)}{f'(x)}, \\ \psi_{8^{th}KT}(x) &= z - \frac{f(x)}{f'(x)}\frac{f(x)f(y)f(z)}{(f(x) - f(y))^2}\frac{f(x)^2 + f(y)(f(y) - f(z))}{(f(x) - f(z))^2(f(y) - f(z))}. \end{aligned} \tag{29}$$

$\psi_{8^{th}LW}$, a method proposed by Liu et al. [8]

$$\begin{aligned} y &= x - \frac{f(x)}{f'(x)}, z = y - \frac{f(x)}{f(x) - 2f(y)}\frac{f(y)}{f'(x)}, \\ \psi_{8^{th}LW}(x) &= z - \frac{f(z)}{f'(x)}\left(\left(\frac{f(x) - f(y)}{f(x) - 2f(y)}\right)^2 + \frac{f(z)}{f(y) - f(z)} + \frac{4f(z)}{f(x) + f(z)}\right). \end{aligned} \tag{30}$$

$\psi_{8^{th}PNPD}$, a method proposed by Petkovic et al. [11]

$$y = x - \frac{f(x)}{f'(x)}, z = x - \left(\left(\frac{f(y)}{f(x)}\right)^2 - \frac{f(x)}{f(y) - f(x)}\right)\frac{f(x)}{f'(x)}, \psi_{8^{th}PNPD}(x) = z - \frac{f(z)}{f'(x)}\left(\varphi(t) + \frac{f(z)}{f(y) - f(z)} + \frac{4f(z)}{f(x)}\right), \tag{31}$$

where $\varphi(t) = 1 + 2t + 2t^2 - t^3$ *and* $t = \dfrac{f(y)}{f(x)}$.

$\psi_{8^{th}SA1}$, a method proposed by Sharma et al. [12]

$$y = x - \frac{f(x)}{f'(x)}, z = y - \left(3 - 2\frac{f[y,x]}{f'(x)}\right)\frac{f(y)}{f'(x)}, \psi_{8^{th}SA1}(x) = z - \frac{f(z)}{f'(x)}\left(\frac{f'(x) - f[y,x] + f[z,y]}{2f[z,y] - f[z,x]}\right). \tag{32}$$

$\psi_{8^{th}SA2}$, a method proposed by Sharma et al. [13]

$$y = x - \frac{f(x)}{f'(x)}, z = y - \frac{f(y)}{2f[y,x] - f'(x)}, \psi_{8^{th}SA2}(x) = z - \frac{f[z,y]}{f[z,x]}\frac{f(z)}{2f[z,y] - f[z,x]} \tag{33}$$

$\psi_{8^{th}CFGT}$, a method proposed by Cordero et al. [6]

$$\begin{aligned} y = x - \frac{f(x)}{f'(x)}, z = y - \frac{f(y)}{f'(x)}\frac{1}{1 - 2t + t^2 - t^3/2}, \psi_{8^{th}CFGT}(x) = \\ z - \frac{1 + 3r}{1 + r}\frac{f(z)}{f[z,y] + f[z,x,x](z - y)}, r = \frac{f(z)}{f(x)}. \end{aligned} \tag{34}$$

$\psi_{8^{th}CTV}$, a method proposed by Cordero et al. [7]

$$y = x - \frac{f(x)}{f'(x)}, z = x - \frac{1 - t}{1 - 2t}\frac{f(x)}{f'(x)}, \psi_{8^{th}CTV}(x) = z - \left(\frac{1 - t}{1 - 2t} - v\right)^2\frac{1}{1 - 3v}\frac{f(z)}{f'(x)}, \ v = \frac{f(z)}{f(y)}. \tag{35}$$

Table 1 shows the efficiency indices of the new methods with some known methods.

Table 1. Comparison of Efficiency Indices.

Methods	p	d	EI
2^{nd}NR	2	2	1.414
4^{th}SB	4	3	1.587
4^{th}CLND	4	3	1.587
4^{th}SJ	4	3	1.587
4^{th}YM	4	3	1.587
8^{th}KT	8	4	1.682
8^{th}LW	8	4	1.682
8^{th}PNPD	8	4	1.682
8^{th}SA1	8	4	1.682
8^{th}SA2	8	4	1.682
8^{th}CFGT	8	4	1.682
8^{th}CTV	8	4	1.682
8^{th}YM	8	4	1.682
16^{th}YM	16	5	1.741

The following test functions and their simple zeros for our study are given below [10]:

$$f_1(x) = \sin(2\cos x) - 1 - x^2 + e^{\sin(x^3)}, \qquad x^* = -0.7848959876612125352...$$
$$f_2(x) = xe^{x^2} - \sin^2 x + 3\cos x + 5, \qquad x^* = -1.2076478271309189270...$$
$$f_3(x) = x^3 + 4x^2 - 10, \qquad x^* = 1.3652300134140968457...$$
$$f_4(x) = sin(x) + cos(x) + x, \qquad x^* = -0.4566247045676308244...$$
$$f_5(x) = \frac{x}{2} - \sin x, \qquad x^* = 1.8954942670339809471...$$
$$f_6(x) = x^2 + sin(\frac{x}{5}) - \frac{1}{4}, \qquad x^* = 0.4099920179891371316...$$

Table 2, shows that corresponding results for $f_1 - f_6$. We observe that proposed method $4^{th}YM$ is converge in a lesser or equal number of iterations and with least error when compare to compared methods. Note that $4^{th}SB$ and $4^{th}SJ$ methods are getting diverge in f_5 function. Hence, the proposed method $4^{th}YM$ can be considered competent enough to existing other compared equivalent methods.

Also, from Tables 3–5 are shows the corresponding results for $f_1 - f_6$. The computational order of convergence agrees with the theoretical order of convergence in all the functions. Note that $8^{th}PNPD$ method is getting diverge in f_5 function and all other compared methods are converges with least error. Also, function f_1 having least error in $8^{th}CFGT$, function f_2 having least error in $8^{th}CTV$, functions f_3 and f_4 having least error in $8^{th}YM$, function f_5 having least error in $8^{th}SA2$, function f_6 having least error in $8^{th}CFGT$. The proposed $16^{th}YM$ method converges less number of iteration with least error in all the tested functions. Hence, the $16^{th}YM$ can be considered competent enough to existing other compared equivalent methods.

Table 2. Numerical results for nonlinear equations.

Methods	$f_1(x), x_0 = -0.9$				$f_2(x), x_0 = -1.6$			
	N	$\|x_1 - x_0\|$	$\|x_N - x_{N-1}\|$	ρ	N	$\|x_1 - x_0\|$	$\|x_N - x_{N-1}\|$	ρ
$2^{nd}NR$ (4)	7	0.1080	7.7326×10^{-74}	1.99	9	0.2044	9.2727×10^{-74}	1.99
$4^{th}SB$ (26)	4	0.1150	9.7275×10^{-64}	3.99	5	0.3343	1.4237×10^{-65}	3.99
$4^{th}CLND$ (27)	4	0.1150	1.4296×10^{-64}	3.99	5	0.3801	1.1080×10^{-72}	3.99
$4^{th}SJ$ (28)	4	0.1150	3.0653×10^{-62}	3.99	5	0.3190	9.9781×10^{-56}	3.99
$4^{th}YM$ (7)	4	0.1150	6.0046×10^{-67}	3.99	5	0.3737	7.2910×10^{-120}	4.00
Methods	$f_3(x), x_0 = 0.9$				$f_4(x), x_0 = -1.9$			
$2^{nd}NR$ (4)	8	0.6263	1.3514×10^{-72}	2.00	8	1.9529	1.6092×10^{-72}	1.99
$4^{th}SB$ (26)	5	0.5018	4.5722×10^{-106}	3.99	5	1.5940	6.0381×10^{-92}	3.99
$4^{th}CLND$ (27)	5	0.5012	4.7331×10^{-108}	3.99	5	1.5894	2.7352×10^{-93}	3.99
$4^{th}SJ$ (28)	5	0.4767	3.0351×10^{-135}	3.99	5	1.5776	9.5025×10^{-95}	3.99
$4^{th}YM$ (7)	5	0.4735	2.6396×10^{-156}	3.99	5	1.5519	1.4400×10^{-102}	3.99
Methods	$f_5(x), x_0 = 1.2$				$f_6(x), x_0 = 0.8$			
$2^{nd}NR$ (4)	9	2.4123	1.3564×10^{-83}	1.99	8	0.3056	3.2094×10^{-72}	1.99
$4^{th}SB$ (26)			Diverge		5	0.3801	2.8269×10^{-122}	3.99
$4^{th}CLND$ (27)	14	0.0566	6.8760×10^{-134}	3.99	5	0.3812	7.8638×10^{-127}	3.99
$4^{th}SJ$ (28)			Diverge		5	0.3780	1.4355×10^{-114}	3.99
$4^{th}YM$ (7)	6	1.2887	2.3155×10^{-149}	3.99	5	0.3840	1.1319×10^{-143}	3.99

Table 3. Numerical results for nonlinear equations.

Methods	$f_1(x), x_0 = -0.9$				$f_2(x), x_0 = -1.6$			
	N	$\|x_1 - x_0\|$	$\|x_N - x_{N-1}\|$	ρ	N	$\|x_1 - x_0\|$	$\|x_N - x_{N-1}\|$	ρ
$8^{th}KT$ (29)	3	0.1151	1.6238×10^{-61}	7.91	4	0.3876	7.2890×10^{-137}	7.99
$8^{th}LW$ (30)	3	0.1151	4.5242×10^{-59}	7.91	4	0.3904	1.1195×10^{-170}	8.00
$8^{th}PNPD$ (31)	3	0.1151	8.8549×10^{-56}	7.87	4	0.3734	2.3461×10^{-85}	7.99
$8^{th}SA1$ (32)	3	0.1151	3.4432×10^{-60}	7.88	4	0.3983	8.4343×10^{-121}	8.00
$8^{th}SA2$ (33)	3	0.1151	6.9371×10^{-67}	7.99	4	0.3927	5.9247×10^{-225}	7.99
$8^{th}CFGT$ (34)	3	0.1151	1.1715×10^{-82}	7.77	5	0.1532	2.0650×10^{-183}	7.99
$8^{th}CTV$ (35)	3	0.1151	4.4923×10^{-61}	7.94	4	0.3925	2.3865×10^{-252}	7.99
$8^{th}YM$ (12)	3	0.1151	1.1416×10^{-70}	7.96	4	0.3896	8.9301×10^{-163}	8.00
$16^{th}YM$ (16)	3	0.1151	0	15.99	3	0.3923	3.5535×10^{-85}	16.20

Table 4. Numerical results for nonlinear equations.

Methods	$f_3(x), x_0 = 0.9$				$f_4(x), x_0 = -1.9$			
	N	$\|x_1 - x_0\|$	$\|x_N - x_{N-1}\|$	ρ	N	$\|x_1 - x_0\|$	$\|x_N - x_{N-1}\|$	ρ
$8^{th}KT$ (29)	4	0.4659	5.0765×10^{-216}	7.99	4	1.4461	5.5095×10^{-204}	8.00
$8^{th}LW$ (30)	4	0.4660	2.7346×10^{-213}	7.99	4	1.4620	3.7210×10^{-146}	8.00
$8^{th}PNPD$ (31)	4	0.3821	9.9119×10^{-71}	8.02	4	1.3858	2.0603×10^{-116}	7.98
$8^{th}SA1$ (32)	4	0.4492	1.5396×10^{-122}	8.00	4	1.4170	2.2735×10^{-136}	7.99
$8^{th}SA2$ (33)	4	0.4652	4.1445×10^{-254}	7.98	4	1.4339	2.5430×10^{-166}	7.99
$8^{th}CFGT$ (34)	4	0.4654	2.4091×10^{-260}	7.99	4	1.4417	4.7007×10^{-224}	7.99
$8^{th}CTV$ (35)	4	0.4652	3.8782×10^{-288}	8.00	4	1.3957	3.7790×10^{-117}	7.99
$8^{th}YM$ (12)	4	0.4653	3.5460×10^{-309}	7.99	4	1.4417	2.9317×10^{-229}	7.99
$16^{th}YM$ (16)	3	0.4652	3.6310×10^{-154}	16.13	3	1.4434	1.8489×10^{-110}	16.36

Table 5. Numerical results for nonlinear equations.

Methods	$f_5(x), x_0 = 1.2$				$f_6(x), x_0 = 0.8$			
	N	$\|x_1 - x_0\|$	$\|x_N - x_{N-1}\|$	ρ	N	$\|x_1 - x_0\|$	$\|x_N - x_{N-1}\|$	ρ
$8^{th}KT$ (29)	5	1.8787	2.6836×10^{-182}	7.99	4	0.3898	6.0701×10^{-234}	7.99
$8^{th}LW$ (30)	6	40.5156	4.6640×10^{-161}	7.99	4	0.3898	6.1410×10^{-228}	7.99
$8^{th}PNPD$ (31)			Diverge		4	0.3894	3.6051×10^{-190}	7.99
$8^{th}SA1$ (32)	7	891.9802	2.1076×10^{-215}	9.00	4	0.3901	5.9608×10^{-245}	8.00
$8^{th}SA2$ (33)	4	0.7161	5.3670×10^{-128}	7.99	4	0.3900	8.3398×10^{-251}	8.61
$8^{th}CFGT$ (34)	5	2.8541	0	7.99	4	0.3900	0	7.99
$8^{th}CTV$ (35)	5	0.6192	1.6474×10^{-219}	9.00	4	0.3901	1.0314×10^{-274}	8.00
$8^{th}YM$ (12)	4	0.7733	1.3183×10^{-87}	7.98	4	0.3900	1.2160×10^{-286}	7.99
$16^{th}YM$ (16)	4	0.6985	0	16.10	3	0.3900	1.1066×10^{-143}	15.73

5. Applications to Some Real World Problems

5.1. Projectile Motion Problem

We consider the classical projectile problem [21,22] in which a projectile is launched from a tower of height $h > 0$, with initial speed v and at an angle θ with respect to the horizontal onto a hill, which is defined by the function ω, called the impact function which is dependent on the horizontal distance, x. We wish to find the optimal launch angle θ_m which maximizes the horizontal distance. In our calculations, we neglect air resistances.

The path function $y = P(x)$ that describes the motion of the projectile is given by

$$P(x) = h + x\tan\theta - \frac{gx^2}{2v^2}\sec^2\theta \tag{36}$$

When the projectile hits the hill, there is a value of x for which $P(x) = \omega(x)$ for each value of x. We wish to find the value of θ that maximize x.

$$\omega(x) = P(x) = h + x\tan\theta - \frac{gx^2}{2v^2}\sec^2\theta \tag{37}$$

Differentiating Equation (37) implicitly w.r.t. θ, we have

$$\omega'(x)\frac{dx}{d\theta} = x\sec^2\theta + \frac{dx}{d\theta}\tan\theta - \frac{g}{v^2}\left(x^2\sec^2\theta\tan\theta + x\frac{dx}{d\theta}\sec^2\theta\right) \tag{38}$$

Setting $\dfrac{dx}{d\theta} = 0$ in Equation (38), we have

$$x_m = \frac{v^2}{g}\cot\theta_m \tag{39}$$

or

$$\theta_m = \arctan\left(\frac{v^2}{g\,x_m}\right) \tag{40}$$

An enveloping parabola is a path that encloses and intersects all possible paths. Henelsmith [23] derived an enveloping parabola by maximizing the height of the projectile fora given horizontal distance x, which will give the path that encloses all possible paths. Let $w = \tan\theta$, then Equation (36) becomes

$$y = P(x) = h + xw - \frac{gx^2}{2v^2}(1 + w^2) \tag{41}$$

Differentiating Equation (41) w.r.t. w and setting $y' = 0$, Henelsmith obtained

$$\begin{aligned} y' &= x - \frac{xg^2}{v^2}(w) = 0 \\ w &= \frac{v^2}{g\,x} \end{aligned} \tag{42}$$

so that the enveloping parabola defined by

$$y_m = \rho(x) = h + \frac{v^2}{2g} - \frac{gx^2}{2v^2} \tag{43}$$

The solution to the projectile problem requires first finding x_m which satisfies $\rho(x) = \omega(x)$ and solving for θ_m in Equation (40) because we want to find the point at which the enveloping parabola ρ intersects the impact function ω, and then find θ that corresponds to this point on the enveloping parabola. We choose a linear impact function $\omega(x) = 0.4x$ with $h = 10$ and $v = 20$. We let $g = 9.8$. Then we apply our IFs starting from $x_0 = 30$ to solve the non-linear equation

$$f(x) = \rho(x) - \omega(x) = h + \frac{v^2}{2g} - \frac{gx^2}{2v^2} - 0.4x$$

whose root is given by $x_m = 36.102990117.....$ and

$$\theta_m = \arctan\left(\frac{v^2}{g\,x_m}\right) = 48.5^\circ.$$

Figure 1 shows the intersection of the path function, the enveloping parabola and the linear impact function for this application. The approximate solutions are calculated correct to 500 significant figures. The stopping criterion $|x_N - x_{N-1}| < \epsilon$, where $\epsilon = 10^{-50}$ is used. Table 6 shows that proposed method 16^{th}YM is converging better than other compared methods. Also, we observe that the computational order of convergence agrees with the theoretical order of convergence.

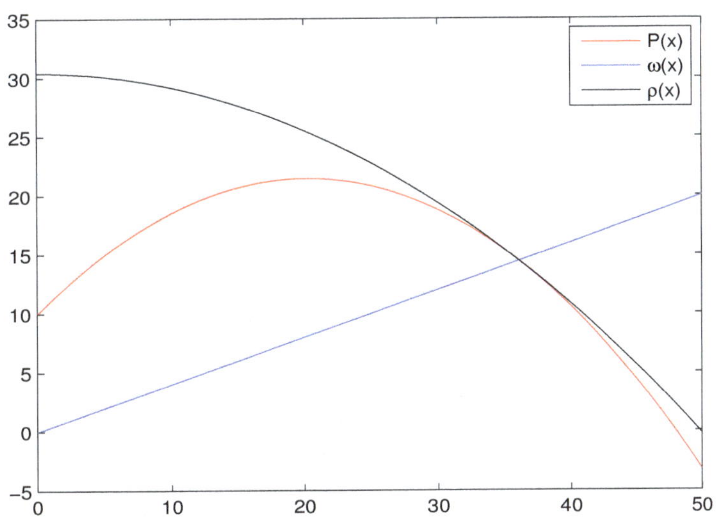

Figure 1. The enveloping parabola with linear impact function.

Table 6. Results of projectile problem.

IF	N	Error	*cpu Time(s)*	ρ
2^{nd}NR	7	4.3980×10^{-76}	1.074036	1.99
4^{th}YM	4	4.3980×10^{-76}	0.902015	3.99
8^{th}KT	3	1.5610×10^{-66}	0.658235	8.03
8^{th}LW	3	7.8416×10^{-66}	0.672524	8.03
8^{th}PNPD	3	4.2702×10^{-57}	0.672042	8.05
8^{th}SA1	3	1.2092×10^{-61}	0.654623	8.06
8^{th}CTV	3	3.5871×10^{-73}	0.689627	8.02
8^{th}YM	3	4.3980×10^{-76}	0.618145	8.02
16^{th}YM	3	0	0.512152	16.01

5.2. Planck's Radiation Law Problem

We consider the following Planck's radiation law problem found in [24]:

$$\varphi(\lambda) = \frac{8\pi ch\lambda^{-5}}{e^{ch/\lambda kT} - 1}, \tag{44}$$

which calculates the energy density within an isothermal blackbody. Here, λ is the wavelength of the radiation, T is the absolute temperature of the blackbody, k is Boltzmann's constant, h is the Planck's constant and c is the speed of light. Suppose, we would like to determine wavelength λ which corresponds to maximum energy density $\varphi(\lambda)$. From (44), we get

$$\varphi'(\lambda) = \Big(\frac{8\pi ch\lambda^{-6}}{e^{ch/\lambda kT} - 1}\Big)\Big(\frac{(ch/\lambda kT)e^{ch/\lambda kT}}{e^{ch/\lambda kT} - 1} - 5\Big) = A \cdot B.$$

It can be checked that a maxima for φ occurs when $B = 0$, that is, when

$$\Big(\frac{(ch/\lambda kT)e^{ch/\lambda kT}}{e^{ch/\lambda kT} - 1}\Big) = 5.$$

Here putting $x = ch/\lambda kT$, the above equation becomes

$$1 - \frac{x}{5} = e^{-x}. \tag{45}$$

Define

$$f(x) = e^{-x} - 1 + \frac{x}{5}. \tag{46}$$

The aim is to find a root of the equation $f(x) = 0$. Obviously, one of the root $x = 0$ is not taken for discussion. As argued in [24], the left-hand side of (45) is zero for $x = 5$ and $e^{-5} \approx 6.74 \times 10^{-3}$. Hence, it is expected that another root of the equation $f(x) = 0$ might occur near $x = 5$. The approximate root of the Equation (46) is given by $x^* \approx 4.96511423174427630369$ with $x_0 = 3$. Consequently, the wavelength of radiation (λ) corresponding to which the energy density is maximum is approximated as

$$\lambda \approx \frac{ch}{(kT)4.96511423174427630369}.$$

Table 7 shows that proposed method 16^{th}YM is converging better than other compared methods. Also, we observe that the computational order of convergence agrees with the theoretical order of convergence.

Table 7. Results of Planck's radiation law problem.

IF	N	Error	$cpu\ Time(s)$	ρ
2^{nd}NR	7	1.8205×10^{-70}	0.991020	2.00
4^{th}YM	5	1.4688×10^{-181}	0.842220	4.00
8^{th}KT	4	4.0810×10^{-288}	0.808787	7.99
8^{th}LW	4	3.1188×10^{-268}	0.801304	7.99
8^{th}PNPD	4	8.0615×10^{-260}	0.800895	7.99
8^{th}SA1	4	1.9335×10^{-298}	0.791706	8.00
8^{th}CTV	4	5.8673×10^{-282}	0.831006	8.00
8^{th}YM	4	2.5197×10^{-322}	0.855137	8.00
16^{th}YM	3	8.3176×10^{-153}	0.828053	16.52

Hereafter, we will study the optimal fourth and eighth order methods along with Newton's method.

6. Corresponding Conjugacy Maps for Quadratic Polynomials

In this section, we discuss the rational map R_p arising from $2^{nd}NR$, proposed methods $4^{th}YM$ and $8^{th}YM$ applied to a generic polynomial with simple roots.

Theorem 3. *($2^{nd}NR$) [18] For a rational map $R_p(z)$ arising from Newton's method (4) applied to $p(z) = (z-a)(z-b)$, $a \neq b$, $R_p(z)$ is conjugate via the a Möbius transformation given by $M(z) = (z-a)/(z-b)$ to*

$$S(z) = z^2.$$

Theorem 4. *($4^{th}YM$) For a rational map $R_p(z)$ arising from Proposed Method (7) applied to $p(z) = (z-a)(z-b)$, $a \neq b$, $R_p(z)$ is conjugate via the a Möbius transformation given by $M(z) = (z-a)/(z-b)$ to*

$$S(z) = z^4.$$

Proof. Let $p(z) = (z-a)(z-b)$, $a \neq b$, and let M be Möbius transformation given by $M(z) = (z-a)/(z-b)$ with its inverse $M^{-1}(z) = \frac{(zb-a)}{(z-1)}$, which may be considered as map from $\mathbb{C} \cup \{\infty\}$. We then have

$$S(z) = M \circ R_p \circ M^{-1}(z) = M\Big(R_p\Big(\frac{zb-a}{z-1}\Big)\Big) = z^4.$$

□

Theorem 5. *($8^{th}YM$) For a rational map $R_p(z)$ arising from Proposed Method (12) applied to $p(z) = (z-a)(z-b)$, $a \neq b$, $R_p(z)$ is conjugate via the a Möbius transformation given by $M(z) = (z-a)/(z-b)$ to*

$$S(z) = z^8.$$

Proof. Let $p(z) = (z-a)(z-b)$, $a \neq b$, and let M be Möbius transformation given by $M(z) = (z-a)/(z-b)$ with its inverse $M^{-1}(z) = \frac{(zb-a)}{(z-1)}$, which may be considered as map from $\mathbb{C} \cup \{\infty\}$. We then have

$$S(z) = M \circ R_p \circ M^{-1}(z) = M\Big(R_p\Big(\frac{zb-a}{z-1}\Big)\Big) = z^8.$$

□

Remark 1. *The methods (29)–(35) are given without proof, which can be worked out with the help of Mathematica.*

Remark 2. *All the maps obtained above are of the form $S(z) = z^p R(z)$, where $R(z)$ is either unity or a rational function and p is the order of the method.*

7. Basins of Attraction

The study of dynamical behavior of the rational function associated to an iterative method gives important information about convergence and stability of the method. The basic definitions and dynamical concepts of rational function which can found in [4,25].

We take a square $\mathbb{R} \times \mathbb{R} = [-2,2] \times [-2,2]$ of 256×256 points and we apply our iterative methods starting in every $z^{(0)}$ in the square. If the sequence generated by the iterative method attempts a zero z_j^* of the polynomial with a tolerance $|f(z^{(k)})| < \times 10^{-4}$ and a maximum of 100 iterations, we decide that $z^{(0)}$ is in the basin of attraction of this zero. If the iterative method starting in $z^{(0)}$ reaches a zero in N iterations ($N \leq 100$), then we mark this point $z^{(0)}$ with colors if $|z^{(N)} - z_j^*| < \times 10^{-4}$. If $N > 50$, we conclude that the starting point has diverged and we assign a dark blue color. Let N_D be a number of diverging points and we count the number of starting points which converge in 1, 2, 3, 4, 5 or above 5 iterations. In the following, we describe the basins of attraction for Newton's method and some higher order Newton type methods for finding complex roots of polynomials $p_1(z) = z^2 - 1$, $p_2(z) = z^3 - 1$ and $p_3(z) = z^5 - 1$.

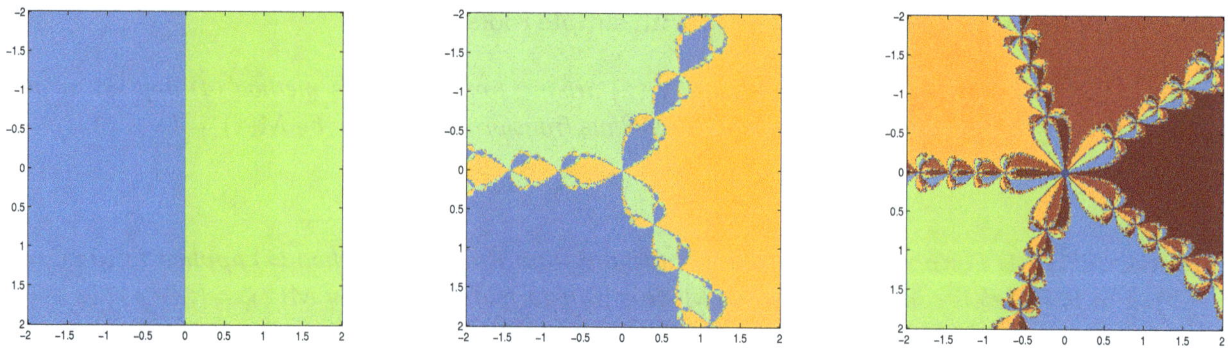

$p_1(z) = z^2 - 1$ $p_2(z) = z^3 - 1$ $p_3(z) = z^5 - 1$

Figure 2. Basins of attraction for $2^{nd}NR$ for the polynomial $p_1(z)$, $p_2(z)$, $p_3(z)$.

Figures 2 and 3 shows the polynomiographs of the methods for the polynomial $p_1(z)$. We can see that the methods 2^{nd}NR, 4^{th}YM, 8^{th}SA2 and 8^{th}YM performed very nicely. The methods 4^{th}SB, 4^{th}SJ, 8^{th}KT, 8^{th}LW, 8^{th}PNPD, 8^{th}SA1, 8^{th}CFGT and 8^{th}CTV are shows some chaotic behavior near the boundary points. The method 4^{th}CLND have sensitive to the choice of initial guess in this case.

Figures 2 and 4 shows the polynomiographs of the methods for the polynomial $p_2(z)$. We can see that the methods 2^{nd}NR, 4^{th}YM, 8^{th}SA2 and 8^{th}YM performed very nicely. The methods 4^{th}SB, 8^{th}KT, 8^{th}LW and 8^{th}CTV are shows some chaotic behavior near the boundary points. The methods 4^{th}CLND, 4^{th}SJ, 8^{th}PNPD, 8^{th}SA1, and 8^{th}CFGT have sensitive to the choice of initial guess in this case.

Figures 2 and 5 shows the polynomiographs of the methods for the polynomial $p_3(z)$. We can see that the methods 4^{th}YM, 8^{th}SA2 and 8^{th}YM are shows some chaotic behavior near the boundary points. The methods 2^{nd}NR, 4^{th}SB, 4^{th}CLND, 4^{th}SJ, 8^{th}KT, 8^{th}LW, 8^{th}PNPD, 8^{th}SA1, 8^{th}CFGT and 8^{th}CTV have sensitive to the choice of initial guess in this case. In Tables 8–10, we classify the number of converging and diverging grid points for each iterative method.

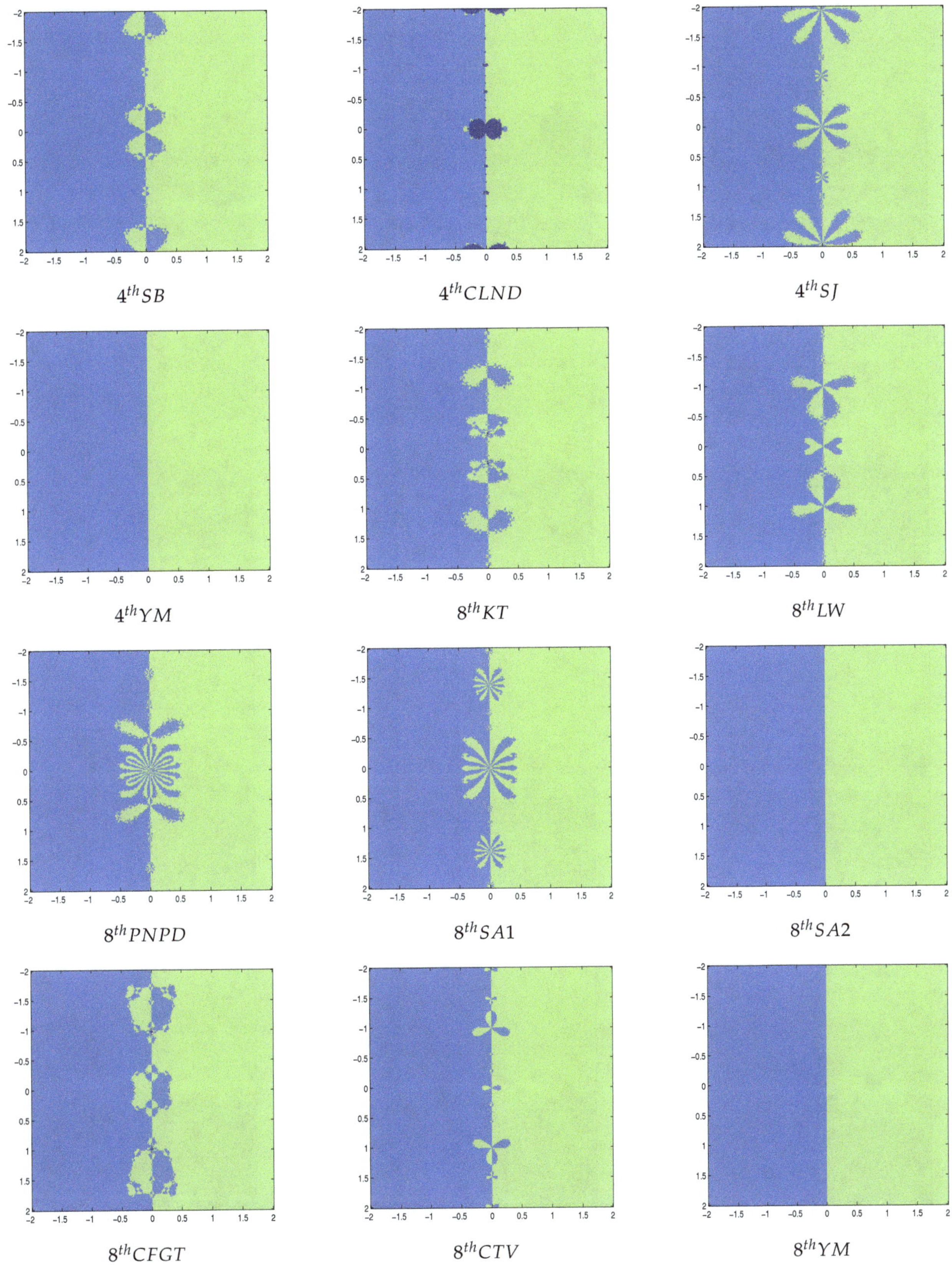

Figure 3. Basins of attraction for $p_1(z) = z^2 - 1$.

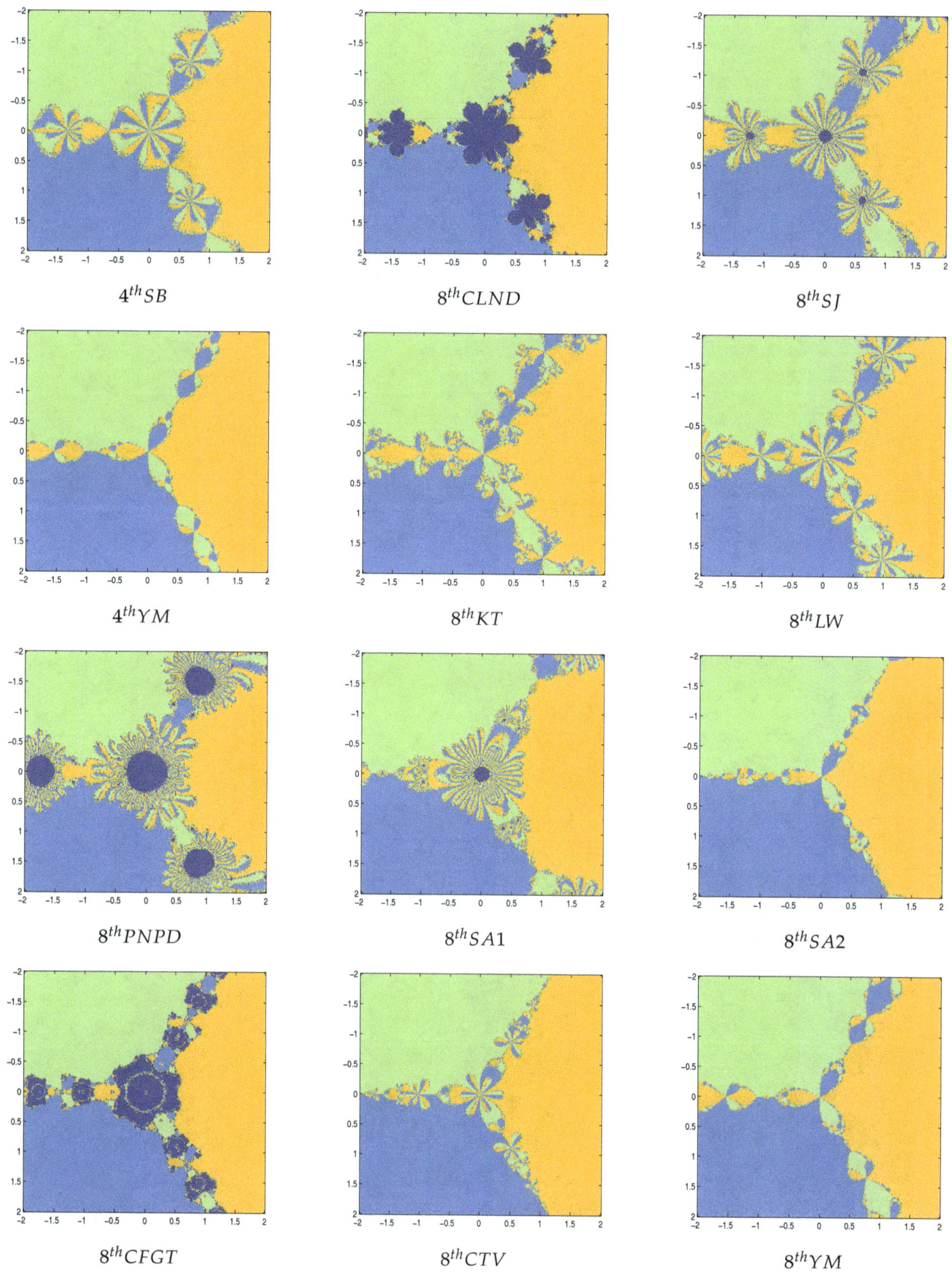

Figure 4. Basins of attraction for $p_2(z) = z^3 - 1$.

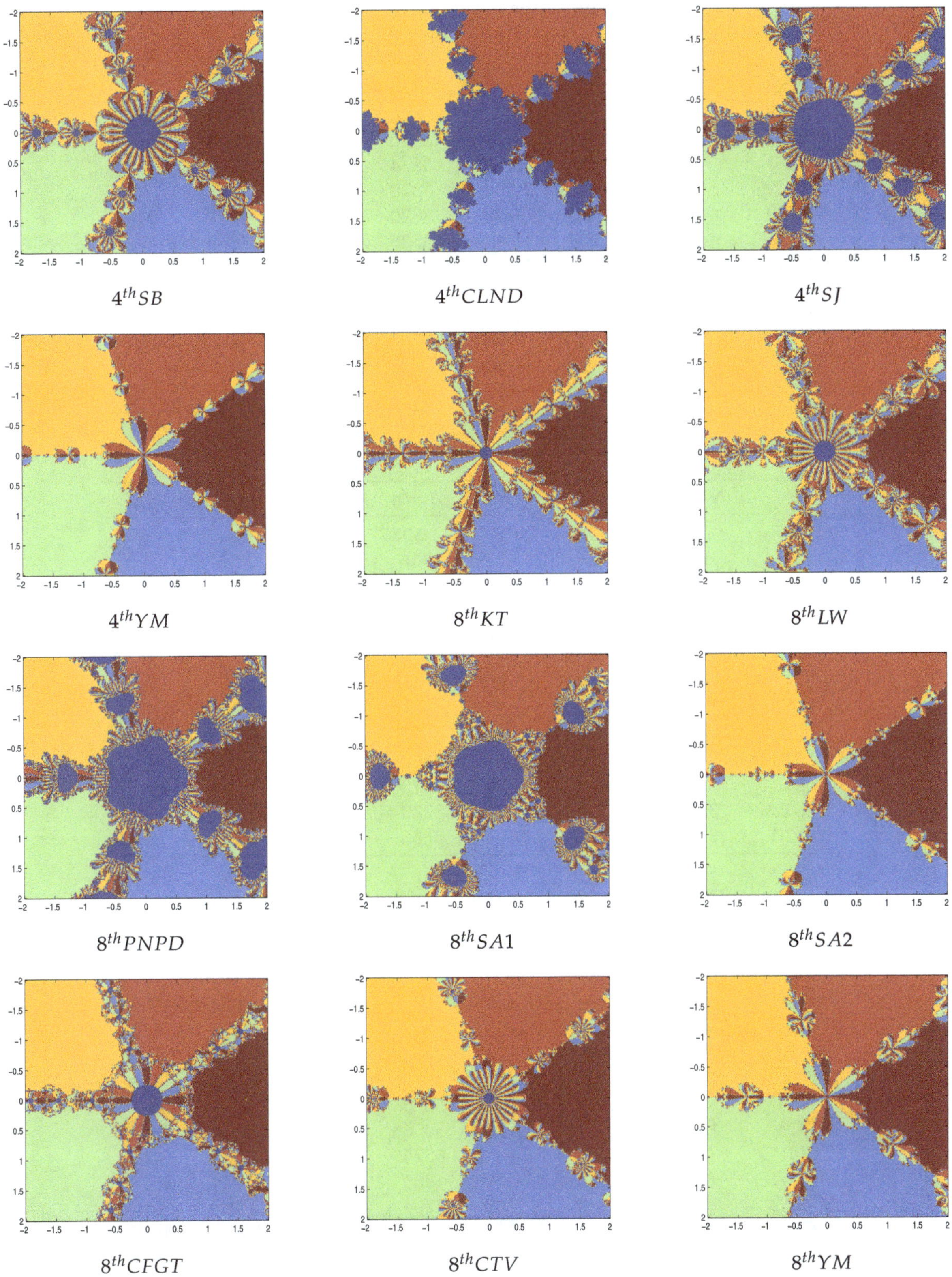

Figure 5. Basins of attraction for $p_3(z) = z^5 - 1$.

Table 8. Results of the polynomials $p_1(z) = z^2 - 1$.

IF	$N = 1$	$N = 2$	$N = 3$	$N = 4$	$N = 5$	$N > 5$	N_D
2^{nd}NR	4	516	7828	23,272	20,548	13,368	0
4^{th}SB	340	22,784	29,056	6836	2928	3592	0
4^{th}CLND	372	24,600	29,140	6512	2224	2688	1076
4^{th}SJ	300	19,816	28,008	5844	2968	8600	0
4^{th}YM	520	31,100	27,520	4828	1208	360	0
8^{th}KT	4684	44,528	9840	3820	1408	1256	24
8^{th}LW	4452	43,236	11,408	3520	1540	1380	0
8^{th}PNPD	2732	39,768	13,112	3480	1568	4876	16
8^{th}SA1	4328	45,824	8136	2564	1484	3200	0
8^{th}SA2	15,680	45,784	3696	376	0	0	0
8^{th}CFGT	9616	43,716	7744	2916	980	564	64
8^{th}CTV	7124	48,232	7464	1892	632	192	0
8^{th}YM	8348	50,792	5572	824	0	0	0

Table 9. Results of the polynomials $p_2(z) = z^3 - 1$.

IF	$N = 1$	$N = 2$	$N = 3$	$N = 4$	$N = 5$	$N > 5$	N_D
2^{nd}NR	0	224	2908	11,302	19,170	31,932	0
4^{th}SB	160	9816	27,438	9346	5452	13,324	6
4^{th}CLND	170	11,242	28,610	9984	4202	11,328	7176
4^{th}SJ	138	7760	25,092	8260	5058	19,228	1576
4^{th}YM	270	18,064	30,374	9862	3688	3278	0
8^{th}KT	2066	34,248	11,752	6130	4478	6862	0
8^{th}LW	2092	33,968	12,180	4830	3030	9436	0
8^{th}PNPD	1106	25,712	11,258	3854	1906	21,700	10,452
8^{th}SA1	1608	36,488	12,486	3718	1780	9456	872
8^{th}SA2	6432	46,850	9120	2230	640	264	0
8^{th}CFGT	3688	40,740	13,696	4278	1390	1744	7395
8^{th}CTV	3530	43,554	11,724	3220	1412	2096	0
8^{th}YM	3816	43,596	12,464	3636	1302	722	0

Table 10. Results of the polynomials $p_3(z) = z^5 - 1$.

IF	$N = 1$	$N = 2$	$N = 3$	$N = 4$	$N = 5$	$N > 5$	N_D
2^{nd}NR	2	100	1222	4106	7918	52,188	638
4^{th}SB	76	3850	15,458	18,026	5532	22,594	5324
4^{th}CLND	86	4476	18,150	17,774	5434	19,616	12,208
4^{th}SJ	62	3094	11,716	16,840	5682	28,142	19,900
4^{th}YM	142	7956	27,428	15,850	5726	8434	0
8^{th}KT	950	17,884	20,892	5675	4024	16,111	217
8^{th}LW	1032	18,764	20,622	5056	3446	16,616	1684
8^{th}PNPD	496	12,770	21,472	6576	2434	21,788	14,236
8^{th}SA1	692	26,212	15,024	4060	1834	17,714	8814
8^{th}SA2	2662	41,400	12,914	4364	1892	2304	0
8^{th}CFGT	2008	21,194	23,734	6180	3958	8462	1953
8^{th}CTV	1802	36,630	13,222	4112	2096	7674	350
8^{th}YM	1736	27,808	21,136	5804	2704	6348	0

We note that a point z_0 belongs to the Julia set if and only if the dynamics in a neighborhood of z_0 displays sensitive dependence on the initial conditions, so that nearby initial conditions lead to wildly different behavior after a number of iterations. For this reason, some of the methods are getting divergent points. The common boundaries of these basins of attraction constitute the Julia set of the iteration function. It is clear that one has to use quantitative measures to distinguish between the methods, since we have a different conclusion when just viewing the basins of attraction.

In order to summarize the results, we have compared mean number of iteration and total number of functional evaluations (TNFE) for each polynomials and each methods in Table 11. The best method based on the comparison in Table 11 is 8^{th}SA2. The method with the fewest number of functional evaluations per point is 8^{th}SA2 followed closely by 4^{th}YM. The fastest method is 8^{th}SA2 followed closely by 8^{th}YM. The method with highest number of functional evaluation and slowest method is 8^{th}PNPD.

Table 11. Mean number of iteration (N_μ) and TNFE for each polynomials and each methods.

IF	N_μ *for* $p_1(z)$	N_μ *for* $p_2(z)$	N_μ *for* $p_3(z)$	Average	TNFE
2^{nd}NR	4.7767	6.4317	9.8531	7.0205	14.0410
4^{th}SB	3.0701	4.5733	9.2701	5.6378	16.9135
4^{th}CLND	3.6644	8.6354	12.8612	8.3870	25.1610
4^{th}SJ	3.7002	7.0909	14.5650	8.4520	25.3561
4^{th}YM	2.6366	3.1733	4.0183	3.2760	9.8282
8^{th}KT	2.3647	3.1270	4.4501	3.3139	13.2557
8^{th}LW	2.3879	3.5209	6.3296	4.0794	16.3178
8^{th}PNPD	2.9959	10.5024	12.3360	8.6114	34.4457
8^{th}SA1	2.5097	4.5787	9.7899	5.6262	22.5044
8^{th}SA2	1.8286	2.1559	2.5732	2.1859	8.7436
8^{th}CFGT	2.1683	2.8029	3.4959	2.8223	11.2894
8^{th}CTV	2.1047	2.4708	3.9573	2.8442	11.3770
8^{th}YM	1.9828	2.3532	3.3617	2.5659	10.2636

8. Concluding Remarks and Future Work

In this work, we have developed optimal fourth, eighth and sixteenth order iterative methods for solving nonlinear equations using the divided difference approximation. The methods require the computations of three functions evaluations reaching order of convergence is four, four functions evaluations reaching order of convergence is eight and five functions evaluations reaching order of convergence is sixteen. In the sense of convergence analysis and numerical examples, the Kung-Traub's conjecture is satisfied. We have tested some examples using the proposed schemes and some known schemes, which illustrate the superiority of the proposed method 16^{th}YM. Also, proposed methods and some existing methods have been applied on the Projectile motion problem and Planck's radiation law problem. The results obtained are interesting and encouraging for the new method 16^{th}YM. The numerical experiments suggests that the new methods would be valuable alternative for solving nonlinear equations. Finally, we have also compared the basins of attraction of various fourth and eighth order methods in the complex plane.

Future work includes:

- Now we are investigating the proposed scheme to develop optimal methods of arbitrarily high order with Newton's method, as in [26].
- Also, we are investigating to develop derivative free methods to study dynamical behavior and local convergence, as in [27,28].

Author Contributions: The contributions of both the authors have been similar. Both of them have worked together to develop the present manuscript.

Acknowledgments: The authors would like to thank the editors and referees for the valuable comments and for the suggestions to improve the readability of the paper.

References

1. Ostrowski, A.M. *Solutions of Equations and System of Equations*; Academic Press: New York, NY, USA, 1960.
2. Kung, H.T.; Traub, J.F. Optimal order of one-point and multipoint iteration. *J. Assoc. Comput. Mach.* **1974**, *21*, 643–651.
3. Amat, S.; Busquier, S.; Plaza, S. Dynamics of a family of third-order iterative methods that do not require using second derivatives. *Appl. Math. Comput.* **2004**, *154*, 735–746.
4. Amat, S.; Busquier, S.; Plaza, S. Review of some iterative root-finding methods from a dynamical point of view. *Scientia* **2004**, *10*, 3–35.
5. Babajee, D.K.R; Madhu, K.; Jayaraman, J. A family of higher order multi-point iterative methods based on power mean for solving nonlinear equations. *Afrika Matematika* **2016**, *27*, 865–876. [CrossRef]
6. Cordero, A.; Fardi, M.; Ghasemi, M.; Torregrosa, J.R. Accelerated iterative methods for finding solutions of nonlinear equations and their dynamical behavior. *Calcolo* **2014**, *51*, 17–30.
7. Cordero, A.; Torregrosa, J.R.; Vasileva, M.P. A family of modified ostrowski's methods with optimal eighth order of convergence. *Appl. Math. Lett.* **2011**, *24*, 2082–2086.
8. Liu, L.; Wang, X. Eighth-order methods with high efficiency index for solving nonlinear equations. *Appl. Math. Comput.* **2010**, *215*, 3449–3454.
9. Madhu, K. Some New Higher Order Multi-Point Iterative Methods and Their Applications to Differential and Integral Equation and Global Positioning System. Ph.D. Thesis, Pndicherry University, Kalapet, India, June 2016.
10. Madhu, K.; Jayaraman, J. Higher order methods for nonlinear equations and their basins of attraction. *Mathematics* **2016**, *4*, 22.
11. Petkovic, M.S.; Neta, B.; Petkovic, L.D.; Dzunic, J. *Multipoint Methods for Solving Nonlinear Equations*; Elsevier: Amsterdam, The Netherlands, 2012.
12. Sharma, J.R.; Arora, H. An efficient family of weighted-newton methods with optimal eighth order convergence. *Appl. Math. Lett.* **2014**, *29*, 1–6.
13. Sharma, J.R.; Arora, H. A new family of optimal eighth order methods with dynamics for nonlinear equations. *Appl. Math. Comput.* **2016**, *273*, 924–933.
14. Soleymani, F.; Khratti, S.K.; Vanani, S.K. Two new classes of optimal Jarratt-type fourth-order methods. *Appl. Math. Lett.* **2011**, *25*, 847–853.
15. Wait, R. *The Numerical Solution of Algebraic Equations*; John Wiley and Sons: Hoboken, NJ, USA, 1979.
16. Khan, Y.; Fardi, M.; Sayevand, K. A new general eighth-order family of iterative methods for solving nonlinear equations. *Appl. Math. Lett.* **2012**, *25*, 2262–2266.
17. Cordero, A.; Torregrosa, J.R. Variants of Newton's method using fifth-order quadrature formulas. *Appl. Math. Comput.* **2007**, *190*, 686–698.
18. Sharma, R.; Bahl, A. An optimal fourth order iterative method for solving nonlinear equations and its dynamics. *J. Complex Anal.* **2015**, *2015*, 259167.
19. Chun, C; Lee, M. Y.; Neta, B; Dzunic, J. On optimal fourth-order iterative methods free from second derivative and their dynamics. *Appl. Math. Comput.* **2012**, *218*, 6427–6438. [CrossRef]
20. Singh, A.; Jaiswal, J.P. Several new third-order and fourth-order iterative methods for solving nonlinear equations. *Int. J. Eng. Math.* **2014**, *2014*, 828409.
21. Babajee, D.K.R.; Madhu, K. Comparing two techniques for developing higher order two-point iterative methods for solving quadratic equations. *SeMA J.* **2018**, 1–22. [CrossRef]
22. Kantrowitz, R.; Neumann, M.M. Some real analysis behind optimization of projectile motion. *Mediterr. J. Math.* **2014**, *11*, 1081–1097.
23. Henelsmith, N. *Finding the Optimal Launch Angle*; Whitman College: Walla Walla, WA, USA, 2016.
24. Bradie, B. *A Friendly Introduction to Numerical Analysis*; Pearson Education Inc.: New Delhi, India, 2006.
25. Scott, M.; Neta, B.; Chun, C. Basin attractors for various methods. *Appl. Math. Comput.* **2011**, *218*, 2584–2599.

26. Cordero, A.; Hueso, J. L.; Martinez, E.; Torregrosa, J.R. Generating optimal derivative free iterative methods for nonlinear equations by using polynomial interpolation. *Appl. Math. Comput.* **2013**, *57* 1950–1956. [CrossRef]
27. Argyros, I.K.; Magrenan, A.A.; Orcos, L. Local convergence and a chemical application of derivative free root finding methods with one parameter based on interpolation. *J. Math. Chem.* **2016**, *54*, 1404–1416.
28. Zafar, F.; Cordero, A.; Torregrosa, J.R. An efficient family of optimal eighth-order multiple root finders. *Mathematics* **2018**, *6*, 310.

9

Hybrid Second Order Method for Orthogonal Projection onto Parametric Curve in n-Dimensional Euclidean Space

Juan Liang [1,2,†], Linke Hou [3,†,*], Xiaowu Li [4,†,*], Feng Pan [4,†], Taixia Cheng [5,†] and Lin Wang [4,†]

1 Data Science and Technology, North University of China, Taiyuan 030051, Shanxi, China; liangjuan76@126.com
2 Department of Science, Taiyuan Institute of Technology, Taiyuan 030008, Shanxi, China
3 Center for Economic Research, Shandong University, Jinan 250100, Shandong, China
4 College of Data Science and Information Engineering, Guizhou Minzu University, Guiyang 550025, Guizhou, China; panf@vip.163.com (F.P.); wanglin@gzmu.edu.cn (L.W.)
5 Graduate School, Guizhou Minzu University, Guiyang 550025, Guizhou, China; lissacheng@163.com
* Correspondence: abram75@163.com (L.H.); lixiaowu002@126.com (X.L.);

† These authors contributed equally to this work.

Abstract: Orthogonal projection a point onto a parametric curve, three classic first order algorithms have been presented by Hartmann (1999), Hoschek, et al. (1993) and Hu, et al. (2000) (hereafter, H-H-H method). In this research, we give a proof of the approach's first order convergence and its non-dependence on the initial value. For some special cases of divergence for the H-H-H method, we combine it with Newton's second order method (hereafter, Newton's method) to create the hybrid second order method for orthogonal projection onto parametric curve in an n-dimensional Euclidean space (hereafter, our method). Our method essentially utilizes hybrid iteration, so it converges faster than current methods with a second order convergence and remains independent from the initial value. We provide some numerical examples to confirm robustness and high efficiency of the method.

Keywords: point projection; intersection; parametric curve; n-dimensional Euclidean space; Newton's second order method; fixed point theorem

---

1. Introduction

In this research, we will discuss the minimum distance problem between a point and a parametric curve in an n-dimensional Euclidean space, and how to gain the closest point (footpoint) on the curve as well as its corresponding parameter, which is termed as the point projection or inversion problem of a parametric curve in an n-dimensional Euclidean space. It is an important issue in the themes such as geometric modeling, computer graphics, computer-aided geometry design (CAGD) and computer vision [1,2]. Both projection and inversion are fundamental for a series of techniques, for instance, the interactive selection of curves and surfaces [1,3], the curve fitting [1,3], reconstructing curves [2,4,5] and projecting a space curve onto a surface [6]. This vital technique is also used in the ICP (iterative closest point) method for shape registration [7].

The Newton-Raphson algorithm is deemed as the most classic one for orthogonal projection onto parametric curve and surface. Searching the root of a polynomial by a Newton-Raphson algorithm can be found in [8–12]. In order to solve the adaptive smoothing for the standard finite unconstrained minimax problems, Polak et al. [13] have presented a extended Newton's algorithm where a new feedback precision-adjustment rule is used in their extended Newton's algorithm.

Once the Newton-Raphson method reaches its convergence, two advantages emerge and it converges very fast with high precision. However, the result relies heavily on a good guess of initial value in the neighborhood of the solution.

Meanwhile, the classic subdivision method consists of several procedures: Firstly, subdivide NURBS curve or surface into a set of Bézier sub-curves or patches and eliminate redundancy or unnecessary Bézier sub-curves or Bézier patches. Then, get the approximation candidate points. Finally, get the closest point through comparing the distances between the test point and candidate points. This technique is reflected in [1]. Using new exclusion criteria within the subdivision strategy, the robustness for the projection of points on NURBS curves and surfaces in [14] has been improved than that in [1], but this criterion is sometimes too critical. Zou et al. [15] use subdivision minimization techniques which rely on the convex hull characteristic of the Bernstein basis to impute the minimum distance between two point sets. They transform the problem into solving of n-dimensional nonlinear equations, where n variables could be represented as the tensor product Bernstein basis. Cohen et al. [16] develop a framework for implementing general successive subdivision schemes for nonuniform B-splines to generate the new vertices and the new knot vectors which are satisfied with derived polygon. Piegl et al. [17] repeatedly subdivide a NURBS surface into four quadrilateral patches and then project the test point onto the closest quadrilateral until it can find the parameter from the closest quadrilateral. Using multivariate rational functions, Elber et al. [11] construct a solver for a set of geometric constraints represented by inequalities. When the dimension of the solver is greater than zero, they subdivide the multivariate function(s) so as to bind the function values within a specified domain. Derived from [11] but with more efficiency, a hybrid parallel method in [18] exploits both the CPU and the GPU multi-core architectures to solve systems under multivariate constraints. Those GPU-based subdivision methods essentially exploit the parallelism inherent in the subdivision of multivariate polynomial. This geometric-based algorithm improves in performance compared to the existing subdivision-based CPU. Two blending schemes in [19] efficiently remove no-root domains, and hence greatly reduce the number of subdivisions. Through a simple linear combination of functions for a given system of nonlinear equations, no-root domain and searching out all control points for its Bernstein-Bézier basic with the same sign must be satisfied with the seek function. During the subdivision process, it can continuously create these kinds of functions to get rid of the no-root domain. As a result, van Sosin et al. [20] efficiently form various complex piecewise polynomial systems with zero or inequality constraints in zero-dimensional or one-dimensional solution spaces. Based on their own works [11,20], Bartoň et al. [21] propose a new solver to solve a non-constrained (piecewise) polynomial system. Two termination criteria are applied in the subdivision-based solver: the no-loop test and the single-component test. Once two termination criteria are satisfied, it then can get the domains which have a single monotone univariate solution. The advantage of these methods is that they can find all solutions, while their disadvantage is that they are computationally expensive and may need many subdivision steps.

The third classic methods for orthogonal projection onto parametric curve and surface are geometry methods. They are mainly classified into eight different types of geometry methods: tangent method [22,23], torus patch approximating method [24], circular or spherical clipping method [25,26], culling technique [27], root-finding problem with Bézier clipping [28,29], curvature information method [6,30], repeated knot insertion method [31] and hybrid geometry method [32]. Johnson et al. [22] use tangent cones to search for regions with satisfaction of distance extrema conditions and then to solve the minimum distance between a point and a curve, but it is not easy to construct tangent cones at any time. A torus patch approximatively approaches for point projection on surfaces in [24]. For the pure geometry method of a torus patch, it is difficult to achieve high precision of the final iterative parametric value. A circular clipping method can remove the curve parts outside a circle with the test point being the circle's center, and the radius of the elimination circle will shrink until it satisfies the criteria to terminate [26]. Similar to the algorithm [26], a spherical clipping technique for computing the minimum distance with clamped B-spline surface is provided by [25]. A culling technique to remove superfluous curves and surfaces containing no projection from the given point is proposed in [27], which is in line with the idea in [1]. Using Newton's method for the last step [1,25–27], the special case of non-convergence may happen. In view of the convex-hull property of Bernstein-Bézier representations, the problem to be solved can be formulated as a univariate root-finding problem. Given a C^1 parametric curve $c(t)$ and a point p, the projection constraint problem can be formulated as a univariate root-finding problem $\langle c'(t), c(t) - p \rangle = 0$ with a metric induced by the Euclidean scalar product in R^n. If the curve is parametrized by a (piece-wise) polynomial, then the fast root-finding schemes as a Bézier clipping [28,29] can be used. The only issue is the C^1 discontinuities that can be checked in a post-process. One advantage of these methods is that they do not need any initial guess on the parameter value. They adopt the key technology of degree reduction via clipping to yield a strip bounded of two quadratic polynomials. Curvature information is found for computing the minimum distance between a point and a parameter curve or surface in [6,30]. However, it needs to consider the second order derivative and the method [30] is not fit for n-dimensional Euclidean space. Hu et al. [6] have not proved the convergence of their two algorithms. Li et al. [33] have strictly proved convergence analysis for orthogonal projection onto planar parametric curve in [6]. Based on repeated knot insertion, Mørken et al. [31] exploit the relationship between a spline and its control polygon and then present a simple and efficient method to compute zeros of spline functions. Li et al. [32] present the hybrid second order algorithm which orthogonally projects onto parametric surface; it actually utilizes the composite technology and hence converges nicely with convergence order being 2. The geometric method can not only solve the problem of point orthogonal projecting onto parametric curve and surface but also compute the minimum distance between parametric curves and parametric surfaces. Li et al. [23] have used the tangent method to compute the intersection between two spatial curves. Based on the methods in [34,35], they have extended to compute the Hausdorff distance between two B-spline curves. Based on matching a surface patch from one model to the other model which is the corresponding nearby surface patch, an algorithm for solving the Hausdorff distance between two freeform surfaces is presented in Kim et al. [36], where a hierarchy of Coons patches and bilinear surfaces that approximate the NURBS surfaces with bounding volume is adopted. Of course, the common feature of geometric methods is that the ultimate solution accuracy is not very high. To sum up, these algorithms have been proposed to exploit diverse techniques such as Newton's iterative method, solving polynomial equation roots methods, subdividing methods, geometry methods. A review of previous algorithms on point projection and inversion problem is obtained in [37].

More specifically, using the tangent line or tangent plane with first order geometric information, a classical simple and efficient first order algorithm which orthogonally project onto parametric curve and surface is proposed in [38–40] (H-H-H method). However, the proof of the convergence for the H-H-H method can not be found in this literature. In this research, we try to give two contributions. Firstly, we give proof that the algorithm is first order convergent and it does not depend on the initial value. We then provide some numerical examples to show its high convergence rate. Secondly, for several special cases where the H-H-H method is not convergent, there are two methods (Newton's method and the H-H-H method) to combine our method. If the H-H-H method's iterative parametric value is satisfied with the convergence condition of the Newton's method, we then go to Newton's method to increase the convergence process. Otherwise, we go on the H-H-H method until its iterative parametric value is satisfied with the convergence condition of the Newton's method, and we then turn to it as above. This algorithm not only ensures the robustness of convergence, but also improves the convergence rate. Our hybrid method can go faster than the existing methods and ensures the independence to the initial value. Some numerical examples verify our conclusion.

The rest of this paper is arranged as follows. In Section 2, convergence analysis of the H-H-H method is presented. In Section 3, for several special cases where the H-H-H method is not convergent, an improved our method is provided. Convergence analysis for our method is also provided in this section. In Section 4, some numerical examples for our method are verified. In Section 5, conclusions are provided.

2. Convergence Analysis of the H-H-H Method

In this part, we will prove that the algorithm defined by Equations (2) or (3) is of first order convergence and its convergence does not rely on the initial value. Suppose a C^2 curve $c(t) = (f_1(t), f_2(t), \ldots, f_n(t))$ in an n-dimensional Euclidean space $\mathrm{R}^n (n \geq 2)$ and a test point $p = (p_1, p_2, \ldots, p_n)$. The first order geometric method to compute the footpoint q of test point p can be implemented as below. Projecting test point p onto the tangent line of the parametric curve $c(t)$ in an n-dimensional Euclidean space at $t = t_m$ gets a point q determined by $c(t_m)$ and its derivative $c'(t_m)$. The footpoint can be approximated as

$$q = c(t_m) + \Delta t c'(t_m). \tag{1}$$

Then,

$$\Delta t = \frac{\langle c'(t_m), p - c(t_m) \rangle}{\langle c'(t_m), c'(t_m) \rangle}, \tag{2}$$

where $\langle x, y \rangle$ is the scalar product of vectors $x, y \in \mathrm{R}^n$. Equation (2) can also be expressed as

$$K_1(t_m) = t_m + \frac{\langle c'(t_m), p - c(t_m) \rangle}{\langle c'(t_m), c'(t_m) \rangle}. \tag{3}$$

Let $t_m \leftarrow K_1(t_m)$, and repeatedly iterate the above process until $|K_1(t_m) - t_m|$ is less than an error tolerance ε. This method is addressed as H-H-H method [38–40]. Furthermore, convergence of this method will not depend on the choice of the initial value. According to many of our test experiments, when the iterative parametric value approaches the target parametric value α, the iteration step size becomes smaller and smaller, while the corresponding number of iterations becomes bigger and bigger.

Theorem 1. *The convergence order of the method defined by Equations (2) or (3) is one, and its convergence does not depend on the initial value.*

Proof. We adopt the numerical analysis method which is equivalent to those in the literature [41,42]. Firstly, we deduce the expression of footpoint q. Suppose that parameter curve $c(t)$ is a C^2 curve in an n-dimensional Euclidean space $\mathrm{R}^n (n \geq 2)$, where the corresponding projecting point with parameter α is orthogonal projecting of the test point $p = (p_1, p_2, \ldots, p_n)$ onto the parametric curve $c(t)$. It is easy to indicate a relational expression

$$\langle p - h, \mathbf{n} \rangle = 0, \tag{4}$$

where $h = c(\alpha)$ and tangent vector $\mathbf{n} = c'(\alpha)$. In order to solve the intersection (footpoint q) between the tangent line, which goes through the parametric curve $c(t)$ at $t = t_m$, and the perpendicular line, which is determined by the test point p, we try to express the equation of the tangent line as:

$$x = c(t_m) + c'(t_m) \cdot s, \tag{5}$$

where $x = (x_1, x_2, \ldots, x_n)$ and s is a parameter. In addition, the vector of line segment both going through the test point p and the point $c(t_m)$ will be

$$y = p - x, \tag{6}$$

where $y = (y_1, y_2, \ldots, y_n)$. Because the vector (6) and the tangent vector $c'(t_m)$ of Equation (5) are orthogonal to each other, the current parameter value s of Equation (5) is

$$s_0 = \frac{\langle p - c(t_m), c'(t_m) \rangle}{\langle c'(t_m), c'(t_m) \rangle}. \tag{7}$$

Substituting (7) into (5), we have

$$q = c(t_m) + c'(t_m) \cdot s_0. \tag{8}$$

Thus, the footpoint $q = (q_1, q_2, \ldots, q_n)$ is determined by Equation (8).

Secondly, we deduce that the convergence order of the method defined by (2) or (3) is first order convergent. Our proof method absorbs the idea of [41,42]. Substituting (8) into (2), and simplifying, we get the relationship,

$$\Delta t = \frac{\langle p - c(t_m), c'(t_m) \rangle}{\langle c'(t_m), c'(t_m) \rangle}. \tag{9}$$

Using Taylor's expansion, we get

$$c(t_m) = \mathrm{B}_0 + \mathrm{B}_1 e_m + \mathrm{B}_2 e_m^2 + o(e_m^3), \tag{10}$$

$$c'(t_m) = \mathrm{B}_1 + 2\mathrm{B}_2 e_m + o(e_m^2), \tag{11}$$

where $e_m = t_m - \alpha$, and $\mathrm{B}_i = (1/i!)c^{(i)}(\alpha), i = 0, 1, 2, \ldots$ From (10) and (11) and combining with (4), the numerator of Equation (9) can be transformed into the following one:

$$\begin{aligned} &\langle p - c(t_m), c'(t_m) \rangle \\ &= L_1 e_m + L_2 e_m^2 + o(e_m^3), \end{aligned} \tag{12}$$

where $L_1 = 2\langle p - \mathrm{B}_0, \mathrm{B}_2 \rangle - \langle \mathrm{B}_1, \mathrm{B}_1 \rangle, L_2 = -3\langle \mathrm{B}_1, \mathrm{B}_2 \rangle$. By (11), the denominator of Equation (9) can be changed as follows:

$$\begin{aligned} &\langle c'(t_m), c'(t_m) \rangle \\ &= M_1 + M_2 e_m + M_3 e_m^2 + o(e_m^3), \end{aligned} \tag{13}$$

where $M_1 = \langle \mathrm{B}_1, \mathrm{B}_1 \rangle, M_2 = 4\langle \mathrm{B}_1, \mathrm{B}_2 \rangle, M_3 = 4\langle \mathrm{B}_2, \mathrm{B}_2 \rangle$. Substituting Equations (12) and (13) into the right-hand side of Equation (9), we get

$$\begin{aligned} \Delta t &= \frac{\langle p - c(t_m), c'(t_m) \rangle}{\langle c'(t_m), c'(t_m) \rangle} \\ &= \frac{L_1 e_m + L_2 e_m^2 + o(e_m^3)}{M_1 + M_2 e_m + M_3 e_m^2 + o(e_m^3)}. \end{aligned} \tag{14}$$

Using Taylor's expansion by Maple 18, and through simplification, we get

$$\begin{aligned} K_1(t_m) &= \alpha + (\frac{L_1}{M_1} + 1)e_m + \frac{L_2 M_1 - L_1 M_2}{M_1^2} e_m^2 + o(e_m^3), \\ &= \alpha + (\frac{L_1}{M_1} + 1)e_m + o(e_m^2), \\ &= \alpha + C_0 e_m + o(e_m^2), \end{aligned} \tag{15}$$

where the symbol C_0 is the coefficient of the first order error e_m of Equation (15). The result implies the iterative Equations (2) or (3) is of first order convergence.

Now, we try to interpret that Equations (2) or (3) do not depend on the initial value.

Our proof method absorbs the idea of references [43,44]. Without loss of generality, we only prove that convergence of Equations (2) or (3) does not depend on the initial value in two-dimensional case. As to convergence of Equations (2) or (3) not being dependent on the initial value in general n-dimensional Euclidean space case, it is completely equivalent to the two-dimensional case.

Firstly, we interpret Figure 1. For a horizontal axis t, there are two points are on the planar parametric curve $c(t)$. For the first point $c(t_m)$ on the horizontal axis, the test point p orthogonal projects it onto the planar parametric curve $c(t)$ and yields the second point and its corresponding parameter value α on the horizontal axis. Then, by the iterative methods (2) or (3), the line segment connected by the point p and the point $c(\alpha)$ is perpendicular to the tangent line of the planar parametric curve $c(t)$ at $t = \alpha$. The footpoint q is determined by the tangent line of the planar parametric curve $c(t)$ through the point $c(t_m)$. Evidently, the parametric value t_{m+1} of footpoint q can be used as the next iterative value. M is the corresponding parametric value of the middle point of the point $c(t_m)$ and the footpoint q.

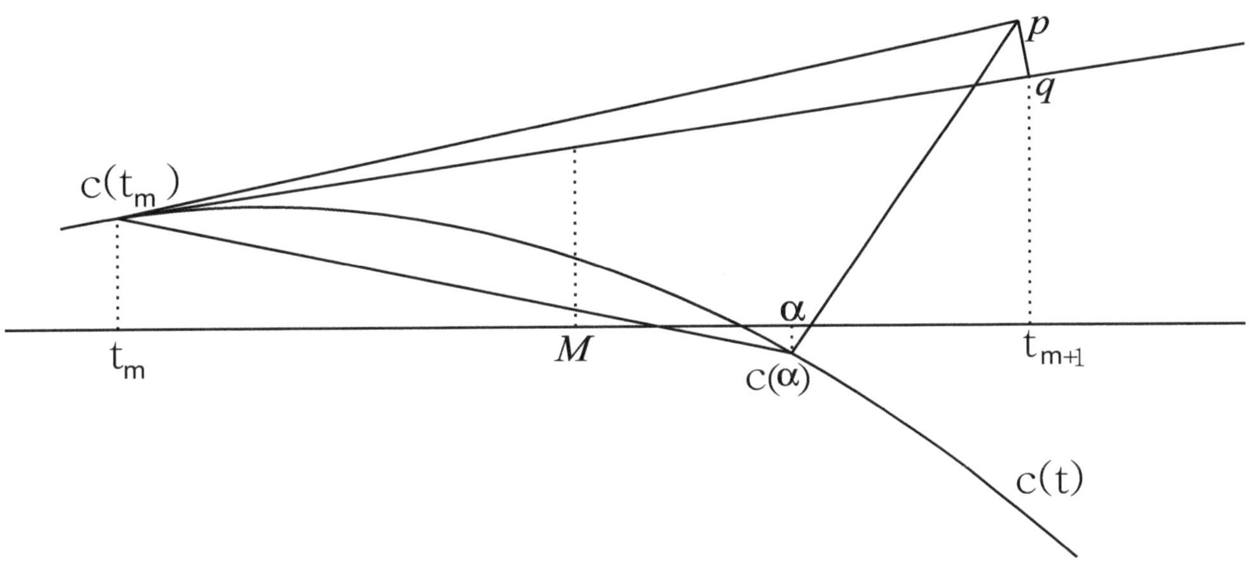

Figure 1. Geometric illustration for convergence analysis.

Secondly, we prove the argument whose convergence of Equations (2) or (3) does not depend on the initial value. It is easy to know that t denotes the corresponding parameter for the first dimensional of the planar parametric curve on the two-dimensional plane. When the iterative Equations (2) or (3) start to run, we suppose that the iterative parameter value is satisfied with the inequality relationship $t_m < \alpha$ and the corresponding parameter of the footpoint q is t_{m+1}, as shown in Figure 1. The middle point of two points $(t_{m+1}, 0)$ and $(t_m, 0)$ is $(M, 0)$, i.e., $M = \frac{t_m + t_{m+1}}{2}$, and, because of $0 < \Delta t = t_{m+1} - t_m$, then there exists an inequality $t_m < M < \alpha$. Equivalently, $t_m - \alpha < t_{m+1} - \alpha < \alpha - t_m = -(t_m - \alpha)$, which can be expressed as $|e_{m+1}| < |e_m|$, where $e_m = t_m - \alpha$. If $t_m > \alpha$, we can get the same result through the same method. Thus, an iterative error expression $|e_{m+1}| < |e_m|$ in a two-dimensional plane is demonstrated. Thus, it is known that convergence of the iterative Equations (2) or (3) does not depend on the initial value in two-dimensional planes (see Figure 1). Furthermore, we could get the argument that convergence of the iterative Equations (2) or (3) does not depend on the initial value in an n-dimensional Euclidean space. The proof is completed. □

3. The Improved Algorithm

3.1. Counterexamples

In Section 2, convergence of the H-H-H method does not depend on the initial value. For special cases with non-convergence by the H-H-H method, we then enumerate nine counterexamples.

Counterexample 1. *There are a parametric curve* $c(t) = (t, 1 + t^2)$ *and a test point* $p = (0,0)$. *The projection point and parametric value of the test point* p *are* $(0,1)$ *and* $\alpha = 0$, *respectively. As to many initial values, the H-H-H method fails to converge to* α. *When the initial values are* $t = -3, -2, -1.5, 1.5, 2, 3$, *respectively, there repeatedly appear alternating oscillatory iteration values of* 0.412415429665, −0.412415429665. *Furthermore, for a parametric curve* $c(t) = (t, 1 + a_1t^2 + a_2t^4 + a_3t^6 + a_4t^8 + a_5t^{10})$, $a_1 \neq 0, a_2 \neq 0, a_3 \neq 0, a_4 \neq 0, a_5 \neq 0$, *about* $p = (0,0)$ *and many initial values, the H-H-H method fails to converge to* α *(see Figure 2).*

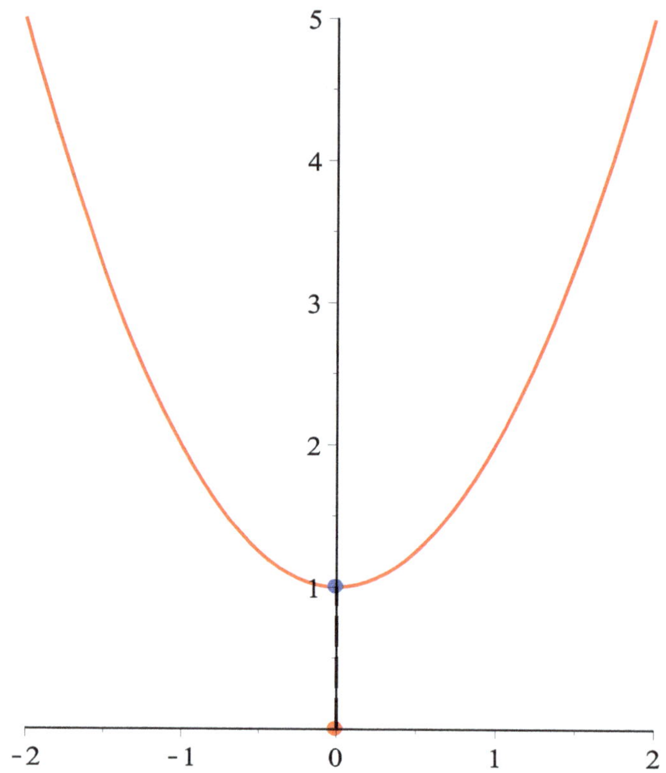

Figure 2. Geometric illustration for counterexample 1.

Counterexample 2. *There are a parametric curve* $c(t) = (t, t^2, t^4, t^6, 1 + t^2 + t^4 + t^6 + t^8)$ *and a test point* $p = (0,0,0,0,0)$. *The projection point and parametric value of the test point* p *are* $(0,0,0,0,1)$ *and* $\alpha = 0$, *respectively. For any initial value, the H-H-H method fails to converge to* α. *When the initial values are* $t = -5, -4, -3, -2, -1, 1, 2, 3, 4, 5$, *respectively, there repeatedly appear alternating oscillatory iteration values of* 0.304949569175, −0.304949569175. *Furthermore, for a parametric curve* $c(t) = (a_0t, a_1t^2, a_2t^4, a_3t^6, 1 + a_4t^2 + a_5t^4 + a_6t^6 + a_7t^8 + a_8t^{10} + a_9t^{28})$, $a_0 \neq 0, a_1 \neq 0, a_2 \neq 0, a_3 \neq 0, a_4 \neq 0, a_5 \neq 0, a_6 \neq 0, a_7 \neq 0, a_8 \neq 0, a_9 \neq 0$, *about point* $p = (0,0,0,0,0)$ *and any initial value, the H-H-H method fails to converge to* α.

Counterexample 3. *There are a parametric curve* $c(t) = (t, \sin(t)), t \in [0,3]$ *and a test point* $p = (4,9)$. *The projection point and parametric value of the test point* p *are* $(1.842576, 0.9632946)$ *and* $\alpha = 1.842576$, *respectively. For point* p *and any initial value, the H-H-H method fails to converge to* α. *When the initial values are* $t = -5, -4, -3, -2, -1, 1, 2, 3, 4, 5$, *respectively, there repeatedly appear alternating oscillatory iteration values of* 2.165320, 0.0778704, 6.505971, 9.609789. *In addition, for a parametric curve* $c(t) = (t, \sin(at)), a \neq 0$, *for any test point* p *and any initial value, the H-H-H method fails to converge to* α *(see Figure 3).*

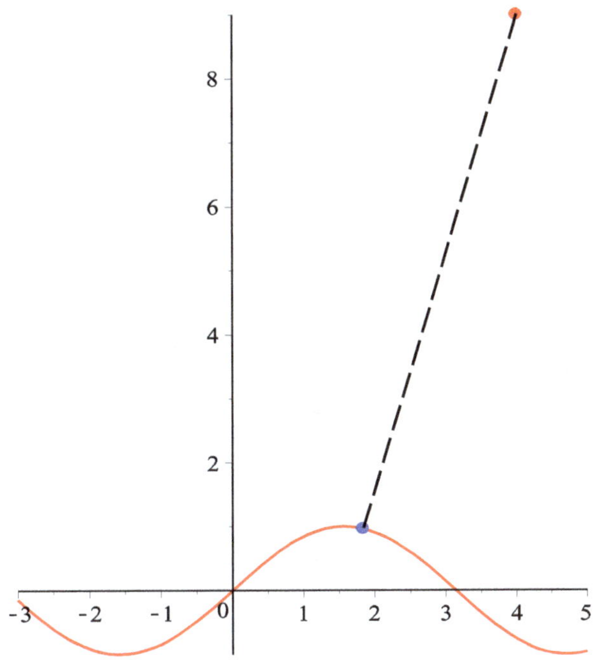

Figure 3. Geometric illustration of counterexample 3.

Counterexample 4. *There are a parametric curve* $c(t) = (t, \cos(t)), t \in [0,3]$ *and a test point* $p = (2,6)$. *The projection point and parametric value of the test point p are* (0.3354892, 0.9442493) *and* $\alpha = 0.3354892$, *respectively. For test point p and any initial value, the H-H-H method fails to converge to* α. *When the initial value is* $t = -5$, *alternating oscillatory iteration values of* 5.18741299662, 3.59425803253, -0.507188248308, 1.6901041247, 3.82746208506 *repeatedly appear. When the initial value is* $t = 2$, *very irregular oscillatory iteration values of* 0.652526561595, -0.720371663877, -2.39555359952, 0.365881194752, 2.06880954777, 3.18725085474, 1.71447110647, *etc. appear In addition, for a parametric curve* $c(t) = (t, \cos(at))$, $a \neq 0$, *for any test point p and any initial value, the H-H-H method fails to converge to* α *(see Figure 4).*

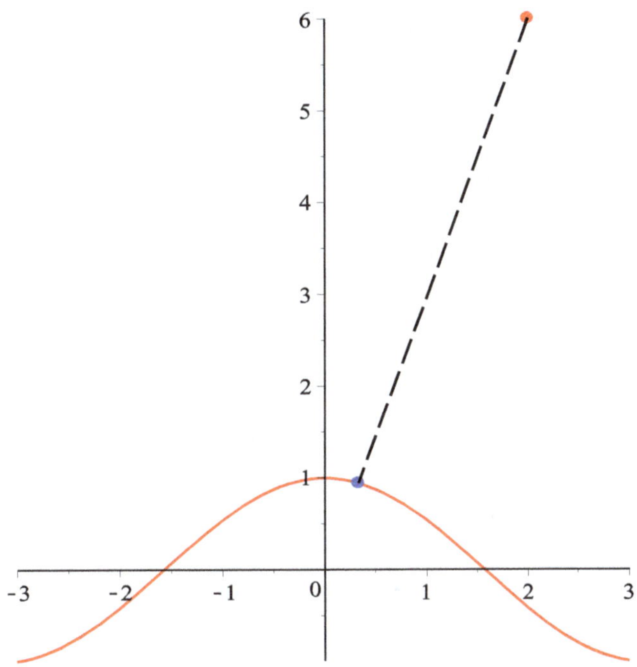

Figure 4. Geometric illustration of counterexample 4.

Counterexample 5. *There are a parametric curve $c(t) = (t,t,t,t,\sin(t)), t \in [6,9]$ and a test point $p = (3,5,7,9,11)$. The projection point and parametric value of the test point p are (7.310786, 7.310786, 7.310786, 7.310786, 0.8560612) and $\alpha = 7.310786$, respectively. For point p and any initial value, the H-H-H method fails to converge to α. When the initial values are $t = -9,-7,-5,6,8$, respectively, there repeatedly appear alternating oscillatory iteration values of 7.24999006346, 6.37363460615. In addition, for a parametric curve $c(t) = (t,t,t,t,\sin(at)), t \in [6,9]$, $a \neq 0$ with a test point $p = (3,5,7,9,11)$, for any initial value, the H-H-H method fails to converge to α.*

Counterexample 6. *There are a parametric curve $c(t) = (t,t,t,t,\cos(t)), t \in [4,8]$ and a test point $p = (2,4,6,8,10)$. The projection point and parametric value of the test point p are (5.883406, 5.883406, 5.883406, 5.883406, 0.9211469) and $\alpha = 5.883406$, respectively. For point p and any initial value, the H-H-H method fails to converge to α. When the initial values are $t =, -4,-3,-2,4,5,6,7$, respectively, there repeatedly appear alternating oscillatory iteration values of 4.17182145828, 7.80116702003. In addition, about a parametric curve $c(t) = (t,t,t,t,\cos(at)), t \in [4,8]$, $a \neq 0$ with a point $p = (2,4,6,8,10)$, for any initial value, the H-H-H method fails to converge. The non-convergence explanation of the three counterexamples below are similar to the preceding six ones and omitted to save space.*

Counterexample 7. *There are a parametric curve $c(t) = (t^4+2t^2+1, t^2+1, t^4+2, t^2, 3t^6+t^4+2t^2)$ in five-dimensional Euclidean space and a test point $p = (0,0,0,0,0)$. The projection point and parametric value of the test point p are $(1,1,2,0,0)$ and $\alpha = 0$, respectively. For any initial value t_0, the H-H-H method fails to converge. We also test many other examples, such as when parametric curve is completely symmetrical and the point is on the symmetrical axis of parametric curve. For any initial value t_0, the same results remain.*

Counterexample 8. *There are a parametric curve $c(t) = (t,\sin(t),t,\sin(t),\sin(t))$, $t \in [-5,5]$ in five-dimensional Euclidean space and a test point $p = (3,4,5,6,7)$. The corresponding orthogonal projection parametric value α are -3.493548, -2.280571, 1.875969, 4.791677, respectively. For any initial value t_0, the H-H-H method fails to converge.*

Counterexample 9. *There is a parametric curve $c(t) = (\sin(t),\cos(t),t, \sin(t),\cos(t))$, $t \in [-5,5]$ in five-dimensional Euclidean space and a test point $p = (3,4,5,6,7)$. The corresponding orthogonal projection parametric value α are -4.833375, -3.058735, 0.9730030, 3.738442, respectively. For any initial value t_0, the H-H-H method fails to converge.*

3.2. The Improved Algorithm

Due to the H-H-H method's non-convergence for some special cases, the improved algorithm is presented to ensure the converge for any parametric curve, test point and initial value. The most classic Newton's method can be expressed as

$$t_{m+1} = t_m - \frac{f(t_m)}{f'(t_m)}, \tag{16}$$

where $f(t) =< T_1, V_1 >= 0$, $T_1 = c'(t), V_1 = p - c(t)$. It converges faster than the H-H-H method. However, the convergence of this depends on the chosen initial value. Only when the local convergence condition for the Newton's method is satisfied, the method can acquire high effectiveness. In order to improve the robustness and rate of convergence, based on the the H-H-H method, our method is proposed. Combining the respective advantage of their two methods, if the iterative parametric value of the H-H-H method is satisfied with the convergence condition of the Newton's method, we then go to the method to increase the convergence process. Otherwise, we continue the H-H-H method until it can generate iterative parametric value while satisfying the convergence condition by the Newton's method, and we then go to the iterative process mentioned above. Thus, we run to the end of the whole process. The procedure not only ensures the robustness of convergence, but also improves the convergence rate. Using a hybrid strategy, our method is faster than current methods and independent from the initial value. Some numerical examples verify our conclusion. Our method can be realized as follows (see Figure 5).

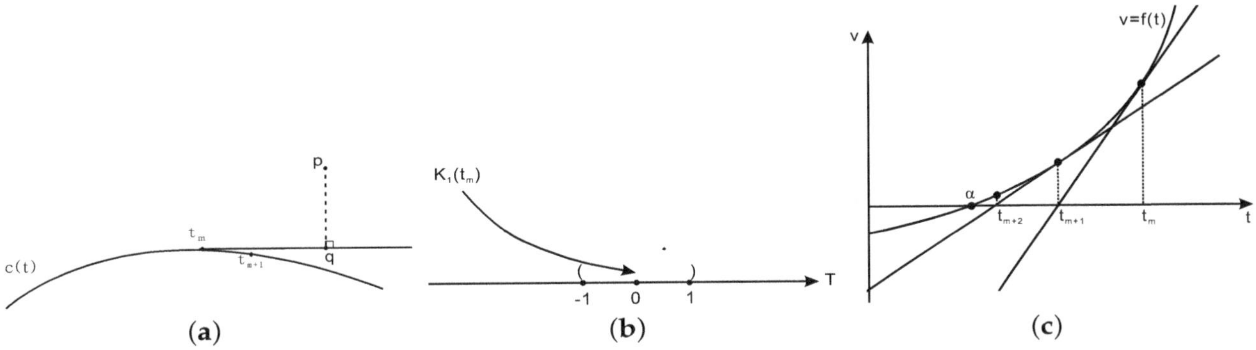

Figure 5. Geometric illustration for our method. (**a**) Running the H-H-H method; (**b**) Judging the H-H-H method whether being satisfied the convergence condition of fixed point theorem for the Newton's iterative method; (**c**) Running the Newton's iterative method.

Hybrid second order method
Input: Initial iterative value t_0, test point p and parametric curve $c(t)$ in an n-dimensional Euclidean space.
Output: The corresponding parameter α determined by orthogonal projection point.
Step 1. Initial iterative parametric value t_0 is input.
Step 2. Using the iterative Equation (3), calculate the parametric value $K_1(t_0)$, and update $K_1(t_0)$ to t_1, namely, $t_1 = K_1(t_0)$.
Step 3. Determine whether absolute value of difference between the current t_0 and the new t_1 is near 0. If so, this algorithm is ended.
Step 4. Substitute the new t_1 into $\left|\frac{f(t)f''(t)}{f'(t)^2}\right|$, determine if $\left|\frac{f(t_1)f''(t_1)}{f'(t_1)^2}\right| < 1$.

If ($\left|\frac{f(t_1)f''(t_1)}{f'(t_1)^2}\right| < 1$) {

Using Newton's iterative Equation (16), compute $t_0 = t_1 - \frac{f(t_1)}{f'(t_1)}$ until absolute value of difference between the current t_1 and the new t_0 is near 0; then, this algorithm ends.

}

Else {

turn to **Step 2.**

}

Remark 1. *Firstly, a geometric illustration of our method in Figure 5 would be presented. Figure 5a illustrates the second step of our method where the next iterative parameter value* $t_{m+1} = K_1(t_m) = t_m + \frac{\langle c'(t_m), p-c(t_m)\rangle}{\langle c'(t_m), c'(t_m)\rangle}$ *is determined by the iterative Equation (3). During the iterative process, the step* Δt *will become smaller and smaller. Thus, the next iterative parameter value* t_{m+1} *comes close to parameter value* t_m *but far from the footpoint* q*. If the third step of our method is not over, then our method goes into the fourth step. Figure 5b is judging condition of a fixed point theorem of the fourth step of our method. If* T = $\left|\frac{f(t)f''(t)}{f'(t)^2}\right| < 1$*, then it turns to the Newton's method in Figure 5c until it runs to the end of the whole process of Newton's second order iteration; otherwise, it goes to the second step in Figure 5a.*

Secondly, we give an interpretation for the singularity case of the iterative Equation (16). As to some special cases where the H-H-H method is not convergent in Section 3.1, our method still converges. We test many examples for arbitrary initial value, arbitrary test point and arbitrary parametric curve and find that our method remains more robust to converge than the H-H-H method. If the first order derivative $f'(t_m)$ *of the iterative Equation (16) develops into 0, i.e.,* $f'(t_m) = 0$ *about some non-negative integer* m*, we use a perturbed method to solve the special problem, which adopts the idea in [23,45]. Namely, the function* $f'(t_m) = 0$ *could be* *increased by a very small positive number ε, i.e., $f'(t_m) = f'(t_m) + \varepsilon$, and then the iteration by Equation (16) is continued in order to calculate the parameter value. On the other hand, if the curve can be parametrized by a (piece-wise) polynomial, then the fast root-finding schemes such as Bézier clipping [28,29] are efficient ones. The only issue is the C^1 discontinuities that can be checked in a post-process. One then does not need any initial guess on the parameter value.*

Thirdly, if the curve is only C^0 continuous, and the closest point can be exactly such a point, then the derivative is not well defined and our method may fail to find such a point. Namely, there are singular points on the parametric curve. We adopt the following technique to solve the problem of singularity. We use the methods [46–48] to find all singular points on the parametric curve and the corresponding parametric value of each singular point as many as possible. Then, the hybrid second order method comes into work. If the current iterative parametric value t_m is the corresponding parametric value of a singular point, we make a very small perturbation ε to the current iterative parametric value t_m, i.e., $t_m = t_m + \varepsilon$. The purpose of this behavior is to enable the hybrid second order method to run normally. Then, from all candidate points (singular points and orthogonal projection points), a corresponding point is selected so that the distance between the corresponding point and the test point is the minimum one. When the entire program terminates, the minimum distance and its corresponding parameter value are found.

3.3. Convergence Analysis of the Improved Algorithm

In this subsection, we prove the convergence analysis of our method.

Theorem 2. In Reference [49] (Fixed Point Theorem)

If $\phi(x) \in C[c,d]$, $\phi(x) \in [c,d]$ for all $x \in [c,d]$; furthermore, if $\phi'(x)$ exists on (c,d) and a positive constant $L < 1$ exists with $|\phi'(x)| \leq L$ for all $x \in (c,d)$, then there exists exactly one fixed point in $[c,d]$.

In addition, if $\phi(t) = t - \dfrac{f(t)}{f'(t)}$, the corresponding fixed point theorem of Newton's method is as follows:

Theorem 3. *Let $f : [c,d] \to [c,d]$ be a differentiable function, if for all $t \in [c,d]$, there is*

$$\left| \frac{f(t)f''(t)}{f'^2(t)} \right| < 1. \tag{17}$$

Then, there is a fixed point $l_0 \in [c,d]$ in Newton's iteration expression (16) such that $l_0 = l_0 - \dfrac{f(l_0)}{f'(l_0)}$. Meanwhile, the iteration sequence $\{t_m\}$ been from expression (16) can converge to the fixed point when $\forall t_0 \in [c,d]$.

Theorem 4. *Our method is second order convergent.*

Proof: Let α be a simple zero for a nonlinear function $f(t) =< T_1, V_1 >= 0$, where $T_1 = c'(t), V_1 = p - c(t)$. Using Taylor's expansion, we have

$$f(t_m) = f'(\alpha)[e_m + b_2 e_m^2 + b_3 e_m^3 + o(e_m^4)], \tag{18}$$

$$f'(t_m) = f'(\alpha)[2b_2 e_m + 3b_3 e_m^2 + o(e_m^3)], \tag{19}$$

where $b_k = \dfrac{f^{(k)}(\alpha)}{k! f'(\alpha)}, k = 2, 3, \ldots,$ and $e_m = t_m - \alpha$. Combining with (15), we then have

$$y_m = \phi(t_m) = t_m - \frac{f(t_m)}{f'(t_m)} = \alpha + b_2 C_0^2 e_m^2 + o(e_m^3). \tag{20}$$

This means that the convergence order of our method is 2. The proof is completed. □

Theorem 5. *Convergence of our method does not depend on the initial value.*

Proof. According to the description of our method, if the iterative parametric value of the H-H-H method is satisfied with the convergence condition of the Newton's method, we then go to the Newton's method. Otherwise, we steadily adopt the H-H-H method until its iterative parametric value is satisfied with the convergence condition of the Newton's method, and we go to Newton's method. Then, we run to the end of the whole process. Theorem 1 ensures that it does not depend on the initial value. If our method goes to the fourth step and if it is appropriate to the condition of the fixed point theorem (Theorem 3), Newton's method is realized by our method. Then, the fourth step of our method being also independent of the initial value can be confirmed by Theorem 3. In brief, convergence of our method does not depend on the initial value via the whole algorithm execution process. The proof is completed. □

4. Numerical Experiments

In order to illustrate the superiority of our method to other algorithms, we provide five numerical examples to confirm its robustness and high efficiency. From Tables 1–14, the iterative termination criteria is satisfied such that $|t_m - \alpha| < 10^{-17} and\, |t_{m+1} - t_n| < 10^{-17}$. All numerical results were computed through g++ in a Fedora Linux 8 environment. The approximate zero α reached up to the 17th decimal place is reflected. These results of our five examples are obtained from computer hardware configuration with T2080 1.73 GHz CPU and 2.5 GB memory.

Example 1. *There is a parametric curve* $c(t) = (f_1(t), f_2(t), f_3(t)) = (6t^7 + t^5, 5t^8 + 3t^6, 10t^{12} + 8t^8 + 6t^6 + 4t^4 + 2t^2 + 3), t \in [-2, 2]$ *in three-dimensional Euclidean space and a test point* $p = (p_1, p_2, p_3) = (2.0, 4.0, 2.0)$. *Using our method, the corresponding orthogonal projection parametric value is* $\alpha = 0.0$, *the initial values* t_0 *are 0,2,4,5,6,8,9,10, respectively. For each initial value, the iteration process runs 10 times and then 10 different iteration times in nanoseconds, respectively. In Table 1, the average run time of our method for eight different initial values are 536,142, 77,622, 101,481, 119,165, 126,502, 142,393, 150,801, 156,413 nanoseconds, respectively. Finally, the overall average running time is 176,315 nanoseconds (see Figure 6). If test point p is (2.0, 2.0, 2.0), the corresponding orthogonal projection parametric value is* $\alpha = 0.0$, *we replicate the procedure using our method and report the results in Table 2. In Table 2, the average running time of our method for 8 different initial values are 627,996, 89,992, 119,241, 139,036, 148,269, 167,364, 167,364, 178,554 nanoseconds, respectively. Finally, the overall average running time is 205,228 nanoseconds (see Figure 7). However, for the above two cases, the H-H-H method does not converge for any initial iterative value.*

Because of a singular point on the parametric curve, we have also added some pre-processing steps before our method. (1) Find the singular point (0,0,3) and the corresponding parametric value 0 by using the methods [21,46–48]. (2) Using our method, the orthogonal projection points of test points (2,4,2) and (2,2,2) and their corresponding parameter values 0 and 0 are calculated, respectively. (3) From all candidate points(singular point and orthogonal projection point), corresponding point is selected so that the distance between the corresponding point and the test point is the minimum one. In Figure 6, the blue point denotes singular point (0,0,3), which is also the orthogonal projecting point of the test point (2,4,2). This is the same for the blue point in Figure 7.

Table 1. Running time for different initial values of Example 1 by our method with test point $p = (2.0, 4.0, 2.0)$.

t_0	0	2	4	5	6	8	9	10
α	0.0	0.0	0.0	0.0	0.0	0.0	0.0	0.0
1	498,454	75,487	105,563	116,470	123,031	134,941	154,253	156,872
2	555,709	81,629	108,064	117,762	125,946	140,940	153,468	155,830
3	509,173	82,824	100,744	111,206	134,367	141,705	150,013	158,715
4	564,222	77,465	96,721	114,757	129,128	173,027	150,320	158,580
5	502,986	81,028	97,142	118,535	120,668	132,856	155,335	149,437
6	553,198	79,520	104,307	120,795	129,351	150,085	151,073	143,065
7	576,814	74,268	100,231	115,002	132,322	139,919	154,754	159,014
8	524,848	81,982	99,604	115,263	122,401	139,345	143,568	175,169
9	528,848	71,228	103,186	140,023	122,040	135,006	145,434	154,016
10	547,161	70,789	99,247	121,834	125,766	136,103	149,790	153,435
Average	536,142	77,622	101,481	119,165	126,502	142,393	150,801	156,413
Total Average	176,315							

Table 2. Running time for different initial values of Example 1 by our method with test point $p = (2.0, 2.0, 2.0)$.

t_0	0	2	4	5	6	8	9	10
α	0.0	0.0	0.0	0.0	0.0	0.0	0.0	0.0
1	595,515	92,371	119,904	135,660	148,751	162,758	171,535	177,355
2	648,825	91,746	119,348	135,284	148,531	162,541	171,333	176,431
3	595,772	91,633	119,248	135,322	148,222	162,240	171,095	176,501
4	648,472	91,565	119,139	135,355	148,165	191,884	171,366	176,395
5	595,856	91,556	119,168	135,406	148,144	162,224	171,417	176,507
6	648,305	91,532	119,018	135,316	148,169	183,342	171,413	176,473
7	647,406	91,587	119,069	135,283	148,197	162,291	171,282	176,397
8	595,423	91,617	119,247	135,140	14,8101	162,116	171,342	196,529
9	646,551	83,167	119,135	172,412	148,149	162,148	171,313	176,390
10	657,838	83,147	119,131	135,179	148,259	162,094	171,609	176,557
Average	627,996	89,992	119,241	139,036	148,269	167,364	171,371	178,554
Total Average	205,228							

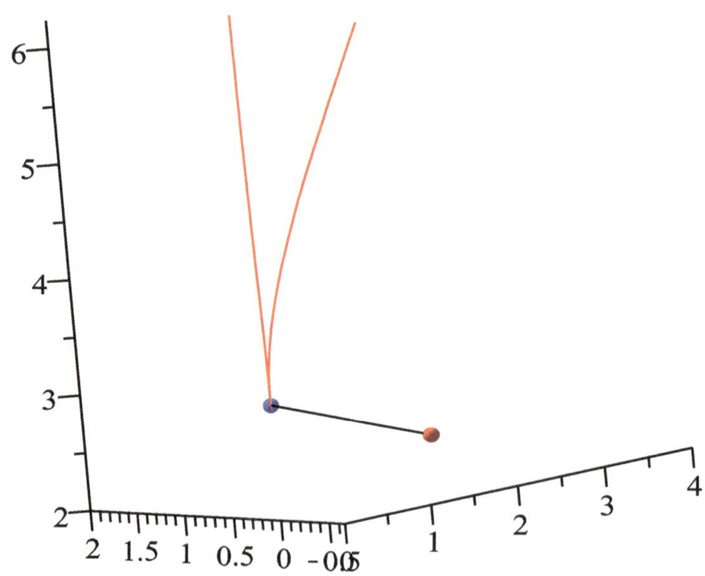

Figure 6. Geometric illustration for the test point $p = (2.0, 4.0, 2.0)$ of Example 1.

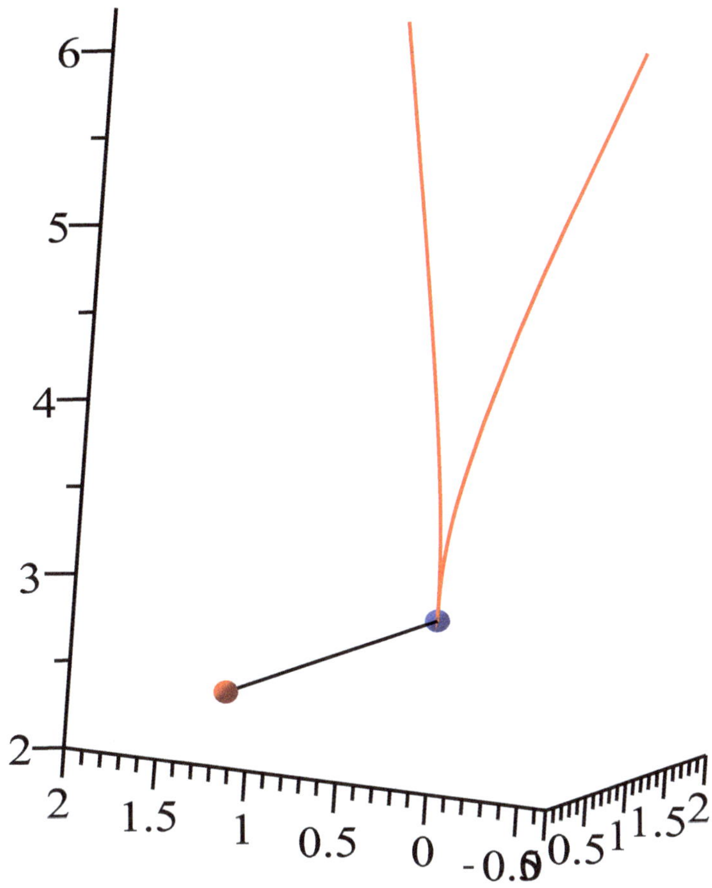

Figure 7. Geometric illustration for the test point $p = (2.0, 2.0, 2.0)$ of Example 1.

Example 2. *There is a spatial quartic quasi-rational Bézier curve* $c(t) = (f_1(t), f_2(t), f_3(t)) = (\frac{u(t)}{a(t)}, \frac{v(t)}{a(t)}, \frac{w(t)}{a(t)})$, *where* $u(t) = 2t^4 + 3t^3 + 3t^2 + 12t + 1$, $v(t) = 4t^4 + 3t^3 + 7t^2 + 7t + 21$, $w(t) = 5t^4 + t^3 + 9t^2 + 11t + 13$, $a(t) = 4t^4 + 8t^3 + 17t^2 + 15t + 6$, $t \in [-2, 2]$ *and a test point* $p = (p_1, p_2, p_3) = (1.0, 3.0, 5.0)$. *The corresponding orthogonal projection parametric value* α *are* -1.4118250062741212, -0.61917136491841674, -0.059335038305820650, 1.8493434997820080, *respectively. Using our method, the initial values* t_0 *are* $-2.4, -2.1, -2.0, -1.8, -1.6, -1.2, -1.0, -0.8$, *respectively. For each initial value, the iteration process runs 10 times and then 10 different iteration times in nanoseconds, respectively. From Table 3, the average running time of our method for eight different initial values are 85,344, 93,936, 79,424, 62,643, 54,482, 22,982, 25,654, 26,868 nanoseconds, respectively. Finally, the overall average running time is 56,417 nanoseconds (see Figure 8). If test point* p *is* $(2.0, 4.0, 8.0)$, *the corresponding orthogonal projection parametric value* α *are* -1.2589948653798823, -0.62724968160147096, -0.14597283439336865, 1.8584532894110559, *respectively. We firstly replicate the procedure using our method and report the results in Table 4. From Table 4, the average running time of our method for eight different initial iterative values are 101,436, 109,001, 95,061, 77,563, 62,366, 27,054, 29,587, 32,501 nanoseconds, respectively. Finally, the overall average running time is 66,821 nanoseconds (see Figure 9). We then replicate the procedure using the algorithm [26] and report the results in Table 5. From Table 5, the average running time of the algorithm [26] for eight different initial values are 619,772, 654,281, 584,653, 467,856, 384,393, 163,225, 183,257, 195,013 nanoseconds, respectively. Finally, the overall average running time is 406,556 nanoseconds. However, for the above two cases, the H-H-H method does not converge for any initial value.*

Table 3. Running time for different initial values of Example 2 by our method with test point $p = (1.0, 3.0, 5.0)$.

t_0	−2.4	−2.1	−2	−1.8	−1.6	−1.2	−1	−0.8
α	−1.4118	0.61917	−1.4118	0.61917	−0.059	−0.059	1.84934	1.84934
1	88,695	90,501	75,137	68,499	52,014	24,731	26,295	28,444
2	89,958	91,254	79,411	64,563	54,321	22,014	26,278	28,024
3	83,956	95,063	79,553	63,237	54,683	22,733	24,813	28,760
4	83,623	96,033	82,022	68,075	51,098	23,270	24,573	26,707
5	83,368	95,700	76,197	63,518	51,752	22,321	24,644	26,586
6	83,631	97,303	80,984	62,608	53,473	21,658	24,009	28,209
7	87,286	94,655	78,483	66,844	52,277	23,502	25,554	28,725
8	87,150	96,316	79,215	64,333	51,554	23,217	26,234	28,295
9	86,300	89,399	94,487	66,665	50,279	23,190	25,791	26,160
10	89,761	96,377	82,362	64,371	50,367	22,332	23,929	27,273
Average	85,344	93,936	79,424	62,643	54,482	22,982	25,654	26,868
Total Average	56,417							

Table 4. Running time for different initial values of Example 2 by our method with test point $p = (2.0, 4.0, 8.0)$.

t_0	−2.4	− −2.1	− −2	−1.8	−1.6	−1.2	−1	−0.8
α	−0.6272	−0.1459	−0.6272	−1.2589	−0.1459	1.858	−1.2589	1.858
1	101,366	109,667	92,799	77,983	62,865	29,460	29,755	32,649
2	102,027	108,844	92,709	77,477	62,269	27,177	29,555	32,458
3	101,526	109,010	92,709	77,587	62,284	26,885	29,619	32,538
4	101,266	108,909	92,724	77,441	62,374	26,785	29,557	32,478
5	101,346	108,944	92,714	77,386	62,214	26,691	29,559	32,505
6	101,315	108,990	92,764	77,557	62,334	26,731	29,564	32,497
7	101,415	108,834	92,614	77,582	62,415	26,720	29,573	32,512
8	101,306	108,945	92,528	77,461	62,309	26,715	29,548	32,493
9	101,562	108,954	116,107	77,542	62,284	26,684	29,549	32,429
10	101,235	108,910	92,939	77,616	62,314	26,690	29,595	32,451
Average	101,436	109,001	95,061	77,563	62,366	27,054	29,587	32,501
Total Average	66,821							

Table 5. Running time for different initial values of Example 2 by the algorithm [26].

t_0	−2.4	−2.1	−2.0	−1.8	−1.6	−1.2	−1.0	−0.8
α	−0.6272	−0.1459	−0.6272	−1.2589	−0.1459	1.858	−1.2589	1.858
1	633,173	660,734	566,675	470,236	391,687	171,352	175,965	198,543
2	597,065	628,741	565,012	485,368	367,539	161,649	185,457	197,798
3	652,494	675,268	600,951	463,899	396,359	163,879	188,682	187,128
4	649,281	653,066	573,597	460,967	385,325	156,876	182,979	195,214
5	622,109	687,282	568,766	472,217	402,669	170,876	189,508	202,540
6	633,737	627,667	562,864	490,735	374,340	165,445	175,457	191,037
7	584,705	637,608	563,523	468,230	395,411	163,631	175,676	187,539
8	607,439	693,001	585,948	449,706	400,728	161,467	189,216	187,433
9	637,036	639,359	671,613	444,834	359,918	157,235	188,119	195,867
10	580,678	640,082	587,577	472,368	369,954	159,834	181,510	207,033
Average	619,772	654,281	584,653	467,856	384,393	163,225	183,257	195,013
Total Average	406,556							

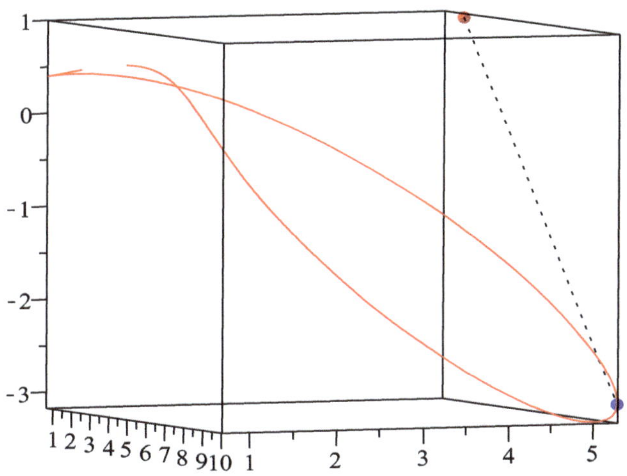

Figure 8. Geometric illustration for the first case of Example 2.

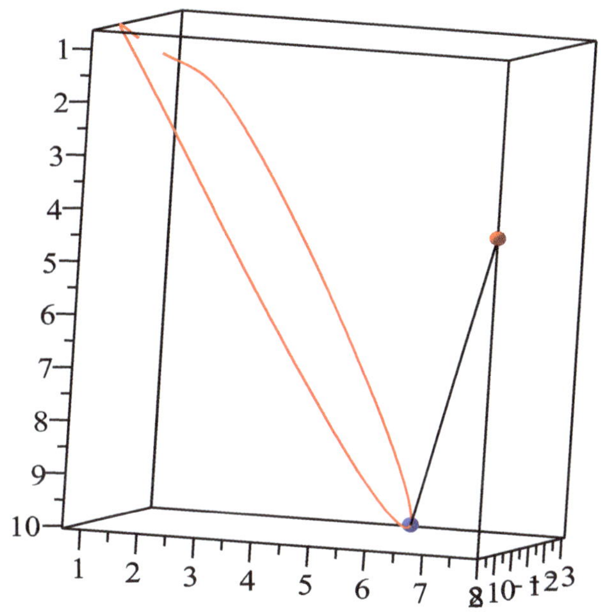

Figure 9. Geometric illustration for the second case of Example 2.

Example 3. *There is a parametric curve* $c(t) = (f_1(t), f_2(t), f_3(t), f_4(t), f_5(t)) = (cos(t), sin(t), t, cos(t), sin(t)), t \in [-2, 2]$ *in five-dimensional Euclidean space and a test point* $p = (p_1, p_2, p_3, p_4, p_5) = (3.0, 4.0, 5.0, 6.0, 7.0)$. *Using our method, the corresponding orthogonal projection parametric value is* $\alpha = 1.1587403612284800$, *the initial values* t_0 *are* $-10, -8, -6, -4, 4, 8, 12, 16$, *respectively. For each initial value, the iteration process runs 10 times and then 10 different iteration times in nanoseconds, respectively. In Table 6, the average running time of our method for eight different initial values are 391,013, 424,444, 391,092, 249,376, 115,617, 170,212, 179,465, 196,912 nanoseconds, respectively. Finally, the overall average running time is 264,766 nanoseconds. If test point* p *is* $(30.0, 40.0, 50.0, 60.0, 70.0)$, *the corresponding orthogonal projection parametric value* α *is* 1.2352898417860202. *We then replicate the procedure using our method and report the results in Table 7. In Table 7, the average running time of our method for eight different initial values are* 577,707, 485,417, 460,913, 289,232, 133,661, 199,470, 211,915, 229,398 *nanoseconds, respectively. Finally, the overall average running time is 323,464 nanoseconds. However, for the above parametric curve and many test points, the H-H-H method does not converge for any initial value.*

Table 6. Running time for different initial values of Example 3 by our method with test point p = (3, 4, 5, 6, 7).

t_0	−10	−8	−6	−4	4	8	12	16
α	1.15874	1.15874	1.15874	1.15874	1.15874	1.15874	1.15874	1.15874
1	407,427	425,388	387,337	306,115	110,887	161,079	187,144	184,119
2	417,729	446,171	398,801	341,895	121,148	169,115	169,954	194,671
3	420,894	390,507	383,308	260,183	115,033	165,103	171,989	198,884
4	383,836	421,365	427,391	242,641	109,521	161,121	179,152	195,714
5	373,696	421,551	373,171	266,584	120,844	187,930	179,184	186,309
6	374,791	445,114	373,974	242,889	119,449	183,082	180,269	201,487
7	381,353	408,011	402,073	216,762	109,054	162,402	172,013	188,206
8	398,662	442,008	373,328	194,821	119,236	192,990	180,472	197,299
9	364,491	417,139	396,843	230,070	110,243	164,273	204,410	196,163
10	387,246	427,188	394,694	191,799	120,759	155,029	170,059	226,270
Average	391,013	424,444	391,092	249,376	115,617	170,212	179,465	196,912
Total Average	264,766							

Table 7. Running time for different initial values of Example 3 by our method with test point p = (30, 40, 50, 60, 70).

t_0	−10	−8	−6	−4	4	8	12	16
α	1.235289	1.235289	1.235289	1.235289	1.235289	1.235289	1.235289	1.235289
1	1,190,730	475,499	453,879	369,551	133,651	191,093	208,202	223,695
2	1,031,760	500,975	486,534	380,881	133,638	190,959	208,490	236,637
3	482,018	475,395	450,480	297,272	133,674	199,528	208,292	223,312
4	428,081	475,588	475,100	277,356	133,635	186,919	208,438	223,802
5	455,282	475,033	448,776	296,510	133,535	220,570	208,139	223,471
6	428,321	499,776	448,617	277,353	133,590	220,625	208,046	223,213
7	428,246	474,978	474,667	247,245	133,620	192,326	208,101	230,791
8	453,374	502,500	448,503	235,415	133,594	220,635	208,087	223,183
9	426,949	474,816	474,167	275,526	133,546	198,204	245,226	223,213
10	452,306	499,605	448,409	235,207	134,128	173,843	208,127	262,661
Average	577,707	485,417	460,913	289,232	133,661	199,470	211,915	229,398
Total Average	323,464							

Example 4. (Reference to [6]) *There is a parametric curve* $c(t) = (f_1(t), f_2(t)) = (t^2, sin(t)), t \in [-3, 3]$ *in two-dimensional Euclidean space and a test point* $p = (p_1, p_2) = (1.0, 2.0)$. *The corresponding orthogonal projection parametric value is* $\alpha = 1.1063055095030472$. *Using our method, the initial values* t_0 *are* $-100, -4, 5, 7, 8, 10, 11, 100$, *respectively. For each initial value, the iteration process runs 10 times and then 10 different iteration times in nanoseconds, respectively. In Table 8, the average running time of our method for eight different initial iterative values are 62,816, 35,042, 27,648, 43,122, 21,625, 38,654, 21,518, 72,917 nanoseconds, respectively. Finally, the overall average running time is 40,418 nanoseconds (see Figure 10). Implementing the same procedure, the overall average running time given by the H-H-H method is 231,613 nanoseconds in Table 9, while the overall average running time given by the second order method [6] is 847,853 nanoseconds in Table 10. Thus, our method is faster than the H-H-H method [38–40] and the second order method [6].*

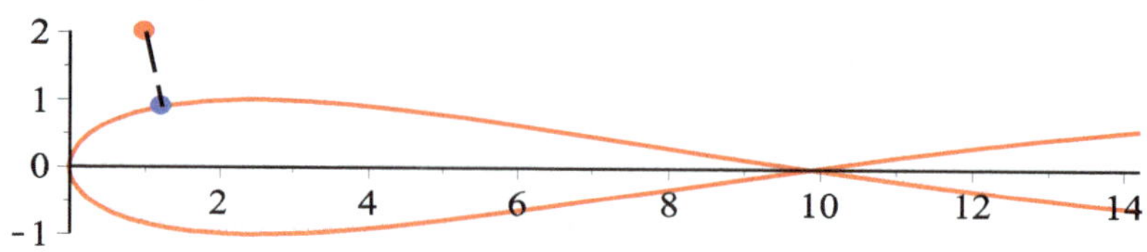

Figure 10. Geometric illustration for Example 4.

Table 8. Running time for different initial values of Example 4 by our method.

t_0	−100	−4	5	7	8	10	11	100
α	1.106305	1.106305	1.106305	1.106305	1.106305	1.106305	1.106305	1.106305
1	63,345	35,580	27,069	41,551	22,304	36,858	21,478	72,257
2	63,192	36,203	28,160	41,733	20,042	38,680	20,338	71,620
3	61,306	33,833	27,400	44,198	23,078	37,704	23,757	73,108
4	66,627	34,502	26,014	44,160	21,147	39,374	22,530	70,154
5	62,583	35,053	29,275	42,800	20,817	39,339	23,046	73,189
6	63,957	34,398	25,650	42,282	22,184	37,376	20,070	75,872
7	60,865	35,929	28,944	42,134	19,964	40,078	21,943	71,608
8	63,522	35,427	27,578	41,688	23,650	39,456	21,076	76,283
9	60,551	35,508	28,563	44,542	20,280	38,463	20,596	71,781
10	62,216	33,987	27,830	46,130	22,781	39,209	20,349	73,296
Average	62,816	35,042	27,648	43,122	21,625	38,654	21,518	72,917
Total Average	40,418							

Table 9. Running time for different initial values of Example 4 by the H-H-H method.

t_0	−100	−4	5	7	8	10	11	100
α	1.106305	1.106305	1.106305	1.106305	1.106305	1.106305	1.106305	1.106305
1	424,579	357,276	443,858	179,583	176,984	175,859	175,249	178,445
2	425,680	358,510	179,137	177,849	182,701	176,665	176,463	207,164
3	359,794	356,912	180,000	180,472	177,867	179,743	178,929	179,372
4	371,119	357,214	179,567	179,804	184,542	177,675	177,854	179,651
5	358,128	358,119	232,337	179,285	179,113	175,632	177,690	181,976
6	358,470	357,893	179,985	179,941	178,600	178,289	178,565	181,868
7	358,083	359,391	178,815	177,857	177,613	178,014	177,385	179,361
8	477,393	357,011	178,029	179,525	175,684	176,000	175,413	180,966
9	356,254	359,356	176,148	178,581	176,351	177,024	185,103	180,013
10	356,801	359,773	213,327	177,252	176,993	178,060	177,655	181,427
Average	384,630	358,146	214,120	179,015	178,645	177,296	178,031	183,024
Total Average	231,613							

Table 10. Running time for different initial values of Example 4 by the Algorithm [6].

t_0	−100	−4	5	7	8	10	11	100
α	1.106305	1.106305	1.106305	1.106305	1.106305	1.106305	1.106305	1.106305
1	681,353	107,102	119,083	120,328	122,504	115,181	113,566	542,116
2	725,571	124,514	136,810	121,111	116,824	111,116	117,466	5,250,481
3	669,249	111,052	122,151	125,261	124,865	116,105	120,309	5,523,805
4	713,982	112,146	131,494	118,104	121,099	111,410	118,658	5,407,166
5	699,433	111,347	118,830	121,003	118,694	115,182	124,917	5,259,412
6	693,396	113,323	116,046	109,176	108,194	111,420	117,342	5,508,049
7	691,375	114,667	115,748	123,330	127,812	118,635	119,208	5,348,517
8	663,125	107,484	127,493	120,134	116,818	111,717	117,079	5,446,703
9	731,148	128,918	122,897	120,947	120,985	113,777	125,463	5,251,580
10	676,286	128,567	130,775	118,031	116,725	111,095	108,275	5,356,125
Average	694,492	115,912	124,133	119,743	119,452	113,564	118,228	5,377,300
Total Average	847,853							

Example 5. (Reference to [6]) *There is a parametric curve* $c(t) = (f_1(t), f_2(t)) = (t, sin(t)), t \in [-3, 3]$ *in two-dimensional Euclidean space and a test point* $p = (p_1, p_2) = (1.0, 2.0)$*, the corresponding orthogonal projection parametric value is* $\alpha = 1.2890239979093887$*. Using our method, the initial values* t_0 *are* $-100, -4, 5, 7, 8, 10, 11, 100$*, respectively. For each initial value, the iteration process runs 10 times and then 10 different iteration time in nanoseconds, respectively. In Table 11, the average running time of our method for eight different initial values are* 50, 579, 28, 238, 22, 687, 34, 974,17, 781, 31, 186, 17, 210, 59, 116 *nanoseconds, respectively. Finally, the overall average running time is 32,721 nanoseconds (see Figure 11). We then replicate the procedure using the second order method [6] and report the results in Table 12. In Table 12, the average running time of the second order method [6] for 8 different initial values are* 320, 035,182, 451, 147, 031, 235, 779, 112, 090, 200, 431, 113, 284, 369, 294 *nanoseconds, respectively. Finally, the overall average*

running time is 210,049 nanoseconds. In addition, we compare the iterations by different methods where the NC denotes non-convergence in Table 13.

Table 11. Running time for different initial values of Example 5 by our method.

t_0	−100	−4	5	7	8	10	11	100
α	1.28902	1.28902	1.28902	1.28902	1.28902	1.28902	1.28902	1.28902
1	52,010	27,426	21,791	33,323	18,399	29,551	16,486	58,995
2	50,335	29,269	23,949	32,810	15,820	30,342	16,066	58,080
3	49,047	26,841	23,061	37,063	19,611	31,569	19,756	57,458
4	52,651	29,124	21,403	33,838	17,472	33,295	18,583	54,566
5	49,871	29,814	25,062	35,870	16,655	32,949	18,304	61,860
6	53,651	28,678	19,550	35,731	18,373	31,429	16,342	59,570
7	47,275	28,115	24,177	35,456	16,933	30,510	18,010	59,042
8	49,982	27,896	22,639	34,292	19,927	30,959	16,449	63,652
9	49,704	29,359	22,502	34,164	17,274	30,391	16,044	61,373
10	51,268	25,859	22,736	37,190	17,342	30,864	16,060	56,564
Average	50,579	28,238	22,687	34,974	17,781	31,186	17,210	59,116
Total Average	32,721							

Table 12. Running time for different initial values of Example 5 by the Algorithm [6].

t_0	−100	−4	5	7	8	10	11	100
α	1.28902	1.28902	1.28902	1.28902	1.28902	1.28902	1.28902	1.28902
1	308,942	191,002	152,199	235,287	114,568	199,404	110,512	379,771
2	348,554	175,800	146,728	232,260	102,698	190,860	101,754	352,834
3	311,680	190,863	148,384	242,131	118,602	207,376	125,517	408,978
4	332,421	166,849	145,131	234,536	102,795	198,956	113,523	370,826
5	319,660	185,059	160,358	235,072	108,557	211,429	119,911	350,188
6	329,882	177,252	132,242	233,702	120,945	199,978	107,366	363,299
7	304,977	200,038	151,398	229,166	102,315	220,162	122,013	354,466
8	326,645	171,624	137,588	228,181	113,627	195,782	108,512	369,899
9	291,369	191,878	156,871	247,614	108,418	189,534	112,319	363,905
10	326,221	174,148	139,415	239,836	128,377	190,831	111,411	378,781
Average	320,035	182,451	147,031	235,779	112,090	200,431	113,284	369,294
Total Average	210,049							

Table 13. Comparison of iterations by different methods in Example 5.

t_0	−100.0	−4.0	5.0	7.0	8.0	10.0	11.0	100.0
α	1.28902	1.28902	1.28902	1.28902	1.28902	1.28902	1.28902	1.28902
H-H-H method [38–40]	NC	NC	NC	NC	NC	NC	NC	NC
Second order method [6]	75	30	32	32	33	29	31	101
Newton's method	NC	NC	NC	NC	NC	NC	NC	NC
Our method	15	19	17	17	15	17	15	23

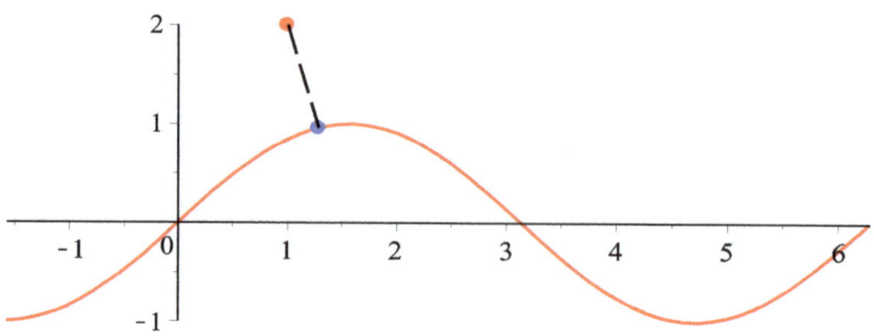

Figure 11. Geometric illustration for Example 5.

Remark 2. *From the results of five examples, the overall average running time of our method is 145.5 μs. From the results of Table 9, the overall average running time of the H-H-H method is 231.6 μs. From results of six examples in [26], the overall average running time of the algorithm [1] is 680.8 μs. From results of* *six examples in [26], the overall average running time of the algorithm [14] is 1270.8 μs. From results of Table 5, the overall average running time of the algorithm [26] is 406.6 μs. From results of Tables 10 and 12, the overall average running time of the algorithm [6] is 528.9 μs. Table 14 displays time comparison for these algorithms. In short, the robustness and efficiency of our method are more superior to those of the existing algorithms [1,6,14,26,38–40].*

Table 14. Time comparison of various algorithms.

Algorithms	Ours	H-H-H	Algorithm [1]	Algorithm [14]	Algorithm [6]	Algorithm [26]
Time (μs)	145.5	231.6	680.8	1270.8	528.9	406.6

Remark 3. *For general parametric curve containing the elementary functions, such as* $sin(t)$*,*$cos(t)$*,* e^t*,* $\ln t$*,* $\arcsin t$*,* $\arccos t$*, etc., it is very difficult to transform general parametric curve into Bézier-type curve. In contrast, our method can deal with the general parametric curve containing the elementary functions. Furthermore, the convergence of our method does not depend on the initial value. From Table 13, only the H-H-H method or the Newton's method can not ensure convergence, while our method can ensure convergence. For multiple solutions of orthogonal projection, our approach works as follows:*
(1) The parameter interval $[a, b]$ *of parametric curve* $c(t)$ *is divided into M identical subintervals.*
(2) An initial value is selected randomly in each interval.
(3) Using our method and using each initial parametric value, do iterations, respectively. Suppose that the iterative parametric values are $\alpha_1, \alpha_2, \ldots, \alpha_M$*, respectively.*
(4) Calculate the local minimum distances $d_1, d_2, \ldots, d_M$*, where* $d_i = \|p - c(\alpha_i)\|$*.*
(5) Seek the global minimum distance $d = \|p - c(\alpha)\|$ *from* $\{\|p - c(a)\|, d_1, d_2, \ldots, d_M, \quad \|p - c(b)\|\}$*.*
If we are to solve all solutions as far as possible, we urge the positive integer M to be as large as possible.

We use Example 2 to illustrate how the procedure works, where, for $t \in [-2, 2]$, three parameter values are -1.4118250062741212, -0.61917136491841674, 1.8493434997820080, respectively. It is easy to find that the projection point with the parameter value -0.61917136491841674 will be the one with minimum distance, whereas other projection points without these parameter values can not be the one with minimum distance. Thus, only the orthogonal projection point with minimum distance remains after the procedure to select multiple orthogonal projection points.

Remark 4. *We have done many test examples including five test examples. In the light of these test results, our method has good convergent properties for different initial values, namely, if initial value is* t_0*, then the corresponding orthogonal projection parametric value* α *for the orthogonal projection point of the test point p is suitable for one inequality relationship*

$$|\langle p - c(\alpha), c'(\alpha) \rangle| < 10^{-17}. \tag{21}$$

This indicates that the inequality relationship satisfies requirements of Equation (4). This shows that convergence of our method does not depend on the initial value. Furthermore, our method is robust and efficient, which is satisfied with the previous two of ten challenges proposed by [50].

5. Conclusions

This paper discusses the problem related to a point orthogonal projection onto a parametric curve in an n-dimensional Euclidean space on the basis of the H-H-H method, combining with a fixed point theorem of Newton's method. Firstly, we run the H-H-H method. If the current iterative parametric value from the H-H-H method is satisfied with the convergence condition of the Newton's method, we then go to the method to increase the convergence rate. Otherwise, we continue the H-H-H method to generate the iterative parametric value with satisfaction of the local convergence condition by the Newton's method, and we then go to the previous step. Then, we run to the end of the whole process. The presented procedures ensure the convergence of our method and it does not depend on the initial value. Analysis of convergence demonstrates that our method is second order convergent. Some numerical examples confirm that our method is more efficient and performs better than other methods, such as the algorithms [1,6,14,26,38–40].

In this paper, our discussion focuses the algorithms in the parametric curve C^2. For the parametric curve being C^0,C^1, piecewise curve or having singular points, we only present a preliminary idea. However, we have not completely implemented an algorithm for this kind of spline with low continuity. In the future, we will try to construct several brand new algorithms to handle the kind of spline with low continuity such that they can ensure very good robustness and efficiency. In addition, we also try to extend this idea to handle point orthogonal projecting onto implicit curves and implicit surfaces that include singularity points. Of course, the realization of these ideas is of great challenge. However, it is of great value and significance in practical engineering applications.

Author Contributions: The contribution of all the authors is the same. All of the authors team up to develop the current draft. J.L. is responsible for investigating, providing methodology, writing, reviewing and editing this work. X.L. is responsible for formal analysis, visualization, writing, reviewing and editing of this work. F.P. is responsible for software, algorithm and program implementation to this work. T.C. is responsible for validation of this work. L.W. is responsible for supervision of this work. L.H. is responsible for providing resources, writing, and the original draft of this work.

Acknowledgments: We take the opportunity to thank the anonymous reviewers for their thoughtful and meaningful comments.

References

1. Ma, Y.L.; Hewitt, W.T. Point inversion and projection for NURBS curve and surface: Control polygon approach. *Comput. Aided Geom. Des.* **2003**, *20*, 79–99. [CrossRef]
2. Piegl, L.; Tiller, W. Parametrization for surface fitting in reverse engineering. *Comput.-Aided Des.* **2001**, *33*, 593–603. [CrossRef]
3. Yang, H.P.; Wang, W.P.; Sun, J.G. Control point adjustment for B-spline curve approximation. *Comput.-Aided Des.* **2004**, *36*, 639–652. [CrossRef]
4. Johnson, D.E.; Cohen, E. A Framework for efficient minimum distance computations. In Proceedings of the IEEE Intemational Conference on Robotics & Automation, Leuven, Belgium, 20 May 1998.
5. Pegna, J.; Wolter, F.E. Surface curve design by orthogonal projection of space curves onto free-form surfaces. *J. Mech. Des. ASME Trans.* **1996**, *118*, 45–52. [CrossRef]
6. Hu, S.M.; Wallner, J. A second order algorithm for orthogonal projection onto curves and surfaces. *Comput. Aided Geom. Des.* **2005**, *22*, 251–260. [CrossRef]
7. Besl, P.J.; McKay, N.D. A method for registration of 3-D shapes. *IEEE Trans. Pattern Anal. Mach. Intell.* **1992**, *14*, 239–256. [CrossRef]
8. Mortenson, M.E. *Geometric Modeling*; Wiley: New York, NY, USA, 1985.
9. Limaien, A.; Trochu, F. Geometric algorithms for the intersection of curves and surfaces. *Comput. Graph.* **1995**, *19*, 391–403. [CrossRef]
10. Press, W.H.; Teukolsky, S.A.; Vetterling, W.T.; Flannery, B.P. Numerical recipes. In *C: The Art of Scientific Computing*, 2nd ed.; Cambridge University Press: New York, NY, USA, 1992.
11. Elber, G.; Kim, M.S. Geometric Constraint solver using multivariate rational spline functions. In Proceedings of the 6th ACM Symposiumon Solid Modeling and Applications, Ann Arbor, MI, USA, 4–8 June 2001; pp. 1–10.
12. Patrikalakis, N.; Maekawa, T. *Shape Interrogation for Computer Aided Design and Manufacturing*; Springer: Berlin, Germany, 2001.
13. Polak, E.; Royset, J.O. Algorithms with adaptive smoothing for finite minimax problems. *J. Optim. Theory Appl.* **2003**, *119*, 459–484. [CrossRef]
14. Selimovic, I. Improved algorithms for the projection of points on NURBS curves and surfaces. *Comput. Aided Geom. Des.* **2006**, 439–445. [CrossRef]

15. Zhou, J.M.; Sherbrooke, E.C.; Patrikalakis, N. Computation of stationary points of distance functions. *Eng. Comput.* **1993**, *9*, 231–246. [CrossRef]
16. Cohen, E.; Lyche, T.; Riesebfeld, R. Discrete B-splines and subdivision techniques in computer-aided geometric design and computer graphics. *Comput. Graph. Image Process.* **1980**, *14*, 87–111. [CrossRef]
17. Piegl, L.; Tiller, W. *The NURBS Book*; Springer: New York, NY, USA, 1995.
18. Park, C.-H.; Elber, G.; Kim, K.-J.; Kim, G.-Y.; Seong, J.-K. A hybrid parallel solver for systems of multivariate polynomials using CPUs and GPUs. *Comput.-Aided Des.* **2011**, *43*, 1360–1369. [CrossRef]
19. Bartoň, M. Solving polynomial systems using no-root elimination blending schemes. *Comput.-Aided Des.* **2011**, *43*, 1870–1878.
20. van Sosin, B.; Elber, G. Solving piecewise polynomial constraint systems with decomposition and a subdivision-based solver. *Comput.-Aided Des.* **2017**, *90*, 37–47. [CrossRef]
21. Bartoň, M.; Elber, G.; Hanniel, I. Topologically guaranteed univariate solutions of underconstrained polynomial systems via no-loop and single-component tests. *Comput.-Aided Des.* **2011**, *43*, 1035–1044.
22. Johnson, D.E.; Cohen, E. Distance extrema for spline models using tangent cones. In Proceedings of the 2005 Conference on Graphics Interface, Victoria, Canada, 9–11 May 2005.
23. Li, X.W.; Xin, Q.; Wu, Z.N.; Zhang, M.S.; Zhang, Q. A geometric strategy for computing intersections of two spatial parametric curves. *Vis. Comput.* **2013**, *29*, 1151–1158. [CrossRef]
24. Liu, X.-M.; Yang, L.; Yong, J.-H.; Gu, H.-J.; Sun, J.-G. A torus patch approximation approach for point projection on surfaces. *Comput. Aided Geom. Des.* **2009**, *26*, 593–598. [CrossRef]
25. Chen, X.-D.; Xu, G.; Yong, J.-H.; Wang, G.Z.; Paul, J.-C. Computing the minimum distance between a point and a clamped B-spline surface. *Graph. Models* **2009**, *71*, 107–112. [CrossRef]
26. Chen, X.-D.; Yong, J.-H.; Wang, G.Z.; Paul, J.-C.; Xu, G. Computing the minimum distance between a point and a NURBS curve. *Comput.-Aided Des.* **2008**, *40*, 1051–1054. [CrossRef]
27. Oh, Y.-T.; Kim, Y.-J.; Lee, J.; Kim, Y.-S. Gershon Elber, Efficient point-projection to freeform curves and surfaces. *Comput. Aided Geom. Des.* **2012**, *29*, 242–254. [CrossRef]
28. Sederberg, T.W.; Nishita, T. Curve intersection using Bézier clipping. *Comput.-Aided Des.* **1990**, *22*, 538–549. [CrossRef]
29. Bartoň, M.; Jüttler, B. Computing roots of polynomials by quadratic clipping. *Comput. Aided Geom. Des.* **2007**, *24*, 125–141.
30. Li, X.W.; Wu, Z.N.; Hou, L.K.; Wang, L.; Yue, C.G.; Xin, Q. A geometric orthogonal pojection strategy for computing the minimum distance between a point and a spatial parametric curve. *Algorithms* **2016**, *9*, 15. [CrossRef]
31. Mørken, K.; Reimers, M. An unconditionally convergent method for computing zeros of splines and polynomials. *Math. Comput.* **2007**, *76*, 845–865. [CrossRef]
32. Li, X.W.; Wang, L.; Wu, Z.N.; Hou, L.K.; Liang, J.; Li, Q.Y. Hybrid second-order iterative algorithm for orthogonal projection onto a parametric surface. *Symmetry* **2017**, *9*, 146. [CrossRef]
33. Li, X.W.; Wang, L.; Wu, Z.N.; Hou, L.K.; Liang, J.; Li, Q.Y. Convergence analysis on a second order algorithm for orthogonal projection onto curves. *Symmetry* **2017**, *9*, 210.
34. Chen, X.-D.; Ma, W.Y.; Xu, G.; Paul, J.-C. Computing the Hausdorff distance between two B-spline curves. *Comput.-Aided Des.* **2010**, *42*, 1197–1206. [CrossRef]
35. Chen, X.-D.; Chen, L.Q.; Wang, Y.G.; Xu, G.; Yong, J.-H.; Paul, J.-C. Computing the minimum distance between two Bézier curves. *J. Comput. Appl. Math.* **2009**, *229*, 294–301. [CrossRef]
36. Kim, Y.J.; Oh, Y.T.; Yoon, S.H.; Kim, M.S.; Elber, G. Efficient Hausdorff distance computation for freeform geometric models in close proximity. *Comput.-Aided Des.* **2013**, *45*, 270–276. [CrossRef]
37. Sundar, B.R.; Chunduru, A.; Tiwari, R.; Gupta, A.; Muthuganapathy, R. Footpoint distance as a measure of distance computation between curves and surfaces. *Comput. Graph.* **2014**, *38*, 300–309. [CrossRef]
38. Hoschek, J.; Lasser, D. *Fundamentals of Computer Aided Geometric Design*; A. K. Peters: Natick, MA, USA, 1993.
39. Hu, S.M.; Sun, J.G.; Jin, T.G.; Wang, G.Z. Computing the parameter of points on NURBS curves and surfaces via moving affine frame method. *J. Softw.* **2000**, *11*, 49–53.
40. Hartmann, E. On the curvature of curves and surfaces defined by normal forms. *Comput. Aided Geom. Des.* **1999**, *16*, 355–376. [CrossRef]
41. Li, X.W.; Mu, C.L.; Ma, J.W.; Wang, C. Sixteenth-order method for nonlinear Equations. *Appl. Math. Comput.* **2010**, *215*, 3754–3758. [CrossRef]

42. Liang, J.; Li, X.W.; Wu, Z.N.; Zhang, M.S.; Wang, L.; Pan, F. Fifth-order iterative method for solving multiple roots of the highest multiplicity of nonlinear equation. *Algorithms* **2015**, *8*, 656–668. [CrossRef]
43. Melmant, A. Geometry and Convergence of Euler's and Halley's Methods. *SIAM Rev.* **1997**, *39*, 728–735. [CrossRef]
44. Traub, J.F. A Class of Globally Convergent Iteration Functions for the Solution of Polynomial Equations. *Math. Comput.* **1966**, *20*, 113–138. [CrossRef]
45. Śmietański, M.J. A perturbed version of an inexact generalized Newton method for solving nonsmooth equations. *Numer. Algorithms* **2013**, *63*, 89–106. [CrossRef]
46. Chen, F.; Wang, W.-P.; Liu, Y. Computing singular points of plane rational curves. *J. Symb. Comput.* **2008**, *43*, 92–117. [CrossRef]
47. Jia, X.-H.; Goldman, R. Using Smith normal forms and μ-bases to compute all the singularities of rational planar curves. *Comput. Aided Geom. Des.* **2012**, *29*, 296–314. [CrossRef]
48. Shi, X.-R.; Jia, X.-H.; Goldman, R. Using a bihomogeneous resultant to find the singularities of rational space curves. *J. Symb. Comput.* **2013**, *53*, 1–25. [CrossRef]
49. Burden, R.L.; Faires, J.D. *Numerical Analysis*, 9th ed.; Brooks/Cole Cengage Learning: Boston, MA, USA, 2011.
50. Piegl, L.A. Ten challenges in computer-aided design. *Comput.-Aided Des.* **2005**, *37*, 461–470. [CrossRef]

10

Higher-Order Derivative-Free Iterative Methods for Solving Nonlinear Equations and their Basins of Attraction

Jian Li [1,*], **Xiaomeng Wang** [1] **and Kalyanasundaram Madhu** [2,*]

[1] Inner Mongolia Vocational College of Chemical Engineering, Hohhot 010070, China; xiaomengw081@gmail.com

[2] Department of Mathematics, Saveetha Engineering College, Chennai 602105, India

* Correspondence: ljian4064@gmail.com (J.L.); kalyan742@pec.edu (K.M.)

Abstract: Based on the Steffensen-type method, we develop fourth-, eighth-, and sixteenth-order algorithms for solving one-variable equations. The new methods are fourth-, eighth-, and sixteenth-order converging and require at each iteration three, four, and five function evaluations, respectively. Therefore, all these algorithms are optimal in the sense of Kung–Traub conjecture; the new schemes have an efficiency index of 1.587, 1.682, and 1.741, respectively. We have given convergence analyses of the proposed methods and also given comparisons with already established known schemes having the same convergence order, demonstrating the efficiency of the present techniques numerically. We also studied basins of attraction to demonstrate their dynamical behavior in the complex plane.

Keywords: Kung–Traub conjecture; multipoint iterations; nonlinear equation; optimal order; basins of attraction

MSC: 65H05; 65D05; 41A25

---

1. Introduction

Finding faster and exact roots of scalar nonlinear equations is the most important problem in engineering, scientific computing, and applied mathematics. In general, this is the problem of solving a nonlinear equation $f(x) = 0$. Analytical methods for finding solutions of such problems are almost nonavailable, so the only way to get appropriate solutions by numerical methods is based on iterative algorithms. Newton's method is one of the well-known and famous methods for finding solutions of nonlinear equations or local minima in problems of optimization. Despite its nice properties, it will often not work efficiently in some real-life practical applications. Ill conditioning of the problems, the computational expense of functional derivative, accurate initial guesses, and a late convergence rate generally lead to difficulties in its use. Nevertheless, many advantages in all of these drawbacks have been found and led to efficient algorithms or codes that can be easily used (see References [1,2] and references therein). Hence, Steffensen developed a derivative-free iterative method ($SM2$) (see References [3]):

$$w^{(n)} = x^{(n)} + f(x^{(n)}),\ x^{(n+1)} = x^{(n)} - \frac{f(x^{(n)})}{f[x^{(n)}, w^{(n)}]}, \tag{1}$$

where $f[x^{(n)}, w^{(n)}] = \frac{f(x^{(n)}) - f(w^{(n)})}{x^{(n)} - w^{(n)}}$, which preserves the convergence order and efficiency index of Newton's method.

The main motivation of this work is to implement efficient derivative-free algorithms for finding the solution of nonlinear equations. We obtained an optimal iterative method that will support the conjecture [4]. Kung–Traub conjectured that multipoint iteration methods without memory based on d functional evaluations could achieve an optimal convergence order 2^{d-1}. Furthermore, we studied the behavior of iterative schemes in the complex plane.

Let us start a short review of the literature with some of the existing methods with or without memory before proceeding to the proposed idea. Behl et al. [5] presented an optimal scheme that does not need any derivative evaluations. In addition, the given scheme is capable of generating new optimal eighth-order methods from the earlier optimal fourth-order schemes in which the first sub-step employs Steffensen's or a Steffensen-type method. Salimi et al. [6] proposed a three-point iterative method for solving nonlinear equations. The purpose of this work is to upgrade a fourth-order iterative method by adding one Newton step and by using a proportional approximation for the last derivative. Salimi et al. [7] constructed two optimal Newton–Secant-like iterative methods for finding solutions of nonlinear equations. The classes have convergence orders of four and eight and cost only three and four function evaluations per iteration, respectively. Matthies et al. [8] proposed a three-point iterative method without memory for solving nonlinear equations with one variable. The method provides a convergence order of eight with four function evaluations per iteration. Sharifi et al. [9] presented an iterative method with memory based on the family of King's methods to solve nonlinear equations. The method has eighth-order convergence and costs only four function evaluations per iteration. An acceleration of the convergence speed is achieved by an appropriate variation of a free parameter in each step. This self-accelerator parameter is estimated using Newton's interpolation fourth degree polynomial. The order of convergence is increased from eight to 12 without any extra function evaluation. Khdhr et al. [10] suggested a variant of Steffensen's iterative method with a convergence order of 3.90057 for solving nonlinear equations that are derivative-free and have memory. Soleymani et al. [11] presented derivative-free iterative methods without memory with convergence orders of eight and sixteen for solving nonlinear equations. Soleimani et al. [12] proposed a optimal family of three-step iterative methods with a convergence order of eight by using a weight function alongside an approximation for the first derivative. Soleymani et al. [13] gave a class of four-step iterative schemes for finding solutions of one-variable equations. The produced methods have better order of convergence and efficiency index in comparison with optimal eighth-order methods. Soleymani et al. [14] constructed a class of three-step eighth order iterative methods by using an interpolatory rational function in the third step. Each method of the class reaches the optimal efficiency index according to the Kung–Traub conjecture concerning multipoint iterative methods without memory. Kanwar et al. [15] suggested two new eighth-order classes of Steffensen–King-type methods for finding solutions of nonlinear equations numerically. Cordero et al. [1] proposed a general procedure to obtain derivative-free iterative methods for finding solutions of nonlinear equations by polynomial interpolation. In addition, many authors have worked with these ideas on different iterative schemes [16–24], describing the basin of attraction of some well-known iterative scheme. In this work, we developed a novel fourth-order iterative scheme, eighth-order iterative scheme, and sixteenth-order iterative scheme, that are without memory, are derivative-free, and are optimal.

The rest of this paper is ordered as follows. In Section 2, we present the proposed fourth-, eighth-, and sixteenth-order methods that are free from derivatives. Section 3 presents the convergence order of the proposed scheme. In Section 4, we discuss some well-known iterative methods for the numerical and effectiveness comparison of the proposed schemes. In Section 5, we display the performance of proposed methods and other compared algorithms described by problems. The respective graphical fractal pictures obtained from each iteration scheme for test problems are given in Section 6 to show the consistency of the proposed methods. Finally, Section 7 gives concluding remarks.

2. Development of Derivative-Free Scheme

2.1. Optimal Fourth-Order Method

Let us start from Steffensen's method and explain the procedure to get optimal methods of increasing order. The idea is to compose a Steffensen's iteration with Newton's step as follows:

$$\begin{cases} w^{(n)} = x^{(n)} + f(x^{(n)})^3, \\ y^{(n)} = x^{(n)} - \frac{f(x^{(n)})^4}{f(w^{(n)})-f(x^{(n)})}, \\ z^{(n)} = y^{(n)} - \frac{f(y^{(n)})}{f'(y^{(n)})}. \end{cases} \tag{2}$$

The resulting iteration has convergence order four, with the composition of two second-order methods, but the method is not optimal because it uses four function evaluations. In order to get an optimality, we need to reduce a function and to preserve the same convergence order, and so, we estimate $f'(y^{(n)})$ by the following polynomial:

$$\mathcal{N}_2(t) = f(y^{(n)}) + (t-y^{(n)})f[y^{(n)}, w^{(n)}] + (t-y^{(n)})(t-w^{(n)})f[y^{(n)}, w^{(n)}, x^{(n)}], \tag{3}$$

where

$$f[x^{(0)}, x^{(1)}, x^{(2)}, ..., x^{(k-1)}, x^{(k)}] = \frac{f[x^{(1)}, x^{(2)}, ..., x^{(k-1)}, x^{(k)}] - f[x^{(0)}, x^{(1)}, x^{(2)}, ..., x^{(k-1)}]}{x^{(k)} - x^{(0)}},\ x^{(k)} \neq x^{(0)},$$

is the generalized divided differences of kth order at $x^{(0)} \leq x^{(1)} \leq x^{(2)} \leq ... \leq x^{(k-1)} \leq x^{(k)}$. It is noted that $\mathcal{N}_2(y^{(n)}) = f(y^{(n)})$. Differentiating Equation (3) and putting $t = y^{(n)}$, we get

$$\mathcal{N}_2'(y^{(n)}) = f[y^{(n)}, w^{(n)}] + (y^{(n)} - w^{(n)})f[y^{(n)}, w^{(n)}, x^{(n)}]. \tag{4}$$

Now, approximating $f'(y^{(n)}) \approx \mathcal{N}_2'(y^{(n)})$ in Equation (2), we get a new derivative-free optimal fourth-order method ($PM4$) given by

$$\begin{cases} w^{(n)} = x^{(n)} + f(x^{(n)})^3, \\ y^{(n)} = x^{(n)} - \frac{f(x^{(n)})^4}{f(w^{(n)})-f(x^{(n)})}, \\ z^{(n)} = y^{(n)} - \frac{f(y^{(n)})}{f[y^{(n)},w^{(n)}]+(y^{(n)}-w^{(n)})f[y^{(n)},w^{(n)},x^{(n)}]}. \end{cases} \tag{5}$$

2.2. Optimal Eighth-Order Method

Next, we attempt to get a new optimal eighth-order method in the following way:

$$\begin{cases} w^{(n)} = x^{(n)} + f(x^{(n)})^3, \\ y^{(n)} = x^{(n)} - \frac{f(x^{(n)})^4}{f(w^{(n)})-f(x^{(n)})}, \\ z^{(n)} = y^{(n)} - \frac{f(y^{(n)})}{f[y^{(n)},w^{(n)}]+(y^{(n)}-w^{(n)})f[y^{(n)},w^{(n)},x^{(n)}]}, \\ p^{(n)} = z^{(n)} - \frac{f(z^{(n)})}{f'(z^{(n)})}. \end{cases} \tag{6}$$

The above has eighth-order convergence with five function evaluations, but this is not an optimal method. To get an optimal, we need to reduce a function and to preserve the same convergence order, and so, we estimate $f'(z^{(n)})$ by the following polynomial:

$$\begin{aligned} \mathcal{N}_3(t) = {} & f(z^{(n)}) + (t-z^{(n)})f[z^{(n)}, y^{(n)}] + (t-z^{(n)})(t-y^{(n)})f[z^{(n)}, y^{(n)}, w^{(n)}] \\ & + (t-z^{(n)})(t-y^{(n)})(t-w^{(n)})f[z^{(n)}, y^{(n)}, w^{(n)}, x^{(n)}]. \end{aligned} \tag{7}$$

It is clear that $\mathcal{N}_3(z^{(n)}) = f(z^{(n)})$. Differentiating Equation (7) and setting $t = z^{(n)}$, we get

$$\mathcal{N}_3'(z^{(n)}) = f[z^{(n)}, y^{(n)}] + (z^{(n)} - y^{(n)})f[z^{(n)}, y^{(n)}, w^{(n)}] + (z^{(n)} - y^{(n)})(z^{(n)} - w^{(n)})f[z^{(n)}, y^{(n)}, w^{(n)}, x^{(n)}]. \quad (8)$$

Now, approximating $f'(z^{(n)}) \approx \mathcal{N}_3'(z^{(n)})$ in (6), we get a new derivative-free optimal eighth-order method ($PM8$) given by

$$\begin{cases} w^{(n)} = x^{(n)} + f(x^{(n)})^3, \\ y^{(n)} = x^{(n)} - \frac{f(x^{(n)})^4}{f(w^{(n)}) - f(x^{(n)})}, \\ z^{(n)} = y^{(n)} - \frac{f(y^{(n)})}{f[y^{(n)}, w^{(n)}] + (y^{(n)} - w^{(n)})f[y^{(n)}, w^{(n)}, x^{(n)}]}, \\ p^{(n)} = z^{(n)} - \frac{f(z^{(n)})}{f[z^{(n)}, y^{(n)}] + (z^{(n)} - y^{(n)})f[z^{(n)}, y^{(n)}, w^{(n)}] + (z^{(n)} - y^{(n)})(z^{(n)} - w^{(n)})f[z^{(n)}, y^{(n)}, w^{(n)}, x^{(n)}]}. \end{cases} \quad (9)$$

2.3. Optimal Sixteenth-Order Method

Next, we attempt to get a new optimal sixteenth-order method in the following way:

$$\begin{cases} w^{(n)} = x^{(n)} + f(x^{(n)})^3, \\ y^{(n)} = x^{(n)} - \frac{f(x^{(n)})^4}{f(w^{(n)}) - f(x^{(n)})}, \\ z^{(n)} = y^{(n)} - \frac{f(y^{(n)})}{f[y^{(n)}, w^{(n)}] + (y^{(n)} - w^{(n)})f[y^{(n)}, w^{(n)}, x^{(n)}]}, \\ p^{(n)} = z^{(n)} - \frac{f(z^{(n)})}{f[z^{(n)}, y^{(n)}] + (z^{(n)} - y^{(n)})f[z^{(n)}, y^{(n)}, w^{(n)}] + (z^{(n)} - y^{(n)})(z^{(n)} - w^{(n)})f[z^{(n)}, y^{(n)}, w^{(n)}, x^{(n)}]}, \\ x^{(n+1)} = p^{(n)} - \frac{f(p^{(n)})}{f'(p^{(n)})}. \end{cases} \quad (10)$$

The above has sixteenth-order convergence with six function evaluations, but this is not an optimal method. To get an optimal, we need to reduce a function and to preserve the same convergence order, and so, we estimate $f'(p^{(n)})$ by the following polynomial:

$$\begin{aligned} \mathcal{N}_4(t) = {} & f(p^{(n)}) + (t - p^{(n)})f[p^{(n)}, z^{(n)}] + (t - p^{(n)})(t - z^{(n)})f[p^{(n)}, z^{(n)}, y^{(n)}] \\ & + (t - p^{(n)})(t - z^{(n)})(t - y^{(n)})f[p^{(n)}, z^{(n)}, y^{(n)}, w^{(n)}] \\ & + (t - p^{(n)})(t - z^{(n)})(t - y^{(n)})(t - w^{(n)})f[p^{(n)}, z^{(n)}, y^{(n)}, w^{(n)}, x^{(n)}]. \end{aligned} \quad (11)$$

It is clear that $\mathcal{N}_4(p^{(n)}) = f(p^{(n)})$. Differentiating Equation (11) and setting $t = p^{(n)}$, we get

$$\begin{aligned} \mathcal{N}_4'(p^{(n)}) = {} & f[p^{(n)}, z^{(n)}] + (p^{(n)} - z^{(n)})f[p^{(n)}, z^{(n)}, y^{(n)}] + (p^{(n)} - z^{(n)})(p^{(n)} - y^{(n)})f[p^{(n)}, z^{(n)}, y^{(n)}, w^{(n)}] \\ & + (p^{(n)} - z^{(n)})(p^{(n)} - y^{(n)})(p^{(n)} - w^{(n)})f[p^{(n)}, z^{(n)}, y^{(n)}, w^{(n)}, x^{(n)}]. \end{aligned} \quad (12)$$

Now, approximating $f'(p^{(n)}) \approx \mathcal{N}_4'(p^{(n)})$ in Equation (10), we get a new derivative-free optimal sixteenth-order iterative method ($PM16$) given by

$$\begin{cases} w^{(n)} = x^{(n)} + f(x^{(n)})^3, \\ y^{(n)} = x^{(n)} - \frac{f(x^{(n)})^4}{f(w^{(n)}) - f(x^{(n)})}, \\ z^{(n)} = y^{(n)} - \frac{f(y^{(n)})}{f[y^{(n)}, w^{(n)}] + (y^{(n)} - w^{(n)})f[y^{(n)}, w^{(n)}, x^{(n)}]}, \\ p^{(n)} = z^{(n)} - \frac{f(z^{(n)})}{f[z^{(n)}, y^{(n)}] + (z^{(n)} - y^{(n)})f[z^{(n)}, y^{(n)}, w^{(n)}] + (z^{(n)} - y^{(n)})(z^{(n)} - w^{(n)})f[z^{(n)}, y^{(n)}, w^{(n)}, x^{(n)}]}, \\ x^{(n+1)} = p^{(n)} - \frac{f(p^{(n)})}{\mathcal{N}_4'(p^{(n)})}, \end{cases} \quad (13)$$

where $\mathcal{N}_4'(p^{(n)})$ given in Equation (12).

3. Convergence Analysis

In this part, we will derive the convergence analysis of the proposed schemes in Equations (5), (9), and (13) with the help of MATHEMATICA software.

Theorem 1. *Let $f : D \subset \mathbb{R} \to \mathbb{R}$ be a sufficiently smooth function having continuous derivatives. If $f(x)$ has a simple root x^* in the open interval D and $x^{(0)}$ is chosen in a sufficiently small neighborhood of x^*, then the method of Equation* (5) *is of local fourth-order convergence and and it satisfies the error equation*

$$e_{n+1} = (c[2]^3 - c[2]c[3])e_n^4 + O(e_n^5).$$

Proof. Let $e_n = x^{(n)} - x^*$ and $c[j] = \frac{f^{(j)}(x^*)}{j!f'(x^*)}, j = 2,3,4,....$ Expanding $f(x^{(n)})$ and $f(w^{(n)})$ about x^* by Taylor's method, we have

$$f(x^{(n)}) = f'(x^*)[e_n + c[2]e_n^2 + c[3]e_n^3 + c[4]e_n^4 + \ldots], \tag{14}$$

$$w^{(n)} = e_n + f'(x^*)^3[e_n + c[2]e_n^2 + c[3]e_n^3 + c[4]e_n^4 + \ldots]^3, \tag{15}$$

$$f(w^{(n)}) = f'(x^*)[e_n + c[2]e_n^2 + (f'(x^*)^3 + c[3])e_n^3 + (5f'(x^*)^3c[2] + c[4])e_n^4 + \ldots]. \tag{16}$$

Then, we have

$$y^{(n)} = x^* + c[2]e_n^2 + (-2c[2]^2 + 2c[3])e_n^3 + (4c[2]^3 - 7c[2]c[3] + 3c[4] + f'(x^*)^3c[2])e_n^4 + \ldots . \tag{17}$$

Expanding $f(y^{(n)})$ about x^*, we have

$$f(y^{(n)}) = f'(x^*)[c[2]e_n^2 - 2(c[2]^2 - c[3])e_n^3 + (5c[2]^3 - 7c[2]c[3] + 3c[4] + f'(x^*)^3c[2])e_n^4 + \ldots]. \tag{18}$$

Now, we get the Taylor's expansion of $f[y^{(n)}, w^{(n)}] = \frac{f(y^{(n)}) - f(w^{(n)})}{y^{(n)} - w^{(n)}}$ by replacing Equation (15)–(18).

$$f[y^{(n)}, w^{(n)}] = f'(x^*)[1 + c[2]e_n + (c[2]^2 + c[3])e_n^2 + (f'(x^*)^3c[2] - 2c[2]^3 + c[2]c[3] + c[4])e_n^3 + \ldots]. \tag{19}$$

Also, we have

$$f[y^{(n)}, w^{(n)}, x^{(n)}] = f'(x^*)[c[2] + 2c[3]e_n + (c[2]c[3] + c[4])e_n^2 + \ldots] \tag{20}$$

Using Equations (14)–(20) in the scheme of Equation (5), we obtain the following error equation:

$$e_{n+1} = (c[2]^3 - c[2]c[3])e_n^4 + \ldots . \tag{21}$$

This reveals that the proposed method $PM4$ attains fourth-order convergence. □

Theorem 2. *Let $f : D \subset \mathbb{R} \to \mathbb{R}$ be a sufficiently smooth function having continuous derivatives. If $f(x)$ has a simple root x^* in the open interval D and $x^{(0)}$ is chosen in a sufficiently small neighborhood of x^*, then the method of Equation* (9) *is of local eighth-order convergence and and it satisfies the error equation*

$$e_{n+1} = c[2]^2(c[2]^2 - c[3])(c[2]^3 - c[2]c[3] + c[4])e_n^8 + O(e_n^9).$$

Theorem 3. *Let $f : D \subset \mathbb{R} \to \mathbb{R}$ be a sufficiently smooth function having continuous derivatives. If $f(x)$ has a simple root x^* in the open interval D and $x^{(0)}$ is chosen in a sufficiently small neighborhood of x^*, then the method of Equation (13) is of local sixteenth-order convergence and and it satisfies the error equation*

$$e_{n+1} = c[2]^4\left(c[2]^2 - c[3]\right)^2\left(c[2]^3 - c[2]c[3] + c[4]\right)\left(c[2]^4 - c[2]^2c[3] + c[2]c[4] - c[5]\right)e_n^{16} + O(e_n^{17}).$$

4. Some Known Derivative-Free Methods

Let us consider the following schemes for the purpose of comparison. Derivative-free Kung–Traub's two-step method (KTM4) [4] is as follows:

$$y^{(n)} = x^{(n)} - \frac{f(x^{(n)})}{f[x^{(n)}, w^{(n)}]},\ w^{(n)} = x^{(n)} + f(x^{(n)}),\ x^{(n+1)} = y^{(n)} - \frac{f(y^{(n)})f(w^{(n)})}{[f(w^{(n)}) - f(y^{(n)})]f[x^{(n)}, y^{(n)}]}. \tag{22}$$

Derivative-free Argyros et al. two-step method (AKKB4) [25] is as follows:

$$y^{(n)} = x^{(n)} - \frac{f(x^{(n)})}{f[x^{(n)}, w^{(n)}]},\ w^{(n)} = x^{(n)} + f(x^{(n)}),\ x^{(n+1)} = y^{(n)} - \frac{f(y^{(n)})}{[f(x^{(n)}) - 2f(y^{(n)})]}\frac{f(x^{(n)})}{f[y^{(n)}, w^{(n)}]}\left(1 - \frac{f(y^{(n)})}{f(x^{(n)})}\right). \tag{23}$$

Derivative-free Zheng et al. two-step method (ZLM4) [26] is as follows:

$$y^{(n)} = x^{(n)} - \frac{f(x^{(n)})}{f[x^{(n)}, w^{(n)}]},\ w^{(n)} = x^{(n)} + f(x^{(n)}),\ x^{(n+1)} = y^{(n)} - \frac{f(y^{(n)})}{f[x^{(n)}, y^{(n)}] + (y^{(n)} - x^{(n)})f[x^{(n)}, w^{(n)}, y^{(n)}]}. \tag{24}$$

Derivative-free Argyros et al. three-step method (AKKB8) [25] is as follows:

$$\begin{cases} y^{(n)} = x^{(n)} - \frac{f(x^{(n)})}{f[x^{(n)},w^{(n)}]}, w^{(n)} = x^{(n)} + f(x^{(n)}),\ z^{(n)} = y^{(n)} - \frac{f(y^{(n)})}{[f(x^{(n)})-2f(y^{(n)})]}\frac{f(x^{(n)})}{f[y^{(n)},w^{(n)}]}\left(1 - \frac{f(y^{(n)})}{f(x^{(n)})}\right), \\ x^{(n+1)} = z^{(n)} - \frac{f(z^{(n)})}{f[z^{(n)},y^{(n)}]+(z^{(n)}-y^{(n)})f[z^{(n)},y^{(n)},x^{(n)}]+(z^{(n)}-y^{(n)})(z^{(n)}-x^{(n)})f[z^{(n)},y^{(n)},x^{(n)},w^{(n)}]}. \end{cases} \tag{25}$$

Derivative-free Kanwar et al. three-step method (KBK8) [15] is as follows:

$$\begin{cases} y^{(n)} = x^{(n)} - \frac{f(x^{(n)})}{f[x^{(n)},w^{(n)}]},\ w^{(n)} = x^{(n)} + f(x^{(n)})^3,\ z^{(n)} = y^{(n)} - \frac{f(y^{(n)})}{2f[y^{(n)},x^{(n)}]-f[x^{(n)},w^{(n)}]}, \\ x^{(n+1)} = z^{(n)} - \frac{f(z^{(n)})}{f[y^{(n)},z^{(n)}]+f[w^{(n)},y^{(n)},z^{(n)}](z^{(n)}-y^{(n)})}\left(1 - \left(\frac{f(y^{(n)})}{f(x^{(n)})}\right)^3 - 8\frac{f(y^{(n)})f(z^{(n)})}{f(x^{(n)})^2} + \frac{f(z^{(n)})}{f(x^{(n)})} + 5\left(\frac{f(z^{(n)})}{f(y^{(n)})}\right)^2\right). \end{cases} \tag{26}$$

Derivative-free Soleymani three-step method (SM8) [2] is as follows:

$$\begin{cases} w^{(n)} = x^{(n)} + f(x^{(n)}), y^{(n)} = x^{(n)} - \frac{f(x^{(n)})}{f[x^{(n)},w^{(n)}]}, z^{(n)} = y^{(n)} - \frac{f(y^{(n)})}{f[x^{(n)},w^{(n)}]}\phi_n, \\ x^{(n+1)} = z^{(n)} - \frac{f(z^{(n)})}{f[x^{(n)},w^{(n)}]}\phi_n\psi_n,\ \textit{where}\ \phi_n = \frac{1}{1-f(y^{(n)})/f(x^{(n)})-f(y^{(n)})/f(w^{(n)})}, \\ \psi_n = 1 + \frac{1}{1+f[x^{(n)},w^{(n)}]}\left(\frac{f(y^{(n)})}{f(x^{(n)})}\right)^2 + \left((1 + f[x^{(n)}, w^{(n)}])(2 + f[x^{(n)}, w^{(n)}])\right)\left(\frac{f(y^{(n)})}{f(w^{(n)})}\right)^3 + \frac{f(z^{(n)})}{f(y^{(n)})} + \frac{f(z^{(n)})}{f(x^{(n)})} + \frac{f(z^{(n)})}{f(w^{(n)})}. \end{cases} \tag{27}$$

Derivative-free Zheng et al. four-step method (ZLM16) [26] is as follows:

$$\begin{cases} y^{(n)} = x^{(n)} - \frac{f(x^{(n)})^2}{f(w^{(n)})-f(x^{(n)})}, w^{(n)} = x^{(n)} + f(x^{(n)}), z^{(n)} = y^{(n)} - \frac{f(y^{(n)})}{f[y^{(n)},w^{(n)}]+(y^{(n)}-w^{(n)})f[y^{(n)},w^{(n)},x^{(n)}]}, \\ p^{(n)} = z^{(n)} - \frac{f(z^{(n)})}{f[z^{(n)},y^{(n)}]+(z^{(n)}-y^{(n)})f[z^{(n)},y^{(n)},w^{(n)}]+(z^{(n)}-y^{(n)})(z^{(n)}-w^{(n)})f[z^{(n)},y^{(n)},w^{(n)},x^{(n)}]}, \\ x^{(n+1)} = p^{(n)} - \frac{f(p^{(n)})}{f'(p^{(n)})}, \\ \textit{where}\ f'(p^{(n)}) \approx f[p^{(n)}, z^{(n)}] + (p^{(n)} - z^{(n)})f[p^{(n)}, z^{(n)}, y^{(n)}] + (p^{(n)} - z^{(n)})(p^{(n)} - y^{(n)})f[p^{(n)}, z^{(n)}, y^{(n)}, w^{(n)}] \\ +(p^{(n)} - z^{(n)})(p^{(n)} - y^{(n)})(p^{(n)} - w^{(n)})f[p^{(n)}, z^{(n)}, y^{(n)}, w^{(n)}, x^{(n)}]. \end{cases} \tag{28}$$

5. Test Problems

We compare the performance of the proposed methods along with some existing methods for test problems by using MATLAB. We use the conditions for stopping criteria for $|f(x^{(N)})| < \epsilon$ where

$\epsilon = 10^{-50}$ and N is the number of iterations needed for convergence. The computational order of convergence (coc) is given by

$$\rho = \frac{\ln|(x^{(N)} - x^{(N-1)})/(x^{(N-1)} - x^{(N-2)})|}{\ln|(x^{(N-1)} - x^{(N-2)})/(x^{(N-2)} - x^{(N-3)})|}.$$

The test problems and their roots are given below:

$$f_1(x) = \sin(2\cos x) - 1 - x^2 + e^{\sin(x^3)}, \qquad x^* = -0.7848959876612125352...$$
$$f_2(x) = x^3 + 4x^2 - 10, \qquad x^* = 1.3652300134140968457...$$
$$f_3(x) = \sqrt{x^2 + 2x + 5} - 2\sin x - x^2 + 3, \qquad x^* = 2.3319676558839640103...$$
$$f_4(x) = e^{-x}\sin x + \log(1 + x^2) - 2, \qquad x^* = 2.4477482864524245021...$$
$$f_5(x) = sin(x) + cos(x) + x, \qquad x^* = -0.4566247045676308244...$$

Tables 1–5 show the results of all the test functions with a given initial point. The computational order of convergence conforms with theoretical order of convergence. If the initial points are close to the zero, then we obtain less number of iterations with least error. If the initial points are away from the zero, then we will not obtained the least error. We observe that the new methods in all the test function have better efficiency as compared to other existing methods of the equivalent methods.

Table 1. Comparisons between different methods for $f_1(x)$ at $x^{(0)} = -0.9$.

Methods	N	$\lvert x^{(1)} - x^{(0)}\rvert$	$\lvert x^{(2)} - x^{(1)}\rvert$	$\lvert x^{(3)} - x^{(2)}\rvert$	$\lvert x^{(N)} - x^{(N-1)}\rvert$	coc
$SM2$ (1)	8	0.0996	0.0149	6.1109×10^{-4}	1.0372×10^{-89}	1.99
$KTM4$ (22)	5	0.1144	6.7948×10^{-4}	3.4668×10^{-12}	5.1591×10^{-178}	4.00
$AKKB4$ (23)	4	0.1147	3.6299×10^{-4}	9.5806×10^{-14}	4.6824×10^{-52}	3.99
$ZLM4$ (24)	5	0.1145	6.1744×10^{-4}	1.5392×10^{-12}	1.3561×10^{-184}	4.00
$PM4$ (5)	4	0.1150	1.3758×10^{-4}	2.6164×10^{-16}	3.4237×10^{-63}	3.99
$AKKB8$ (25)	3	0.1151	1.2852×10^{-8}	3.7394×10^{-62}	3.7394×10^{-62}	7.70
$KBK8$ (26)	3	0.1151	8.1491×10^{-8}	1.5121×10^{-56}	1.5121×10^{-56}	7.92
$SM8$ (27)	4	0.1151	1.8511×10^{-6}	1.0266×10^{-43}	0	7.99
$PM8$ (9)	3	0.1151	7.1154×10^{-9}	9.3865×10^{-67}	9.3865×10^{-67}	8.02
$ZLM16$ (28)	3	0.1151	5.6508×10^{-15}	1.4548×10^{-225}	1.4548×10^{-225}	15.82
$PM16$ (13)	3	0.1151	5.3284×10^{-17}	1.2610×10^{-262}	1.2610×10^{-262}	16.01

Table 2. Comparisons between different methods for $f_2(x)$ at $x^{(0)} = 1.6$.

Methods	N	$\lvert x^{(1)} - x^{(0)}\rvert$	$\lvert x^{(2)} - x^{(1)}\rvert$	$\lvert x^{(3)} - x^{(2)}\rvert$	$\lvert x^{(N)} - x^{(N-1)}\rvert$	coc
$SM2$ (1)	12	0.0560	0.0558	0.0520	1.7507×10^{-83}	1.99
$KTM4$ (22)	5	0.2184	0.0163	3.4822×10^{-6}	4.7027×10^{-79}	3.99
$AKKB4$ (23)	33	0.0336	0.0268	0.0171	2.4368×10^{-52}	0.99
$ZLM4$ (24)	5	0.2230	0.0117	4.4907×10^{-7}	3.9499×10^{-95}	3.99
$PM4$ (5)	5	0.2123	0.0224	2.3433×10^{-7}	4.3969×10^{-112}	4.00
$AKKB8$ (25)	4	0.2175	0.0173	1.2720×10^{-9}	1.0905×10^{-66}	8.00
$KBK8$ (26)	D	D	D	D	D	D
$SM8$ (27)	4	0.2344	4.1548×10^{-4}	9.5789×10^{-24}	7.7650×10^{-181}	7.89
$PM8$ (9)	4	0.2345	2.4307×10^{-4}	4.6428×10^{-32}	8.2233×10^{-254}	8.00
$ZLM16$ (28)	3	0.2348	2.2048×10^{-7}	1.9633×10^{-124}	1.9633×10^{-124}	15.57
$PM16$ (13)	3	0.2348	2.8960×10^{-8}	1.7409×10^{-126}	1.7409×10^{-126}	17.11

Table 3. Comparisons between different methods for $f_3(x)$ at $x^{(0)} = 2.7$.

Methods	N	$\lvert x^{(1)} - x^{(0)} \rvert$	$\lvert x^{(2)} - x^{(1)} \rvert$	$\lvert x^{(3)} - x^{(2)} \rvert$	$\lvert x^{(N)} - x^{(N-1)} \rvert$	coc
$SM2$ (1)	7	0.3861	0.0180	4.6738×10^{-05}	1.0220×10^{-82}	1.99
$KTM4$ (22)	4	0.3683	2.8791×10^{-4}	1.0873×10^{-16}	2.2112×10^{-66}	3.99
$AKKB4$ (23)	4	0.3683	2.5241×10^{-4}	5.2544×10^{-17}	9.8687×10^{-68}	3.99
$ZLM4$ (24)	4	0.3683	3.1466×10^{-4}	1.7488×10^{-16}	1.6686×10^{-65}	4.00
$PM4$ (5)	4	0.3683	2.2816×10^{-4}	2.3732×10^{-17}	2.7789×10^{-69}	3.99
$AKKB8$ (25)	3	0.3680	1.7343×10^{-8}	3.8447×10^{-67}	3.8447×10^{-67}	8.00
$KBK8$ (26)	4	0.3680	4.2864×10^{-5}	1.8700×10^{-38}	2.4555×10^{-305}	7.99
$SM8$ (27)	3	0.3680	7.8469×10^{-8}	2.9581×10^{-61}	2.9581×10^{-61}	8.00
$PM8$ (9)	3	0.3680	9.7434×10^{-9}	1.0977×10^{-69}	1.0977×10^{-69}	8.04
$ZLM16$ (28)	3	0.3680	1.4143×10^{-16}	6.3422×10^{-240}	6.3422×10^{-240}	16.03
$PM16$ (13)	3	0.3680	3.6568×10^{-17}	7.4439×10^{-274}	7.4439×10^{-274}	16.04

Table 4. Comparisons between different methods for $f_4(x)$ at $x^{(0)} = 1.9$.

Methods	N	$\lvert x^{(1)} - x^{(0)} \rvert$	$\lvert x^{(2)} - x^{(1)} \rvert$	$\lvert x^{(3)} - x^{(2)} \rvert$	$\lvert x^{(N)} - x^{(N-1)} \rvert$	coc
$SM2$ (1)	7	0.4975	0.0500	2.5378×10^{-4}	1.9405×10^{-73}	2.00
$KTM4$ (22)	4	0.2522	1.7586×10^{-6}	1.5651×10^{-26}	9.8198×10^{-107}	3.99
$AKKB4$ (23)	4	0.5489	0.0011	3.8305×10^{-15}	5.5011×10^{-61}	3.99
$ZLM4$ (24)	4	0.5487	9.0366×10^{-4}	1.4751×10^{-15}	1.0504×10^{-62}	3.99
$PM4$ (5)	4	0.5481	3.0864×10^{-4}	8.0745×10^{-18}	3.7852×10^{-72}	3.99
$AKKB8$ (25)	3	0.5477	5.4938×10^{-7}	4.9628×10^{-56}	4.9628×10^{-56}	8.17
$KBK8$ (26)	3	0.5477	4.1748×10^{-7}	5.8518×10^{-59}	5.8518×10^{-59}	8.47
$SM8$ (27)	3	0.5477	5.4298×10^{-7}	4.1081×10^{-56}	4.1081×10^{-56}	8.18
$PM8$ (9)	3	0.5477	5.8222×10^{-8}	1.1144×10^{-64}	1.1144×10^{-64}	8.13
$ZLM16$ (28)	3	0.5477	2.7363×10^{-14}	7.2982×10^{-229}	7.2982×10^{-229}	16.13
$PM16$ (13)	3	0.5477	5.6240×10^{-16}	1.9216×10^{-257}	1.9216×10^{-257}	16.11

Table 5. Comparisons between different methods for $f_5(x)$ at $x^{(0)} = -0.2$.

Methods	N	$\lvert x^{(1)} - x^{(0)} \rvert$	$\lvert x^{(2)} - x^{(1)} \rvert$	$\lvert x^{(3)} - x^{(2)} \rvert$	$\lvert x^{(N)} - x^{(N-1)} \rvert$	coc
$SM2$ (1)	7	0.3072	0.0499	6.4255×10^{-4}	4.1197×10^{-59}	2.00
$KTM4$ (22)	5	0.2585	0.0019	1.5538×10^{-12}	3.4601×10^{-194}	4.00
$AKKB4$ (23)	4	0.2571	4.4142×10^{-4}	3.4097×10^{-15}	1.2154×10^{-59}	3.99
$ZLM4$ (24)	4	0.2580	0.0013	3.5840×10^{-13}	1.8839×10^{-51}	3.99
$PM4$ (5)	4	0.2569	2.8004×10^{-4}	6.2960×10^{-17}	1.6097×10^{-67}	3.99
$AKKB8$ (25)	3	0.2566	4.1915×10^{-8}	6.3444×10^{-65}	6.3444×10^{-65}	8.37
$KBK8$ (26)	4	0.2566	4.0069×10^{-6}	5.1459×10^{-47}	0	7.99
$SM8$ (27)	4	0.2566	2.9339×10^{-6}	1.0924×10^{-46}	0	7.99
$PM8$ (9)	3	0.2566	3.7923×10^{-11}	9.0207×10^{-90}	9.0207×10^{-90}	7.99
$ZLM16$ (28)	3	0.2566	5.3695×10^{-16}	7.0920×10^{-252}	7.0920×10^{-252}	16.06
$PM16$ (13)	3	0.2566	1.1732×10^{-19}	1.2394×10^{-314}	1.2394×10^{-314}	16.08

6. Basins of Attraction

The iterative scheme gives information about convergence and stability by studying basins of attraction of the rational function. The basic definitions and dynamical concepts of rational function can found in References [17,27,28]. Let us consider a region $\mathbb{R} \times \mathbb{R} = [-2,2] \times [-2,2]$ with 256×256 grids. We test iterative methods in all the grid point $z^{(0)}$ in the square. The iterative algorithms attempt roots z_j^* of the equation with condition $|f(z^{(k)})| < \times 10^{-4}$ and a maximum of 100 iterations; we conclude that $z^{(0)}$ is in the basin of attraction of this zero. If the iterative method starting in $z^{(0)}$ reaches a zero in N iterations, then we mark this point $z^{(0)}$ with colors if $|z^{(N)} - z_j^*| < \times 10^{-4}$. If $N > 50$, then we assign a dark blue color for diverging grid points. We describe the basins of attraction for finding complex roots of $p_1(z) = z^2 - 1$, $p_2(z) = z^3 - 1$, $p_3(z) = (z^2 + 1)(z^2 - 1)$, and $p_4(z) = z^5 - 1$ for proposed methods and some higher-order iterative methods.

In Figures 1–5, we have given the basins of attraction for new methods with some existing methods. We confirm that a point z_0 containing the Julia set whenever the dynamics of point shows sensitivity to the conditions. The neighbourhood of initial points leads to the slight variation in behavior after some iterations. Therefore, some of the compared algorithms obtain more divergent initial conditions.

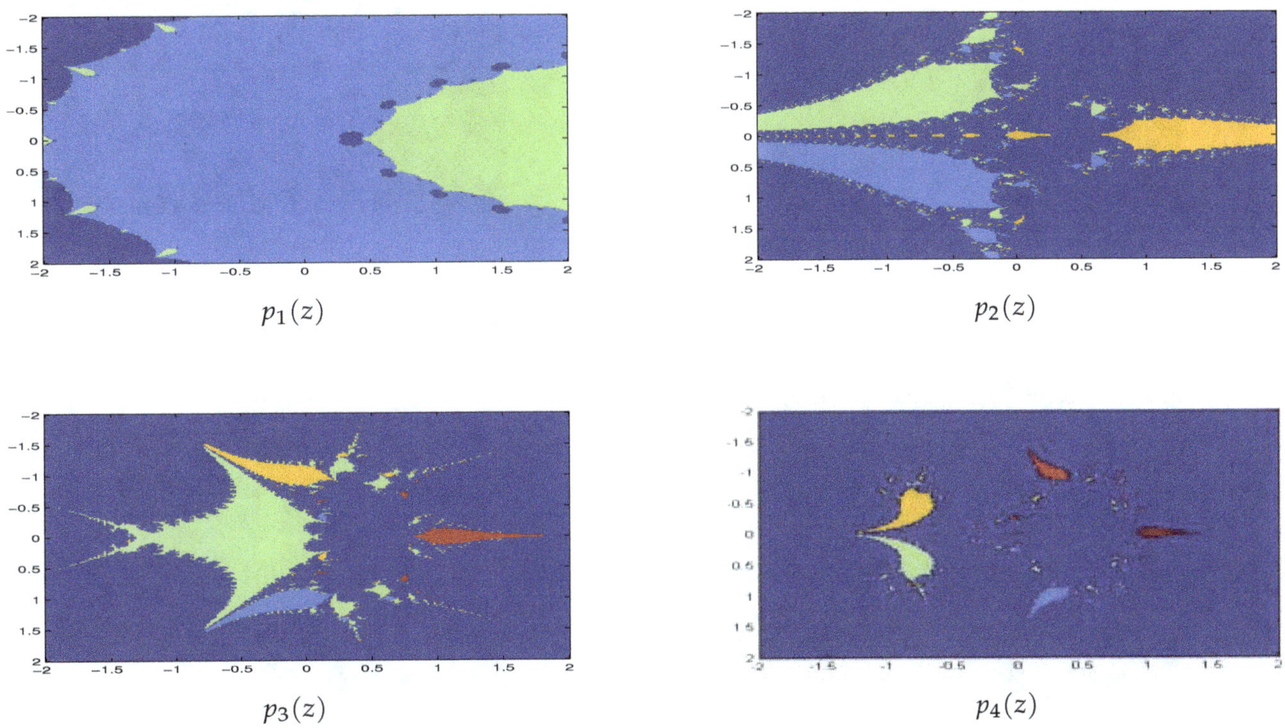

Figure 1. Basins of attraction for $SM2$ for the polynomial.

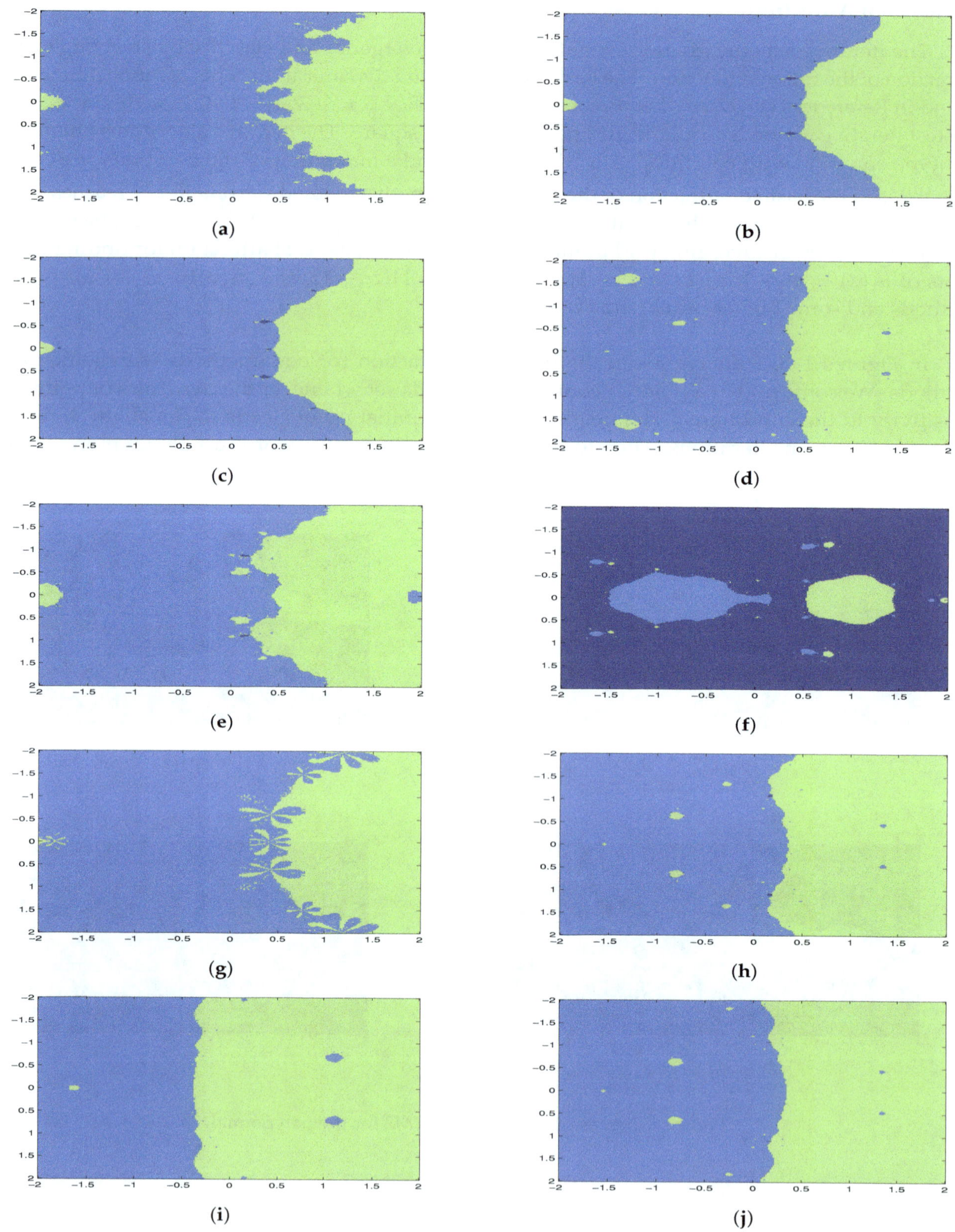

Figure 2. Polynomiographs of $p_1(z)$: (**a**) $KTM4$; (**b**) $AKKB4$; (**c**) $ZLM4$; (**d**) $PM4$; (**e**) $AKKB8$; (**f**) $KBK8$; (**g**) $SM8$; (**h**) $PM8$; (**i**) $ZLM16$; and (**j**) $PM16$.

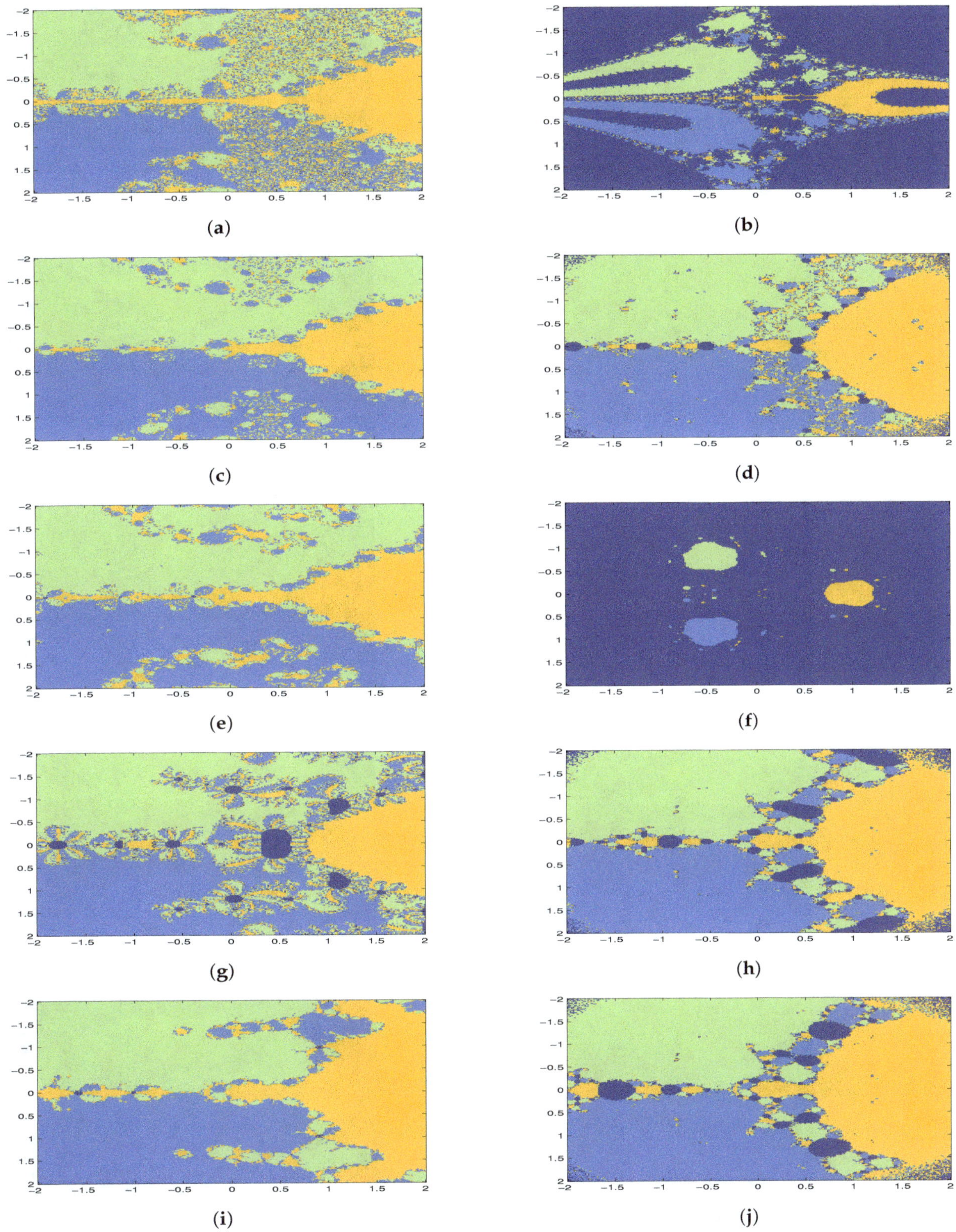

Figure 3. Polynomiographs of $p_2(z)$: (**a**) $KTM4$; (**b**) $AKKB4$; (**c**) $ZLM4$; (**d**) $PM4$; (**e**) $AKKB8$; (**f**) $KBK8$; (**g**) $SM8$; (**h**) $PM8$; (**i**) $ZLM16$; and (**j**) $PM16$.

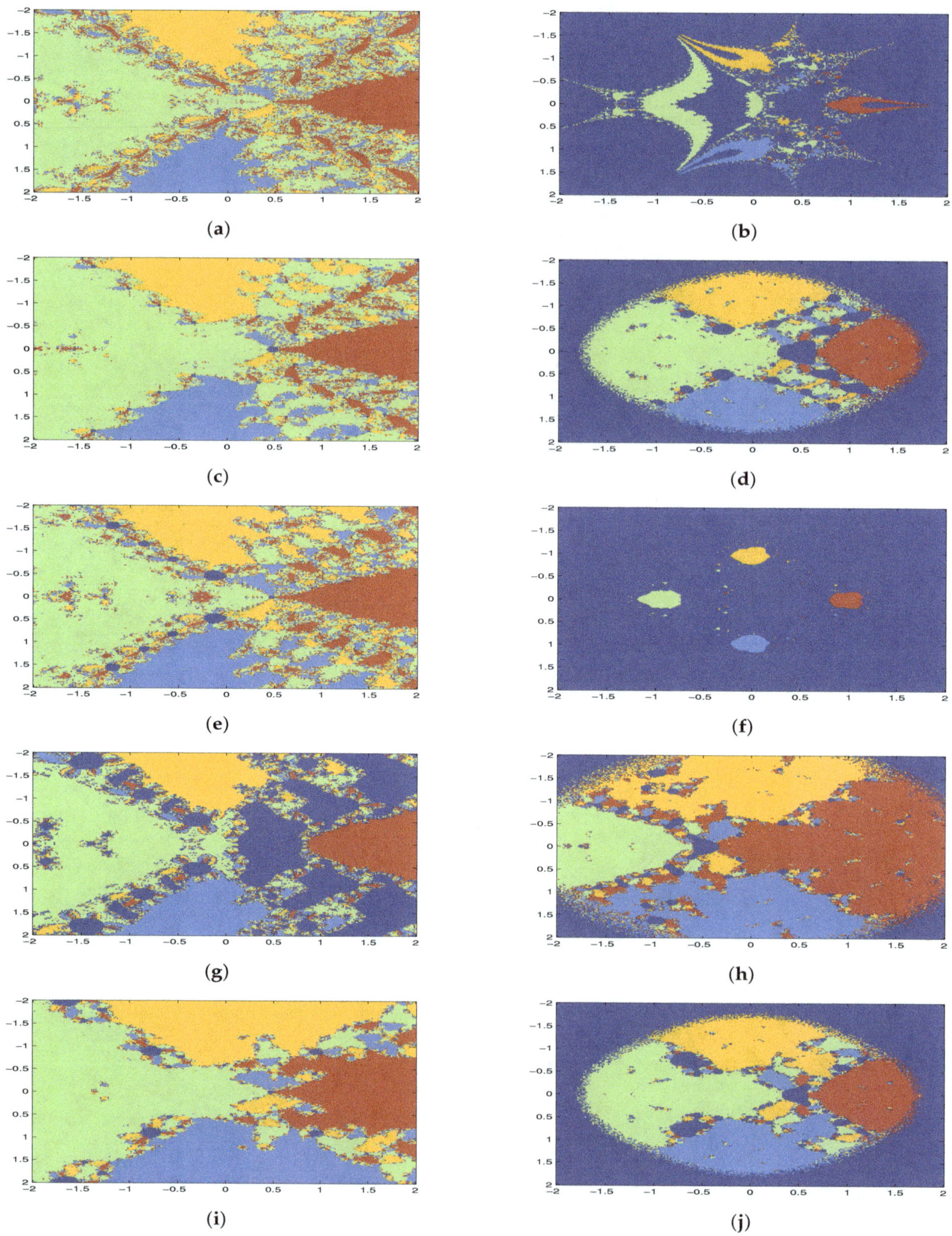

Figure 4. Polynomiographs of $p_3(z)$: (**a**) $KTM4$; (**b**) $AKKB4$; (**c**)$ZLM4$; (**d**) $PM4$; (**e**) $AKKB8$; (**f**) $KBK8$; (**g**) $SM8$; (**h**) $PM8$; (**i**) $ZLM16$; and (**j**) $PM16$.

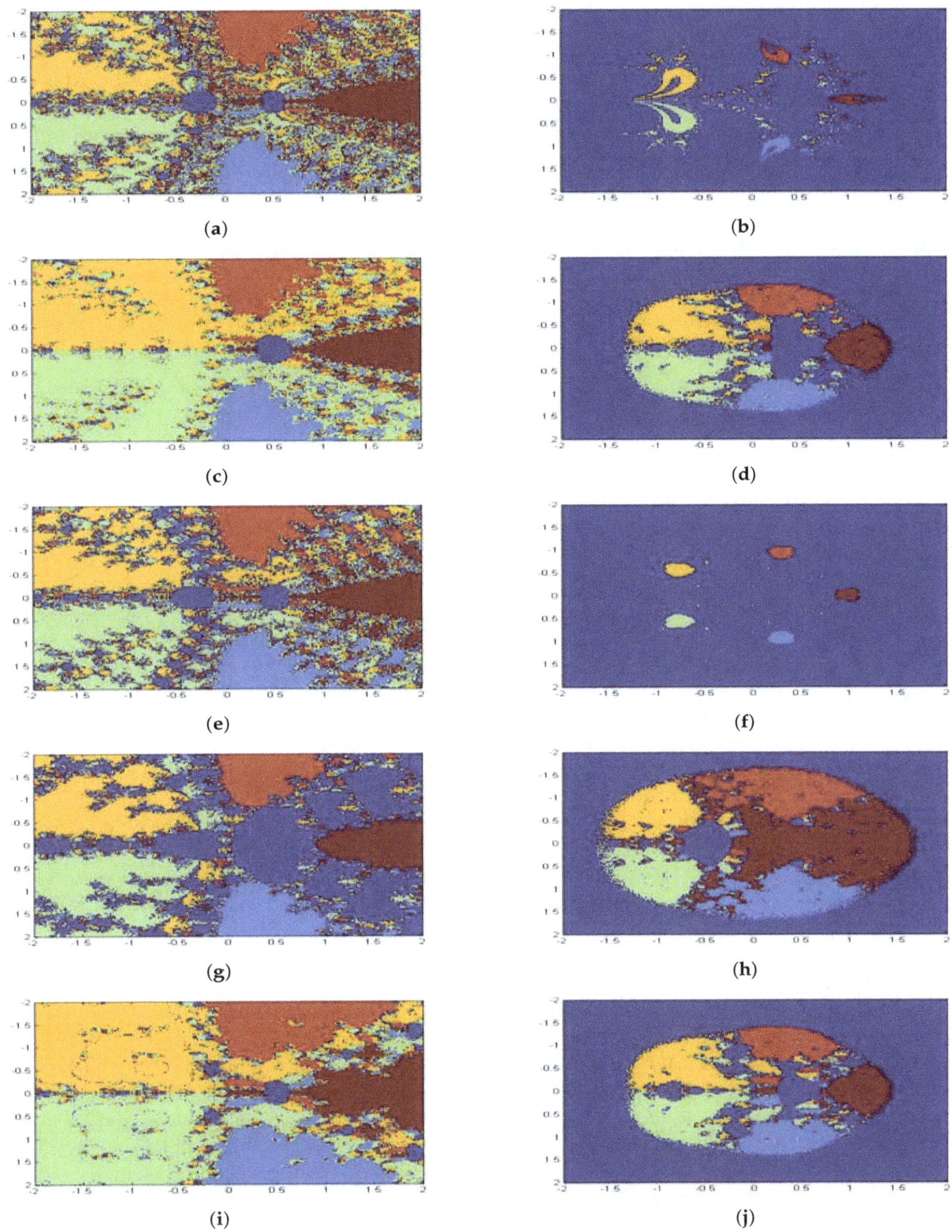

Figure 5. Polynomiographs of $p_4(z)$: (**a**) $KTM4$; (**b**) $AKKB4$; (**c**)$ZLM4$; (**d**) $PM4$; (**e**) $AKKB8$; (**f**) $KBK8$; (**g**) $SM8$; (**h**) $PM8$; (**i**) $ZLM16$; and (**j**) $PM16$.

7. Concluding Remarks

We have proposed fourth-, eighth-, and sixteenth-order methods using finite difference approximations. Our proposed new methods requires 3 functions to get the 4th-order method, 4 functions to obtain the 8th-order method, and 5 functions to get the 16th-order one. We have increased the convergence order of the proposed method, respectively, to four, eight, and sixteen with efficiency indices 1.587, 1.565, and 1.644 respectively. Our new proposed schemes are better than the Steffensen method in terms of efficiency index (1.414). Numerical solutions are tested to show the performance of the proposed algorithms. Also, we have analyzed on the complex region for iterative methods to study their basins of attraction. Hence, we conclude that the proposed methods are comparable to other well-known existing equivalent methods.

Author Contributions: Conceptualization, K.M.; Funding acquisition, J.L. and X.W.; Methodology, K.M.; Project administration, J.L. and X.W.; Resources, J.L. and X.W.; Writing—original draft, K.M.

Acknowledgments: The authors would like to thank the editors and referees for the valuable comments and for the suggestions to improve the readability of the paper.

References

1. Cordero, A.; Hueso, J.L.; Martinez, E.; Torregrosa, J.R. Generating optimal derivative free iterative methods for nonlinear equations by using polynomial interpolation. *Appl. Math. Comp.* **2013**, *57*, 1950–1956. [CrossRef]
2. Soleymani, F. Efficient optimal eighth-order derivative-free methods for nonlinear equations. *Jpn. J. Ind. Appl. Math.* **2013**, *30*, 287–306. [CrossRef]
3. Steffensen, J.F. Remarks on iteration. *Scand. Aktuarietidskr.* **1933**, *16*, 64–72. [CrossRef]
4. Kung, H.; Traub, J. Optimal order of one-point and multi-point iteration. *J. Assoc. Comput. Math.* **1974**, *21*, 643–651. [CrossRef]
5. Behl, R.; Salimi, M.; Ferrara, M.; Sharifi, S.; Alharbi, S.K. Some Real-Life Applications of a Newly Constructed Derivative Free Iterative Scheme. *Symmetry* **2019**, *11*, 239. [CrossRef]
6. Salimi, M.; Long, N.M.A.N.; Sharifi, S.; Pansera, B.A. A multi-point iterative method for solving nonlinear equations with optimal order of convergence. *Jpn. J. Ind. Appl. Math.* **2018**, *35*, 497–509. [CrossRef]
7. Salimi, M.; Lotfi, T.; Sharifi, S.; Siegmund, S. Optimal Newton-Secant like methods without memory for solving nonlinear equations with its dynamics. *Int. J. Comput. Math.* **2017**, *94*, 1759–1777. [CrossRef]
8. Matthies, G.; Salimi, M.; Sharifi, S.; Varona, J.L. An optimal three-point eighth-order iterative method without memory for solving nonlinear equations with its dynamics. *Jpn. J. Ind. Appl. Math.* **2016**, *33*, 751–766. [CrossRef]
9. Sharifi, S.; Siegmund, S.; Salimi, M. Solving nonlinear equations by a derivative-free form of the King's family with memory. *Calcolo* **2016**, *53*, 201–215. [CrossRef]
10. Khdhr, F.W.; Saeed, R.K.; Soleymani, F. Improving the Computational Efficiency of a Variant of Steffensen's Method for Nonlinear Equations. *Mathematics* **2019**, *7*, 306. [CrossRef]
11. Soleymani, F.; Babajee, D.K.R.; Shateyi, S.; Motsa, S.S. Construction of Optimal Derivative-Free Techniques without Memory. *J. Appl. Math.* **2012**, *2012*, 24. [CrossRef]
12. Soleimani, F.; Soleymani, F.; Shateyi, S. Some Iterative Methods Free from Derivatives and Their Basins of Attraction for Nonlinear Equations. *Discret. Dyn. Nat. Soc.* **2013**, *2013*, 10. [CrossRef]
13. Soleymani, F.; Sharifi, M. On a General Efficient Class of Four-Step Root-Finding Methods. *Int. J. Math. Comp. Simul.* **2011**, *5*, 181–189.
14. Soleymani, F.; Vanani, S.K.; Paghaleh, M.J. A Class of Three-Step Derivative-Free Root Solvers with Optimal Convergence Order. *J. Appl. Math.* **2012**, *2012*, 15. [CrossRef]
15. Kanwar, V.; Bala, R.; Kansal, M. Some new weighted eighth-order variants of Steffensen-King's type family for solving nonlinear equations and its dynamics. *SeMA J.* **2016**. [CrossRef]
16. Amat, S.; Busquier, S.; Plaza, S. Dynamics of a family of third-order iterative methods that do not require using second derivatives. *Appl. Math. Comput.* **2004**, *154*, 735–746. [CrossRef]

17. Amat, S.; Busquier, S.; Plaza, S. Review of some iterative root-finding methods from a dynamical point of view. *SCIENTIA Ser. A Math. Sci.* **2004**, *10*, 3–35.
18. Babajee, D.K.R.; Madhu, K. Comparing two techniques for developing higher order two-point iterative methods for solving quadratic equations. *SeMA J.* **2019**, *76*, 227–248. [CrossRef]
19. Cordero, A.; Hueso, J.L.; Martinez, E.; Torregrosa, J.R. A family of iterative methods with sixth and seventh order convergence for nonlinear equations. *Math. Comput. Model.* **2010**, *52*, 1490–1496. [CrossRef]
20. Curry, J.H.; Garnett, L.; Sullivan, D. On the iteration of a rational function: computer experiments with Newton's method. *Commun. Math. Phys.* **1983**, *91*, 267–277. [CrossRef]
21. Soleymani, F.; Babajee, D.K.R.; Sharifi, M. Modified Jarratt Method Without Memory With Twelfth-Order Convergence. *Ann. Univ. Craiova Math. Comput. Sci. Ser.* **2012**, *39*, 21–34.
22. Tao, Y.; Madhu, K. Optimal Fourth, Eighth and Sixteenth Order Methods by Using Divided Difference Techniques and Their Basins of Attraction and Its Application. *Mathematics* **2019**, *7*, 322. [CrossRef]
23. Vrscay, E.R. Julia sets and mandelbrot-like sets associated with higher order Schroder rational iteration functions: a computer assisted study. *Math. Comput.* **1986**, *46*, 151–169.
24. Vrscay, E.R.; Gilbert, W.J. Extraneous fxed points, basin boundaries and chaotic dynamics for Schroder and Konig rational iteration functions. *Numer. Math.* **1987**, *52*, 1–16. [CrossRef]
25. Argyros, I.K.; Kansal, M.; Kanwar, V.; Bajaj, S. Higher-order derivative-free families of Chebyshev-Halley type methods with or without memory for solving nonlinear equations. *Appl. Math. Comput.* **2017**, *315*, 224–245. [CrossRef]
26. Zheng, Q.; Li, J.; Huang, F. An optimal Steffensen-type family for solving nonlinear equations. *Appl. Math. Comput.* **2011**, *217*, 9592–9597. [CrossRef]
27. Madhu, K. Some New Higher Order Multi-Point Iterative Methods and Their Applications to Differential and Integral Equation and Global Positioning System. Ph.D. Thesis, Pndicherry University, Pondicherry, India, June 2016.
28. Scott, M.; Neta, B.; Chun, C. Basin attractors for various methods. *Appl. Math. Comput.* **2011**, *218*, 2584–2599. [CrossRef]

11

Numerical Solution of Heston-Hull-White Three-Dimensional PDE with a High Order FD Scheme

Malik Zaka Ullah

Department of Mathematics, King Abdulaziz University, Jeddah 21589, Saudi Arabia; zmalek@kau.edu.sa

Abstract: A new numerical method for tackling the three-dimensional Heston–Hull–White partial differential equation (PDE) is proposed. This PDE has an application in pricing options when not only the asset price and the volatility but also the risk-free rate of interest are coming from stochastic nature. To solve this time-dependent three-dimensional PDE as efficiently as possible, high order adaptive finite difference (FD) methods are applied for the application of method of lines. It is derived that the new estimates have fourth order of convergence on non-uniform grids. In addition, it is proved that the overall procedure is conditionally time-stable. The results are upheld via several numerical tests.

Keywords: heston model; Hull–White; option pricing; PDE; finite difference (FD)

MSC: 41A25; 65M22

---

1. Introduction

To model different types of derivatives in finance, a common approach is to investigate the connections of these factors to each other, formulated as a stochastic differential equation (SDEs). The factors could be the underlying asset, the volatility, domestic and foreign interest rates, etc., [1,2]. As such, the important action of pricing option under different payoffs can be modeled and simulated via the SDEs or their corresponding partial differential equation (PDE) formulation.

However, a frequently occurring issue is that whatever the model becomes complicated and more realistic, the procedure of having and representing its exact solution becomes harder, see, e.g., [3–5].

To discuss more and from the beginning, the classical model of Black–Scholes in pricing contracts does not cover and illustrate all the aspects of an option in a complete market, such as market risks, stochastic volatility (SV), and asymmetries seen in data of market, [6]. Some remedies to this well-known model are via non-lognormal hypothesis for a SDE, that indicates some modifications of the volatility and the underlying asset. We recall that Heston in [7] extended and improved the behavior of the Black–Scholes model by involving more risky factor into the model, i.e., by considering the volatility to be stochastic as well. Further discussions can be found at [6,8].

On the other hand, as long as the foreign exchange (FX) products are involved and a trader encounters a situation in which the interest rate is not anymore constant during the lifetime of an option, then investigating and proposing an improved model, having stochastic rates of interest, such as the power-reverse dual-currency and the Heston–Cox–Ingersoll–Ross (HCIR) problems (refer to [9] and the references therein for more background).

1.1. Problem Formulation

The option pricing problem under the 3D Heston–Hull–White (HHW) model as a PDE model is defined by [10]:

$$\begin{aligned}\frac{\partial u(s,v,r,t)}{\partial t} =& \frac{1}{2}s^2 v\frac{\partial^2 u(s,v,r,t)}{\partial s^2} + \frac{1}{2}\sigma_1^2 v\frac{\partial^2 u(s,v,r,t)}{\partial v^2} + \frac{1}{2}\sigma_2^2\frac{\partial^2 u(s,v,r,t)}{\partial r^2}\\ &+ \rho_{12}\sigma_1 s v\frac{\partial^2 u(s,v,r,t)}{\partial s\partial v} + \rho_{13}\sigma_2 s\sqrt{v}\frac{\partial^2 u(s,v,r,t)}{\partial s\partial r}\\ &+ \rho_{23}\sigma_1\sigma_2\sqrt{v}\frac{\partial^2 u(s,v,r,t)}{\partial v\partial r}\\ &+ rs\frac{\partial u(s,v,r,t)}{\partial s} + \kappa(\eta - v)\frac{\partial u(s,v,r,t)}{\partial v}\\ &+ a(b(T-t)-r)\frac{\partial u(s,v,r,t)}{\partial r}\\ &- ru(s,v,r,t).\end{aligned} \tag{1}$$

Here $\kappa > 0$ shows the volatility adjustment speed to the analytical mean $\eta > 0$, while σ_1, σ_2, a are some parameters. In addition, the correlation parameters are $\rho_{12}, \rho_{13}, \rho_{23} \in [-1,1]$, b is a time function.

In pricing under call options, the (terminal/)initial condition is given by [11,12]:

$$u(s,v,r,0) = (s-E)^+, \tag{2}$$

where the strike price is E. In a similar way, for a put option, it is given as follows:

$$u(s,v,r,0) = (E-s)^+. \tag{3}$$

As discussed in [13,14], the fair pricing procedure should be carried out by computational schemes since the corresponding high-dimensional PDEs, constructed for such options, do not admit any analytical or semi-analytical solutions, see [15,16] for further background.

1.2. Novelties and Motivation

The contribution of this article reads in proposing a solution method via an un-equally spaced grid having a focus on the hot area in option pricing under the HHW PDE problem. Studying and coding multi dimensional problems with discretization methods while the grid of points are non-uniform is a challenging and intensive task, but could clearly increase the accuracy of the approximate solution by applying fewer numbers of grid nodes in contrasts to the uniform discretization. This reduces the size of the discretized problem and is useful in practice.

To this aim, (1) is tackled by employing high order fourth-order finite difference approximations. We apply fourth order discretizations on a stencil having five and six non-equidistant nodes. Derivation and construction of fourth-order compact FD method for HHW PDE is new and useful in practice.

In fact, the method-of-lines technique is considered to build a set of ODEs with time-varying system matrix. All the side conditions are imposed therein as well. Thence, a method to march along time for the set of ODEs is provided in Section 3 and it is analytically illustrated that the presented numerical procedure is conditionally time-stable when b is not changing by time.

Recalling that here adaptive FD formulas are constructed to hit some features simultaneously, viz., to be effective, results in sparse operators and being able to handling non-uniform grids.

Motivated by recent works in this field (see e.g., [17]), we aim at proposing higher order schemes for the HHW equation on non-uniform meshes so as to increase the accuracy of obtained option prices without increasing the computational load so much. The novelties and contributions of our work are given below:

- We propose fourth-order adaptive discretizations for the spatial variables.
- The beauty of our scheme is the use of non-uniform grid of nodes with an adaptation on the hotzone.
- We provide a new stability bound for the resulting fully discretized set of equations when pricing under HHW PDE using high order discretization methods along the spatial as well as the temporal variables.

1.3. Grid Generation

The option pricing problem (1) is considered in the unbounded area

$$(s, v, r, t) \in \Omega \times (0, T], \tag{4}$$

wherein $\Omega = [0, +\infty) \times [0, +\infty) \times [0, +\infty)$. For tackling the financial model numerically, one can take into account the following domain [18]:

$$\Omega = [0, s_{\max}] \times [0, v_{\max}] \times [-r_{\max}, r_{\max}], \tag{5}$$

wherein $s_{\max}, v_{\max}, r_{\max}$ are three positive real constants and assumed to be large enough.

Since the PDE model is coercive (sometimes called degenerate) at $v = 0$, its payoff is non-smooth at $s = E$, and the working domain has large width, thus it is requisite to use non-uniform meshes, at which the location of the nodes are not equally-spaced. This helps in producing results of higher accuracy with adapting to the hotzone of the problem.

Let $\{s_i\}_{i=1}^{m}$ be a set of non-uniform nodes along s as follows [13,19]:

$$s_i = \varphi(\xi_i), \qquad 1 \le i \le m, \tag{6}$$

where $m > 1$ and $\xi_{\min} = \xi_1 < \xi_2 < \cdots < \xi_m = \xi_{\max}$ are m equi–distant points with the following characteristics: $\xi_{\min} = \sinh^{-1}\left(\frac{s_{\min}-s_{\text{left}}}{d_1}\right)$, $\xi_{\text{int}} = \frac{s_{\text{right}}-s_{\text{left}}}{d_1}$, $\xi_{\max} = \xi_{\text{int}} + \sinh^{-1}\left(\frac{s_{\max}-s_{\text{right}}}{d_1}\right)$, wherein $s_{\min} = 0$. Here $d_1 > 0$ controls the density of the nodes around $s = E$. We also have:

$$\varphi(\xi) = \begin{cases} s_{\text{left}} + d_1 \sinh(\xi), & \xi_{\min} \le \xi < 0, \\ s_{\text{left}} + d_1 \xi, & 0 \le \xi \le \xi_{\text{int}}, \\ s_{\text{right}} + d_1 \sinh(\xi - \xi_{\text{int}}), & \xi_{\text{int}} < \xi \le \xi_{\max}. \end{cases} \tag{7}$$

Throughout this work, we used the same value for $d_1 = \frac{E}{20}$ while $s_{\text{left}} = \max\{0.5, \exp\{-0.25T\}\} \times E$, $[s_{\text{left}}, s_{\text{right}}] \subset [0, s_{\max}]$, $s_{\text{right}} = E$ and $s_{\max} = 14E$.

The nodes along v, i.e., $\{v_j\}_{j=1}^{n}$ are defined by:

$$v_j = d_2 \sinh(\varsigma_j), \qquad 1 \le j \le n, \tag{8}$$

where $d_2 > 0$ gives the concentration around $v = 0$. In this work, we used $d_2 = \frac{v_{\max}}{500}$, where $v_{\max} = 10$. In addition, ς_j are equally spaced points given by:

$$\varsigma_j = (j-1)\Delta\varsigma, \Delta\varsigma = \frac{1}{n-1} \sinh^{-1}\left(\frac{v_{\max}}{d_2}\right), \tag{9}$$

for any $1 \leq j \leq n$. The non-uniform nodes along r are defined as follows:

$$r_k = d_3 \sinh(\zeta_k), \qquad 1 \leq k \leq o, \tag{10}$$

whereas $d_3 = \frac{r_{\max}}{500}$ is a positive parameter and $r_{\max} = 1$. We also have $\zeta_k = (k-1)\Delta\zeta$, $\Delta\zeta = \frac{1}{o-1} \sinh^{-1}\left(\frac{r_{\max}}{d_3}\right)$. Note that denser mesh points in the important area could circumvent the problems happening in solving (1), like non-smoothness of payoffs (2) and (3) at $s = E$, and the degeneracy at $v = r = 0$.

We state that a detailed study into possibly better choices for the involved parameters in mesh generating may be interesting, but this is beyond the scope of the current research. Furthermore, the non-smoothness arising in the payoff would ruin the convergence rate of most derivative approximation particularly on uniform meshes and due to this, the application of non-uniform nodes is indispensable for efficient numerical solution of (1).

1.4. Manuscript Organization

The remaining parts of this work are organized as follows. In Section 2, the weights of the FD scheme over non-uniform grids (here we also call adaptive grids with special emphasis on the hot zone) are derived to attain the higher rate of convergence four.

Section 3 is devoted to the application of a sixth order Runge–Kutta time stepping method to advance along time when semi-discretize the HHW PDE. We prove that the new procedure is time-stable conditionally based on the largest eigenvalue of the system matrix. Section 4 shows that numerical performances are more useful than the earlier schemes with quicker convergence behavior. Finally, some conclusions are drawn in Section 5.

2. Calculating the Weights of the High Order FD Scheme

In this section, by applying a methodology as in ([20], Chapters 3–4) or [21], but with more Taylor expansion terms, we can construct fourth-order FD approximations on (non-uniform) grids.

Five points are required in estimating the first derivative as well as six points in approximating the second derivative in order to obtain a consistent fourth-order scheme throughout the discretized mesh of points.

Without losing the generality, let us construct the weights in the one dimensional case. Then, the concept of tensors using Kronecker product may be applied easily to transfer the weights to the appropriate dimensions. To this objective, consider a sufficiently smooth function $g(s)$ and a grid as follows:

$$\{s_1, s_2, \cdots, s_{m-1}, s_m\}. \tag{11}$$

Consider the following five adjacent nodes:

$$\{\{s_{i-2}, g(s_{i-2})\}, \{s_{i-1}, g(s_{i-1})\}, \{s_i, g(s_i)\}, \{s_{i+1}, g(s_{i+1})\}, \{s_{i+2}, g(s_{i+2})\}\}, \tag{12}$$

and calculate the interpolation polynomial $p(z)$ going via the nodes and then its first derivative $p'(z)$.

At this moment, by employing a computer algebra system to do some symbolic computations and setting $z = s_i$, we attain the FD estimate for the first derivative as follows:

$$g'(s_i) = \alpha_{i-2} g(s_{i-2}) + \alpha_{i-1} g(s_{i-1}) + \alpha_i g(s_i) + \alpha_{i+1} g(s_{i+1}) + \alpha_{i+2} g(s_{i+2}) + \mathcal{O}\left(h^4\right), \tag{13}$$

where the maximum local grid spacing is h and we have

$$\begin{aligned}\alpha_{i-2} &= -\frac{\Gamma_{i-1,i}\Gamma_{i,i+1}\Gamma_{i,i+2}}{\Gamma_{i-2,i-1}\Gamma_{i-2,i}\Gamma_{i-2,i+1}\Gamma_{i-2,i+2}},\\ \alpha_{i-1} &= \frac{\Gamma_{i-2,i}\Gamma_{i,i+1}\Gamma_{i,i+2}}{\Gamma_{i-2,i-1}\Gamma_{i-1,i}\Gamma_{i-1,i+1}\Gamma_{i-1,i+2}},\\ \alpha_{i} &= \frac{1}{\Gamma_{i-2,i}\Gamma_{i,i-1}\Gamma_{i,i+1}\Gamma_{i,i+2}}\Xi_1,\\ \alpha_{i+1} &= \frac{\Gamma_{i-2,i}\Gamma_{i,i-1}\Gamma_{i,i+2}}{\Gamma_{i-2,i+1}\Gamma_{i+1,i-1}\Gamma_{i+1,i}\Gamma_{i+1,i+2}},\\ \alpha_{i+2} &= \frac{\Gamma_{i-2,i}\Gamma_{i,i-1}\Gamma_{i,i+1}}{\Gamma_{i-2,i+2}\Gamma_{i+2,i-1}\Gamma_{i+2,i}\Gamma_{i+2,i+1}},\end{aligned} \tag{14}$$

using $\Gamma_{l,q} = s_l - s_q$ and

$$\begin{aligned}\Xi_1 =& s_{i-2}(s_{i-1}(\Gamma_{i+1,i} + \Gamma_{i+2,i})\\ &+ 3s_i^2 - 2(s_{i+1} + s_{i+2})s_i + s_{i+1}s_{i+2}) + s_i(-4s_i^2 + 3(s_{i+1} + s_{i+2})s_i\\ &- 2s_{i+1}s_{i+2}) + s_{i-1}(3s_i^2 - 2(s_{i+1} + s_{i+2})s_i + s_{i+1}s_{i+2}).\end{aligned} \tag{15}$$

Recalling that the above procedure should be similarly done for the nodes $\{s_1, s_2, s_{m-1}, s_m\}$, viz, to find the weighting coefficients with fourth order of convergence for such nodes, we should consider the five adjacent points and then calculate the interpolating polynomial at that specific point. In this way, the sided FD formulas are constructed and used.

Similarly, FD estimates for the second derivative terms can be obtained applying a similar methodology as above. To this objective, we consider a set of points as follows:

$$\begin{aligned}&\{\{s_{i-3}, g(s_{i-3})\}, \{s_{i-2}, g(s_{i-2})\}, \{s_{i-1}, g(s_{i-1})\},\\ &\{s_i, g(s_i)\}, \{s_{i+1}, g(s_{i+1})\}, \{s_{i+2}, g(s_{i+2})\}\},\end{aligned} \tag{16}$$

and compute the second-derivative interpolating polynomial $p''(z)$ based on z. Now by taking into account $z = s_i$ in Mathematica [22], one obtains that:

$$\begin{aligned}g''(s_i) =& \beta_{i-3}g(s_{i-3}) + \beta_{i-2}g(s_{i-2}) + \beta_{i-1}g(s_{i-1})\\ &+ \beta_i g(s_i) + \beta_{i+1}g(s_{i+1}) + \beta_{i+2}g(s_{i+2}) + \mathcal{O}\left(h^4\right),\end{aligned} \tag{17}$$

where

$$\begin{aligned}\beta_{i-3} &= \frac{\Xi_2}{\Gamma_{i-3,i-2}\Gamma_{i-3,i-1}\Gamma_{i-3,i}\Gamma_{i-3,i+1}\Gamma_{i-3,i+2}},\\ \beta_{i-2} &= \frac{\Xi_3}{\Gamma_{i-3,i-2}\Gamma_{i-2,i-1}\Gamma_{i-2,i}\Gamma_{i-2,i+1}\Gamma_{i-2,i+2}},\\ \beta_{i-1} &= \frac{\Xi_4}{\Gamma_{i-2,i-1}\Gamma_{i-1,i-3}\Gamma_{i-1,i}\Gamma_{i-1,i+1}\Gamma_{i-1,i+2}},\end{aligned} \tag{18}$$

$$\begin{aligned}\beta_{i} &= \frac{\Xi_5}{\Gamma_{i-3,i}\Gamma_{i,i-2}\Gamma_{i,i-1}\Gamma_{i,i+1}\Gamma_{i,i+2}},\\ \beta_{i+1} &= \frac{\Xi_6}{\Gamma_{i-3,i+1}\Gamma_{i+1,i-2}\Gamma_{i+1,i-1}\Gamma_{i+1,i}\Gamma_{i+1,i+2}},\\ \beta_{i+2} &= \frac{\Xi_7}{\Gamma_{i-3,i+2}\Gamma_{i+2,i-2}\Gamma_{i+2,i-1}\Gamma_{i+2,i}\Gamma_{i+2,i+1}}.\end{aligned}$$

Here, we have

$$\begin{aligned}\Xi_2 =&s_{i-1}(-6s_i^2+4(s_{i+1}+s_{i+2})s_i-2s_{i+1}s_{i+2})+2s_i(4s_i^2\\&-3(s_{i+1}+s_{i+2})s_i+2s_{i+1}s_{i+2})+s_{i-2}(-6s_i^2\\&+4(s_{i+1}+s_{i+2})s_i-2s_{i+1}s_{i+2}+s_{i-1}(4s_i-2(s_{i+1}+s_{i+2}))),\\\Xi_3 =&2(s_{i-3}(s_{i-1}(\Gamma_{i+1,i}+\Gamma_{i+2,i})+3s_i^2-2(s_{i+1}\\&+s_{i+2})s_i+s_{i+1}s_{i+2})+s_i(-4s_i^2+3(s_{i+1}+s_{i+2})s_i-2s_{i+1}s_{i+2})\\&+s_{i-1}(3s_i^2-2(s_{i+1}+s_{i+2})s_i+s_{i+1}s_{i+2})),\\\Xi_4 =&2(s_{i-3}(s_{i-2}(\Gamma_{i+1,i}+\Gamma_{i+2,i})+3s_i^2-2(s_{i+1}\\&+s_{i+2})s_i+s_{i+1}s_{i+2})+s_i(-4s_i^2+3(s_{i+1}+s_{i+2})s_i-2s_{i+1}s_{i+2})\\&+s_{i-2}(3s_i^2-2(s_{i+1}+s_{i+2})s_i+s_{i+1}s_{i+2})),\end{aligned} \tag{19}$$

$$\begin{aligned}\Xi_5 =&2(s_{i-2}(s_{i-1}(\Gamma_{i+1,i}+\Gamma_{i+2,i}-s_i)+6s_i^2-3(s_{i+1}+s_{i+2})s_i+s_{i+1}s_{i+2})+s_{i-3}(s_{i-1}(\Gamma_{i+1,i}\\&+\Gamma_{i+2,i}-s_i)+s_{i-2}(\Gamma_{i+1,i}+\Gamma_{i+2,i}+s_{i-1}-s_i)+6s_i^2-3s_{i+1}s_i-3s_{i+2}s_i+s_{i+1}s_{i+2})\\&+s_i(2s_i(3s_{i+1}-5s_i)+s_{i-1}(6s_i-3s_{i+1}))+(3s_i(2s_i-s_{i+1})+s_{i-1}(s_{i+1}-3s_i))s_{i+2}),\end{aligned}$$

$$\begin{aligned}\Xi_6 =&2(s_i(s_{i-1}(2\Gamma_{i,i+2}+s_i)+s_i(3s_{i+2}-4s_i))+s_{i-2}(s_i(2\Gamma_{i,i+2}+s_i)\\&+s_{i-1}(\Gamma_{i+2,i}-s_i))+s_{i-3}(2s_i\Gamma_{i,i+2}\\&+s_{i-1}(\Gamma_{i+2,i}-s_i)+s_{i-2}(\Gamma_{i-1,i}+\Gamma_{i+2,i})+s_i^2)),\\\Xi_7 =&2(s_i(s_{i-1}(2\Gamma_{i,i+1}+s_i)+s_i(3s_{i+1}-4s_i))+s_{i-2}(s_i(2\Gamma_{i,i+1}\\&+s_i)+s_{i-1}(\Gamma_{i+1,i}-s_i))+s_{i-3}(2s_i\Gamma_{i,i+1}\\&+s_{i-1}(\Gamma_{i+1,i}-s_i)+s_{i-2}(\Gamma_{i-1,i}+\Gamma_{i+1,i})+s_i^2)).\end{aligned}$$

Summarizing the following theorem has been established.

Theorem 1. *As long as the function g is sufficiently smooth, the first and second derivative of the this function can be approximated by five and six adjacent points respectively on non-uniform meshes, via the formulas (13) and (17).*

Proof. The proof can be investigated by Taylor expansions as in the derivation in this section. It is hence omitted. □

The procedure for obtaining the weights for the points $\{s_1, s_2, s_3, s_{m-1}, s_m\}$ to keep the fourth convergence order should be investigated by the six adjacent points as described above but for that specific node.

It is noted that the formulations derived in (13) and (17) can be used for both uniform and nonuniform distribution of the discretization nodes, and can be simplified to more simpler formulations if the nodes are equidistant.

3. Application to Option Pricing under 3D HHW PDE

Considering the non-uniform nodes discussed in Section 1 along with the high order FD formulations calculated in Section 2, one is able to derive the differentiation matrices corresponding to the first and second derivatives of the function. These derivative matrices contains the weights of the fourth order approximations and are sparse in general since they are banded matrices whose zero elements are much more than their non-zero elements. These feature would help us in solving the financial model (1) as would be observed later.

For multi-dimensional derivatives, a matrix is constructed such that this is done on the flattened data, and subsequently the Kronecker product of the matrices for the derivatives (in one-dimension) are being considered.

One way for imposing the impact of (13)–(17) is with matrices including the weights of (13)–(17), i.e., the non-equidistant second-order FD weights, as their elements. A matrix which shows an estimation to the differential operator is called as a matrix of differentiation [20]. Forming and implementing the proposed scheme based on these matrices are invaluable aids for analysis.

Taking all the weights into consideration, the PDE (1) can be semi discretized to obtain:

$$\frac{\partial U(t)}{\partial t} = A(t)U(t), \qquad 0 \leq t \leq T, \tag{20}$$

at which $U(t) = \underbrace{(u_{1,1,1}(t), u_{1,1,2}(t), \ldots, u_{m,n,o-1}(t), u_{m,n,o}(t))^*}_{N \text{ elements}}$, is the unknowns vector and $N = m \times n \times o$. Noting that $A_{N \times N}(t)$ is the coefficient of the problem (1) at which the boundaries have not yet been imposed inside.

Here the boundaries along s are defined as follows [13]:

$$u(s, v, r, t) = 0, \qquad s = 0, \tag{21}$$

$$u_s(s, v, r, t) = 1, \qquad s = s_{\max}. \tag{22}$$

For $v = v_{\max}$, the following Dirichlet condition is prescribed:

$$u(s, v, r, t) = s, \qquad v = v_{\max}. \tag{23}$$

Remarking that the nodes which are located on the boundary $v = 0$ are considered as interior nodes and we take a fact into consideration that they must read the PDE model. That is to say, we incorporate the semi-discretized equations at this boundary.

At last, for $r = \pm r_{\max}$, we impose:

$$u_r(s, v, r, t) = 0, \qquad r = r_{\max}, \tag{24}$$

$$u_r(s, v, r, t) = 0, \qquad r = -r_{\max}. \tag{25}$$

By incorporating the above mentioned conditions, we obtain the following system of semi-discretized ODEs as follows:

$$\dot{U}(t) = \bar{A}(t)U(t), \tag{26}$$

where $\bar{A}(t)$ is the coefficient matrix including the boundaries.

Integrator

For discretizing in temporal variable t, many schemes are existing, for example refer to [23]. Explicit methods are basically straightforward to implement, but suffer from stability problems. Implicit schemes are unconditionally stable, but only exhibit low convergence or very time-consuming because of solving nonlinear system of algebraic equations per step.

Consider $\mathbf{u}_\iota$ to be the computational solution for the exact solution $U(t_\iota)$ and choose $k+1$ temporal nodes with the step size $\Delta t = \frac{T}{k}$.

At the moment, we use the δ–stage Runge–Kutta scheme [23] at $t_{\iota+1} = t_\iota + \Delta t$, $(0 \leq \iota \leq k)$ by:

$$\begin{aligned} g_i &= \mathbf{u}_\iota + \Delta t \sum_{j=1}^{\delta} k_j a_{i,j}, \\ k_i &= f(\Delta t c_i + t_\iota, g_i), \\ \mathbf{u}_{\iota+1} &= \mathbf{u}_\iota + \Delta t \sum_{i=1}^{\delta} b_i k_i, \end{aligned} \tag{27}$$

wherein f is defined based on the right hand side of (26). It is generally assumed that the row-sum conditions hold:

$$c_i = \sum_{j=1}^{\delta} a_{i,j}, \qquad i = 1, 2, \ldots, \delta. \tag{28}$$

Now we consider a sixth-order explicit Runge–Kutta scheme (RK6) below [24]:

$$\Lambda = \begin{pmatrix} 0 & 0 & 0 & 0 & 0 & 0 & 0 \\ 1 & 0 & 0 & 0 & 0 & 0 & 0 \\ \frac{3}{8} & \frac{1}{8} & 0 & 0 & 0 & 0 & 0 \\ \frac{8}{27} & \frac{2}{27} & \frac{8}{27} & 0 & 0 & 0 & 0 \\ \frac{3(3p_1-7)}{392} & \frac{(p_1-7)}{49} & \frac{-6(p_1-7)}{49} & \frac{3(p_1-21)}{392} & 0 & 0 & 0 \\ \frac{-3(17p_1+77)}{392} & \frac{(-p_1-7)}{49} & -\frac{(8p_1)}{49} & \frac{3(121p_1+21)}{1960} & \frac{(p_1+6)}{5} & 0 & 0 \\ \frac{(7p_1+22)}{12} & \frac{2}{3} & \frac{2(7p_1-5)}{9} & \frac{-7(3p_1-2)}{20} & \frac{-7(9p_1+49)}{90} & \frac{-7(p_1-7)}{18} & 0 \end{pmatrix}, \tag{29}$$

with $b = (9/180, 0, 64/180, 0, 49/180, 49/180, 9/180)$, $C = (1, 1/2, 2/3, (7-p_1)/14, (7+p_1)/14, 1)$, and $p_1 = 21^{1/2}$.

Notice that a consequence of explicitness is $c_1 = 0$ in (28), so that the function is sampled at the beginning of the current integration step. Here, the sixth-order time-stepping solver consists of seven stages and reaches sixth order of convergence. The sixth order says that the error of local truncation is on the order of $\mathcal{O}(\Delta t^7)$, while the total accumulated error is on the order of $\mathcal{O}(\Delta t^6)$.

In the sequel, we study that under what criteria the numerical discretized solution does not blow up. The following theorem is one of the contributions of this work. This is given for the time-independent case, i.e., when $\bar{A}(t) = \bar{A}$.

Theorem 2. *If the system of ODEs (26) reads the condition of Lipschitz, then the time-stepping method (27)–(29) has conditional stability.*

Proof. To find a stability conditions, we proceed as follows. Incorporating the time-stepping solver (27) on the system of ODEs (26) yields:

$$\begin{aligned} \mathbf{u}_{\iota+1} = & \left(I + \Delta t\bar{A} + \frac{(\Delta t\bar{A})^2}{2!} \right. \\ & + \frac{(\Delta t\bar{A})^3}{3!} + \frac{(\Delta t\bar{A})^4}{4!} \\ & + \frac{(\Delta t\bar{A})^5}{5!} + \frac{(\Delta t\bar{A})^6}{6!} \\ & \left. - \frac{(\Delta t\bar{A})^7}{2160} \right) \mathbf{u}_{\iota}. \end{aligned} \tag{30}$$

Thus, the numerical stability is reduced to:

$$\begin{vmatrix} 1 + \Delta t\omega_i + \frac{(\Delta t\omega_i)^2}{2} + \frac{(\Delta t\omega_i)^3}{6} + \frac{(\Delta t\omega_i)^4}{24} \\ + \frac{(\Delta t\omega_i)^5}{120} + \frac{(\Delta t\omega_i)^6}{720} - \frac{(\Delta t\omega_i)^7}{2126} \end{vmatrix} \leq 1, \tag{31}$$

which is due to (30) for any ω_i as the eigenvalue of $\bar{A}$. Now by considering: $\omega_{\max}(\bar{A})$, the stability condition can now be represented as follows:

$$\left|1+\Delta t\omega_{\max}+\frac{(\Delta t\omega_{\max})^2}{2}+\frac{(\Delta t\omega_{\max})^3}{6}+\frac{(\Delta t\omega_{\max})^4}{24}+\frac{(\Delta t\omega_{\max})^5}{120}+\frac{(\Delta t\omega_{\max})^6}{720}-\frac{(\Delta t\omega_{\max})^7}{2126}\right|\leq 1. \tag{32}$$

Noting that the negative semi-definiteness of $\bar{A}$ makes all its eigenvalues to have negative real parts. Thus, the proposed scheme has numerical stability if the temporal step size Δt satisfy (32). Noting that this can be computed in the language Mathematica [22] via the command

$$\texttt{Eigenvalues}[\texttt{matrix}, 1]. \tag{33}$$

The proof is ended. □

4. Experiments

In this section, some tests were given for our proposed method showed via Adaptive Finite Difference Method (AFDM) to price at the money call options, when $T = 1$ year and $E = 100\$$. A comparison was done by the standard uniform FD scheme [4], which by second order FD approximations and the Euler's scheme as a temporal solver shown by FDM. We also compare with the method provided in [13] shown by Haentjens-In't Method (HIM).

Mathematica 11.0 is used for the simulations [25]. Time is also reported in second while we employ the following stopping condition:

$$\text{Error} = \left|\frac{u_{\text{approx}}(s,v,r,t) - u_{\text{ref}}(s,v,r,t)}{u_{\text{ref}}(s,v,r,t)}\right|, \tag{34}$$

wherein u_{ref} and u_{approx} are the exact and numerical results.

To increase the computational efficiency for very large scale semi-discrete systems that we are dealing with, here we set `AccuracyGoal` → 5, `PrecisionGoal` → 5.

Here, we consider more number of discretization nodes along s rather than v and r, since its working interval is larger than the others and the non-smoothness of the initial condition occurs along this spatial variable.

The non-constant b is defined as follows:

$$b(\tau) = c_1 - c_2 \exp(-c_3\tau), \qquad \tau \geq 0, \tag{35}$$

where c_1, c_2, c_3 are constants, and $\tau = T - t$. The following two test cases are considered:

1. $\kappa = 3.0$, $\eta = 0.12$, $a = 0.20$, $\sigma_1 = 0.80$, $\sigma_2 = 0.03$, $\rho_{12} = 0.6$, $\rho_{13} = 0.2$, $\rho_{23} = 0.4$, $c_1 = 0.05$, $c_2 = 0$, $c_3 = 0$, where the reference value is $u_{\text{ref}}(100, 0.04, 0.1, 1) \simeq 16.176$.
2. $\kappa = 0.5$, $\eta = 0.8$, $a = 0.16$, $\sigma_1 = 0.90$, $\sigma_2 = 0.03$, $\rho_{12} = -0.5$, $\rho_{13} = 0.2$, $\rho_{23} = 0.1$, $c_1 = 0.055$, $c_2 = 0$, $c_3 = 0$, where the reference value is $u_{\text{ref}}(100, 0.04, 0.1, 1) \simeq 20.994$.

The results are brought forward in Tables 1 and 2 showing the stable and efficient valuations of options under HHW PDE via the new high-order procedure. Furthermore, to reveal the positivity and stability of the numerical results, in Experiment 2, and by considering $m = 30$, $n = 18$ and $o = 18$ discretization nodes, the results based on AFDM are plotted in Figures 1 and 2.

Table 1. Error results in Heston–Hull–White (HHW) Option 1.

Procedure	m	n	o	Size	Δt	Price	Error	CPU Timing
FDM								
	10	8	6	480	0.002	25.492	5.7×10^{-1}	0.39
	14	10	10	1400	0.001	11.098	3.1×10^{-1}	0.62
	18	12	12	2592	0.0005	17.203	6.3×10^{-2}	1.22
	24	14	14	4704	0.00025	18.731	1.5×10^{-1}	3.39
	28	16	16	7168	0.0002	13.329	1.7×10^{-1}	6.07
	45	22	22	21,780	0.00005	14.636	9.4×10^{-2}	70.54
HIM								
	10	8	6	480	0.001	14.472	1.0×10^{-1}	0.44
	14	10	10	1400	0.0005	15.300	5.3×10^{-2}	0.85
	18	12	12	2592	0.00025	15.615	3.4×10^{-2}	2.09
	24	14	14	4704	0.0001	15.806	2.2×10^{-2}	8.04
	28	16	16	7168	0.0001	15.871	1.8×10^{-2}	11.84
	50	22	22	24,200	0.000025	16.006	5.9×10^{-3}	186.77
AFDM								
	10	8	6	480	0.005	15.123	6.5×10^{-2}	0.41
	14	10	10	1400	0.002	15.986	1.1×10^{-2}	0.82
	18	12	12	2592	0.001	16.059	7.2×10^{-3}	1.99
	24	14	14	4704	0.000625	16.136	2.4×10^{-3}	5.56
	28	16	16	7168	0.0005	16.160	9.8×10^{-4}	9.02
	50	22	22	24,200	0.0001	16.179	1.8×10^{-4}	103.26

Table 2. Error results in HHW option 2.

Procedure	m	n	o	Size	Δt	Price	Error	CPU Timing
FDM								
	20	10	10	2000	0.00025	22.022	4.9×10^{-2}	1.77
	24	12	12	3456	0.0002	21.436	2.1×10^{-2}	3.14
	26	14	14	5096	0.0001	19.678	6.1×10^{-2}	8.67
	28	16	16	7168	0.0001	17.376	1.7×10^{-1}	11.72
	30	18	18	9720	0.00005	17.404	1.7×10^{-1}	35.50
	36	20	20	14,400	0.000025	20.510	2.2×10^{-2}	107.89
	38	22	22	18,392	0.000025	20.275	3.3×10^{-2}	161.16
	42	22	22	20,328	0.00002	18.370	1.2×10^{-1}	244.32
HIM								
	20	10	10	2000	0.00025	20.631	1.6×10^{-2}	1.69
	24	12	12	3456	0.0002	20.709	1.2×10^{-2}	3.40
	26	14	14	5096	0.0001	20.729	1.1×10^{-2}	8.64
	28	16	16	7168	0.0001	20.748	1.0×10^{-2}	12.36
	30	18	18	9720	0.00005	20.767	9.9×10^{-3}	36.54
	36	20	20	14,400	0.000025	20.810	7.8×10^{-3}	108.92
	38	22	22	18,392	0.000025	20.818	7.5×10^{-3}	166.08
	42	22	22	20,328	0.00002	20.833	6.7×10^{-3}	250.67
AFDM								
	20	10	10	2000	0.0005	20.832	7.7×10^{-3}	0.89
	24	12	12	3456	0.0004	20.899	4.5×10^{-3}	3.36
	26	14	14	5096	0.00025	20.910	4.0×10^{-3}	6.62
	28	16	16	7168	0.0002	20.926	3.2×10^{-3}	11.27
	30	18	18	9720	0.0001	20.951	2.0×10^{-3}	34.55
	36	20	20	14,400	0.0000625	20.972	1.0×10^{-3}	98.22
	38	22	22	18,392	0.00005	20.980	6.6×10^{-4}	165.27
	42	22	22	20,328	0.00004	20.999	2.3×10^{-4}	240.25

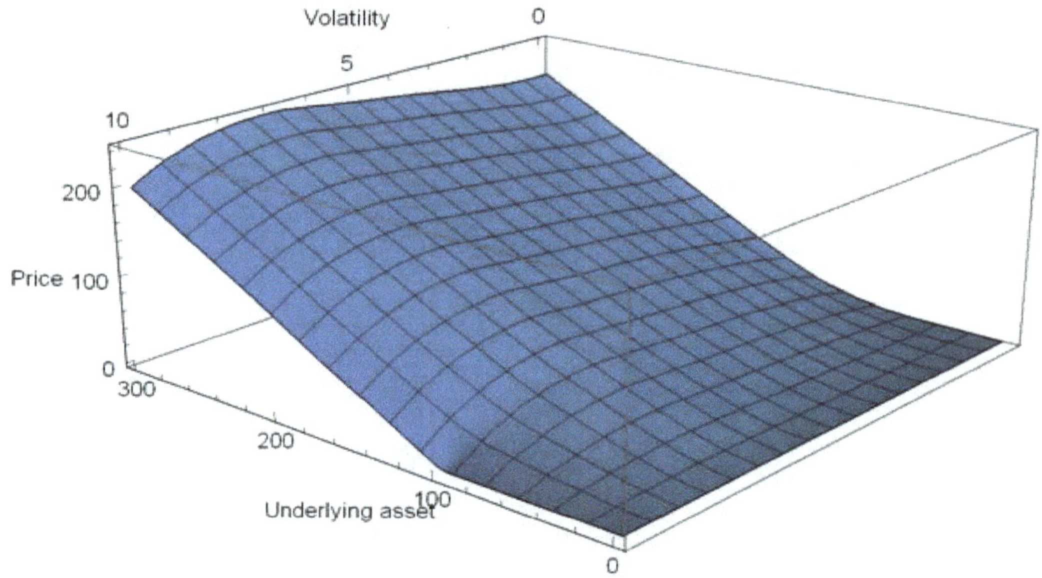

Figure 1. A numerical solution based on AFDM in Heston–Hull–White (HHW) option 2.

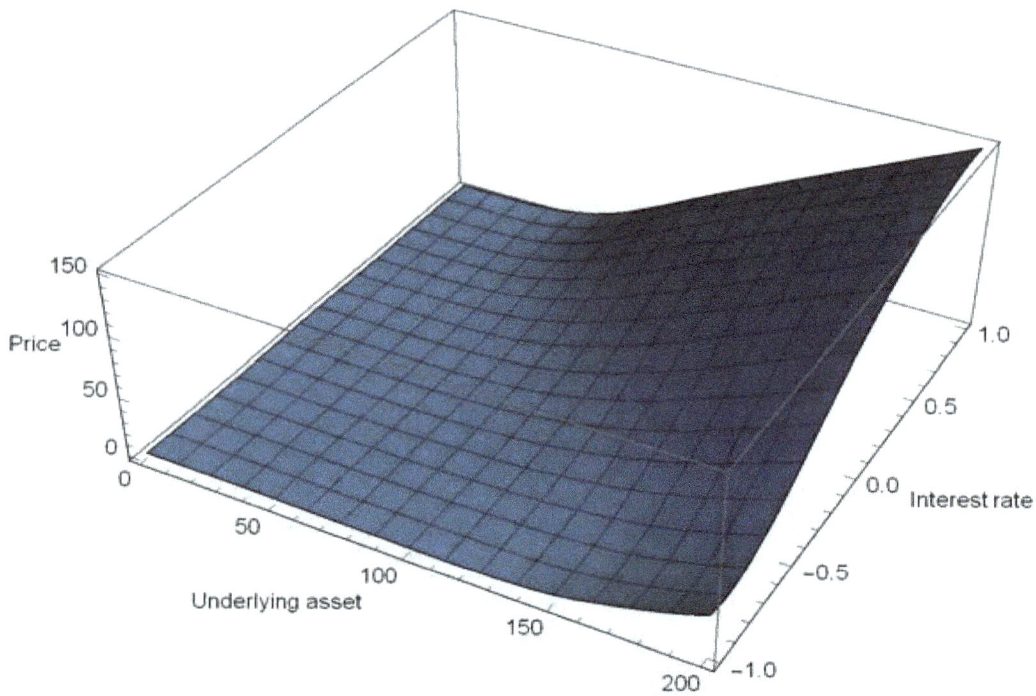

Figure 2. A numerical solution based on AFDM in HHW option 2.

5. Ending Comments

In financial engineering, it is famous that the Black–Scholes PDE could not be useful in real application due to several restrictions. Several ideas to observe the market's reality are models based upon the stochastic volatility and interest rate models. The resulted PDE problem in this way is hard to be solved theoretically due to higher involved dimensions and so numerical methods are required.

In this paper, we have proposed a new discretized numerical method based on adaptive FD methodology on non-uniform grids in order to tackle an important problem in computational finance known as HHW PDE (1). It was proved that the new procedure has conditional stability and shown to be efficient in practice.

Further discussions can be investigated to extend the results of this work for other types of options defined on HHW model such as digital (binary) options, at which the initial condition is not only non-smooth at the strike but also discontinues.

Acknowledgments: This work was supported by the Deanship of Scientific Research (DSR), King Abdulaziz University, Jeddah, under grant No. (D-235-130-1439). The authors, therefore, gratefully acknowledge the DSR technical and financial support.

References

1. Brigo, D.; Mercurio, F. *Interest Rate Models-Theory and Practice: With Smile, Inflation and Credit*, 2nd ed.; Springer Finance: Berlin, Germany, 2007.
2. Cakici, N.; Chatterjee, S.; Chen, R.-R. Default risk and cross section of returns. *J. Risk Financ. Manag.* **2019**, *12*, 95. [CrossRef]
3. Ballestra, L.V.; Cecere, L. A numerical method to estimate the parameters of the CEV model implied by American option prices: Evidence from NYSE. *Chaos Solitons Fractals* **2016**, *88*, 100–106. [CrossRef]
4. Duffy, D.J. *Finite Difference Methods in Financial Engineering: A Partial Differential Equation Approach*; Wiley: Chichester, UK, 2006.
5. Magoulès, F.; Gbikpi-Benissan, G.; Zou, Q. Asynchronous iterations of parareal algorithm for option pricing models. *Mathematics* **2018**, *6*, 45. [CrossRef]
6. Fouque, J.-P.; Papanicolaou, G.; Sircar, K.R. *Derivatives in Financial Markets with Stochastic Volatility*; Cambridge Univ. Press: Cambridge, UK, 2000.
7. Heston, S.L. A closed-form solution for options with stochastic volatility with applications to bond and currency options. *Rev. Finan. Stud.* **1993**, *6*, 327–343. [CrossRef]
8. Hull, J.; White, A. Using Hull-White interest rate trees. *J. Deriv.* **1996**, *4*, 26–36. [CrossRef]
9. Schöbel, R.; Zhu, J. Stochastic volatility with an Ornstein-Uhlenbeck process: An extension. *Eur. Financ. Rev.* **1999**, *3*, 23–46. [CrossRef]
10. Guo, S.; Grzelak, L.A.; Oosterlee, C.W. Analysis of an affine version of the Heston-Hull-White option pricing partial differential equation. *Appl. Numer. Math.* **2013**, *72*, 143–159. [CrossRef]
11. Sargolzaei, P.; Soleymani, F. A new finite difference method for numerical solution of Black-Scholes PDE. *Adv. Diff. Equat. Control Process.* **2010**, *6*, 49–55.
12. Soleymani, F.; Barfeie, M. Pricing options under stochastic volatility jump model: A stable adaptive scheme. *Appl. Numer. Math.* **2019**, *145*, 69–89. [CrossRef]
13. Haentjens, T.; In't Hout, K.J. Alternating direction implicit finite difference schemes for the Heston-Hull-White partial differential equation. *J. Comput. Fin.* **2012**, *16*, 83–110. [CrossRef]
14. Soleymani, F.; Akgül, A. Asset pricing for an affine jump-diffusion model using an FD method of lines on non-uniform meshes. *Math. Meth. Appl. Sci.* **2019**, *42*, 578–591. [CrossRef]
15. Itkin, A.; Carr, P. Jumps without tears: A new splitting technology for Barrier options. *Int. J. Numer. Anal. Model.* **2011**, *8*, 667–704.
16. Sumei, Z.; Jieqiong, Z. Efficient simulation for pricing barrier options with two-factor stochastic volatility and stochastic interest rate. *Math. Prob. Eng.* **2017**, *2017*, 3912036. [CrossRef]
17. Soleymani, F.; Saray, B.N. Pricing the financial Heston-Hull-White model with arbitrary correlation factors via an adaptive FDM. *Comput. Math. Appl.* **2019**, *77*, 1107–1123. [CrossRef]
18. Kwok, Y.K. *Mathematical Models of Financial Derivatives*, 2nd ed.; Springer: Heidelberg, Germany, 2008.
19. Ballestra, L.V.; Sgarra, C. he evaluation of American options in a stochastic volatility model with jumps: An efficient finite element approach. *Comput. Math. Appl.* **2010**, *60*, 1571–1590. [CrossRef]
20. Fornberg, B. *A Practical Guide to Pseudospectral Methods*; Cambridge University Press: Cambridge , UK, 1996.
21. Soleymani, F.; Barfeie, M.; Khaksar Haghani, F. Inverse multi-quadric RBF for computing the weights of FD method: Application to American options. *Commun. Nonlinear Sci. Numer. Simul.* **2018**, *64*, 74–88. [CrossRef]
22. Wellin, P.R.; Gaylord, R.J.; Kamin, S.N. *An Introduction to Programming with Mathematica*, 3rd ed.; Cambridge University Press: Cambridge, UK, 2005.

23. Sofroniou, M.; Knapp, R. *Advanced Numerical Differential Equation Solving in Mathematica, Wolfram Mathematica, Tutorial Collection;* Wolfram Research, Inc.: Champaign, IL, USA, 2008.
24. Luther, H.A. An explicit sixth-order Runge-Kutta formula. *Math. Comput.* **1967**, *22*, 434–436. [CrossRef]
25. Mangano, S. *Mathematica Cookbook*; O'Reilly Media: Sevastopol, CA, USA, 2010.

12

A Unified Convergence Analysis for Some Two-Point Type Methods for Nonsmooth Operators

Sergio Amat [1,*,†], Ioannis Argyros [2,†], Sonia Busquier [1,†], Miguel Ángel Hernández-Verón [3,†] and María Jesús Rubio [3,†]

1 Departamentode Matemática Aplicada y Estadística, Universidad Politécnica de Cartagena, 11003 Cádiz, Spain
2 Department of Mathematical Sciences, Cameron University, Lawton, OK 73505, USA
3 Departamento de Matemáticas y Computación, Universidad de La Rioja, Calle Madre de Dios, 53, 26006 Logrono, Spain
* Correspondence: sergio.amat@upct.es.
† These authors contributed equally to this work.

Abstract: The aim of this paper is the approximation of nonlinear equations using iterative methods. We present a unified convergence analysis for some two-point type methods. This way we compare specializations of our method using not necessarily the same convergence criteria. We consider both semilocal and local analysis. In the first one, the hypotheses are imposed on the initial guess and in the second on the solution. The results can be applied for smooth and nonsmooth operators.

Keywords: iterative methods; nonlinear equations; Newton-type methods; smooth and nonsmooth operators

---

1. Introduction

One of the most important techniques in order to approximate nonlinear equations are iterative methods [1–6]. In this paper, we present a unified approach for two-point Newton-type methods for smooth and nonsmooth operators [7–10]. We will consider two types of convergence. The semilocal convergence is where the hypotheses are imposed on the initial guess; and local convergence is where the hypotheses are imposed on the solution. Our family includes a great variety of methods. We are interested also in the application of these methods in practice (nonlinear systems, boundary problems and image processing).

For a greater generality, in this study, let X and Y be two Banach spaces and D a nonempty, open, and convex set; let $F_1 : D \subset X \to Y$ and $F_2 : D \subset X \to Y$ be continuous operators. Moreover, we assume that the operator F_1 has a continuous Fréchet derivative and F_2 is a continuous operator whose differentiability is not assumed. We consider the equation

$$F(x) = F_1(x) + F_2(x) = 0. \tag{1}$$

To solve this equation, we use the two-point Newton-type methods defined by

$$x_{k+1} = x_k - L_{k-1,k}^{-1}(F_1(x_k) + F_2(x_k)) \tag{2}$$

for each $k = 0, 1, 2, \ldots$, where $x_{-1}, x_0 \in D$ are the initial points, $L(.,.) : D \times D \to \mathfrak{L}(X, Y)$ and $\mathfrak{L}(X, Y)$ is the space of bounded linear operators from X into Y. We have denoted by $L_{k-1,k} = L(x_{k-1}, x_k)$.

If $F_2(x) \neq 0$, we have that the operator F is not Fréchet differentiable. In general, to approximate a solution of (1) in this situation, derivative-free iterative methods are used [11–14]. To obtain this type of iterative processes, it is common to approximate derivatives by difference divided. Remember that, given an operator $H : D \subset X \to Y$, we call $[x, y; H] \in \mathfrak{L}(X, Y)$ a first order divided differences operator for H on the points x and y $(x \neq y)$ in D if

$$[x, y; H](x - y) = H(x) - H(y). \tag{3}$$

So, to solve (1) with iterative methods given from (2), we can consider at least two different procedures. Firstly, we have the Zincenko method [15], given by the following algorithm:

$$\begin{cases} \text{Given } x_{-1},\ x_0 \in D, \\ x_{k+1} = x_k - \left[F_1'(x_k)\right]^{-1} F(x_k), \quad n \geq 0, \end{cases} \tag{4}$$

where we directly eliminate the nondifferentiable part of F, i.e., F_2. So, in this case, $L_{k-1,k} = F_1'(x_k)$ in (2). Secondly, we can consider an approximation of F' by divided differences, the secant-type methods [16,17]:

$$\begin{cases} \text{Given } x_{-1},\ x_0 \in D, \\ y_k = \lambda x_k + (1 - \lambda) x_{k-1}, \quad \lambda \in [0, 1), \\ x_{k+1} = x_k - [y_k, x_k; F]^{-1} F(x_k), \quad n \geq 0, \end{cases} \tag{5}$$

where the secant method, for $\lambda = 0$, is obtained. So, in this case, $L_{k-1,k} = [y_k, x_k; F]$ in (2). But, if we consider a better approximation of the derivative of F, an approximation of second order, we have the Kurchatov method [18]:

$$\begin{cases} \text{Given } x_{-1},\ x_0 \in D, \\ x_{k+1} = x_k - [x_{k-1}, 2x_k - x_{k-1}; F]^{-1} F(x_n), \quad n \geq 0, \end{cases} \tag{6}$$

in this case, $L_{k-1,k} = [x_{k-1}, 2x_k - x_{k-1}; F]$ in (2).

By using this procedure of decomposition for operator F, we see that we can also consider the application of iterative methods that use derivatives when F is nondifferentiable. So, if we consider decomposition of F given in (1), we can use the Newton-secant-type algorithm:

$$\begin{cases} \text{Given } x_{-1},\ x_0 \in D, \\ y_k = \lambda x_k + (1 - \lambda) x_{k-1}, \quad \lambda \in [0, 1), \\ x_{k+1} = x_k - \left(F_1'(x_k) + [y_k, x_k; F]\right)^{-1} F(x_k), \quad n \geq 0, \end{cases} \tag{7}$$

where $L_{k-1,k} = F_1'(x_k) + [y_k, x_k; F]$ in (2). The other possibility, from the decomposition method, is to consider the Newton–Kurchatov [19] algorithm:

$$\begin{cases} \text{Given } x_{-1},\ x_0 \in D, \\ x_{k+1} = x_k - \left(F_1'(x_k) + [x_{k-1}, 2x_k - x_{k-1}; F]\right)^{-1} F(x_k), \quad n \geq 0, \end{cases} \tag{8}$$

where $L_{k-1,k} = F_1'(x_k) + [x_{k-1}, 2x_k - x_{k-1}; F]$ in (2). Another possibility is to consider Steffensen-type methods, that is, the methods associated to divided differences like $[x_k, x_k + F(x_k); F]$.

As we can see, there are a lot of iterative methods that can be written as algorithms (2).

The main aim of this paper is to obtain a general study for the convergence, local and semilocal, for these Newton-type of iterative methods given in (2).

2. Convergence Analysis for Two-Point Newton-Type Methods

In this section, we present both semilocal and local convergence analysis. In the first one, the hypotheses are imposed on the initial guess; and in the second, on the solution. The results can be applied for smooth and nonsmooth operators.

2.1. Local Convergence Analysis

We start the local analysis of method (2). Let $v_0 : [0,+\infty) \times [0,+\infty) \to [0,+\infty)$ be a nondecreasing continuous function. Assume that the equation

$$v_0(t,t) = 1 \tag{9}$$

has at least one positive root r_0. Let also $v : [0,r_0) \times [0,r_0) \to [0,+\infty)$ be a nondecreasing continuous function. Define function $\bar{v}$ on the interval $[0,r_0)$ by

$$\bar{v}(t) = \frac{v(t,t)}{1 - v_0(t,t)} - 1. \tag{10}$$

Assume that the equation

$$\bar{v}(t) = 0 \tag{11}$$

has a minimal positive solution r. It follows that for each $t \in [0,r)$

$$0 \leq v_0(t,t) < 1 \tag{12}$$

and

$$0 \leq \bar{v}(t) < 1. \tag{13}$$

Our analysis of method (2) will use the conditions (A):

- (a_1) There exist a solution $x^* \in D$ of Equation (1), and $B \in \mathfrak{L}(X,Y)$ such that $B^{-1} \in \mathfrak{L}(Y,X)$.
- (a_2) Condition (9) holds and for each $x,u \in D$

$$\|B^{-1}(L(x,u) - B)\| \leq v_0(\|x - x^*\|, \|u - y^*\|),$$

where v_0 is defined previously, and r_0 is given in (9).

Set $D_0 = D \cap \bar{U}(x^*, r_0)$.

- (a_3) For each $x,z \in D_0$, and any solution $y \in D$ of Equation (1)

$$\|B^{-1}(F_1(x) + F_2(x) - L(z,x)y)\| \leq v(\|z - y\|, \|x - y\|)\|x - y\|,$$

where v is defined previously, and $L(\cdot,\cdot) : D_0 \times D_0 \to \mathfrak{L}(X,Y)$.

- (a_4) $\bar{U}(x^*,r) \subset D$, where r is given in (10).
- (a_5)

$$\xi := \frac{v(r,r)}{1 - v_0(r,r)} \in [0,1).$$

We are able to perform our local analysis of method (2) based on the aforementioned conditions (A).

Theorem 1. *Assume that the conditions (A) hold. Then, sequence x_k, defined by method (2) for $x_{-1}, x_0 \in U(x^*,r) - x^*$, is well defined in $U(x^*,r)$; remains in $U(x^*,r)$; and converges to x^*. Finally, the following estimates hold.*

$$\|x_{k+1} - x^*\| \leq \frac{v(\|x_{k-1} - x^*\|, \|x_k - x^*\|)}{1 - v_0(\|x_{k-1} - x^*\|, \|x_k - x^*\|)}\|x_k - x^*\| \leq \|x_k - x^*\| < r. \tag{14}$$

The vector x^* is the only solution of Equation (1) in $\bar{U}(x^*,r)$.

Proof. We will use mathematical induction on k.

Let $x, u \in U(x^*,r)$.

Using (2), (a_1) and (a_2), we obtain

$$\|B^{-1}(L(x,u)-B)\| \le v_0(\|x-x^*\|,\|u-x^*\|) \le v_0(r,r) < 1. \tag{15}$$

Using the Banach lemma on invertible operators [20] and (15), we deduce that $L(x,u)^{-1} \in \mathfrak{L}(Y,X)$, and

$$\|L(x,u)^{-1}B\| \le \frac{1}{1-v_0(\|x-x^*\|,\|u-x^*\|)}. \tag{16}$$

In particular, estimate (16) holds for $x = x_0$, so x_1 is well defined by method (2) for $k=0$.

Using the definition of the method (2) (for $k=0$); (a_1), (a_3), (13), and (16) (for $k=0$) that

$$\begin{aligned}
\|x_1-x^*\| &= \|x_0-x^*-L(x_{-1},x_0)^{-1}(x_{-1},x_0)(F_1(x_0)+F_2(x_0))\| \\
&= \|[L(x_{-1},x_0)^{-1}B][B^{-1}(F_1(x_0)+F_2(x_0)-L(x_{-1},x_0)(x_0-x^*))]\| \\
&\le \|L(x_{-1},x_0)^{-1}B\|\|B^{-1}(F_1(x_0)+F_2(x_0)-L(x_{-1},x_0)(x_0-x^*))\| \\
&\le \frac{v(\|x_{-1}-x^*\|,\|x_0-x^*\|)}{1-v_0(\|x_{-1}-x^*\|,\|x_0-x^*\|)}\|x_0-x^*\| \le \|x_0-x^*\| < r,
\end{aligned} \tag{17}$$

which shows estimate (14) for $k=0$ and $x_1 \in U(x^*,r)$.

Replace x_0, x_1 by x_i, x_{i+1} in the preceding estimates to complete the induction for estimate (14). Then, from the estimate

$$\|x_{i+1}-x^*\| \le \mu\|x_i-x^*\| < r, \tag{18}$$

where

$$\mu = \frac{v(\|x_{-1}-x\|,\|x_0-x^*\|)}{1-v_0(\|x_{-1}-x^*\|,\|x_0-x^*\|)} \in [0,1),$$

thus, $\lim_{i\to+\infty} x_i = x^*$ and $x_{i+1} \in U(x^*,r)$. Moreover, for the uniqueness part, let $y^* \in \bar{U}(x^*,r)$ with $F_1(y^*)+F_2(y^*)=0$. Using $(a3)$, $(a5)$, and estimate (17), we obtain in turn that

$$\begin{aligned}
\|x_{i+1}-y^*\| &\le \|L(x_{i-1},x_i)^{-1}B\|\|B^{-1}(F_1(x_i)+F_2(x_i)-L(x_{i-1},x_i)(x_i-y^*))\| \\
&\le \frac{v(\|x_{i-1}-y^*\|,\|x_i-y^*\|)}{1-v_0(\|x_{i-1}-y^*\|,\|x_i-y^*\|)}\|x_i-y^*\| \\
&\le \mu\|x_i-y^*\| < \mu^{i+1}\|x_0-y^*\|,
\end{aligned} \tag{19}$$

which shows $\lim_{i\to+\infty} x_i = y^*$—but, we showed $\lim_{i\to+\infty} x_i = x^*$. Hence, we conclude that $x^* = y^*$. □

Remark 1.

- *Condition (a_3) can be replaced by the stronger: for each $x, y \in D_0$*

$$\|B^{-1}(F_1(x)+F_2(x)-L(x)(x-y))\| \le v_1(\|x-y\|)\|x-y\|,$$

where function v_1 is as v. However, for each $t, t' \ge 0$

$$v(t,t') \le v_1(t,t').$$

- *Linear operator B does not necessarily depend on q, where $q = x^*$ or $q = x_0$. It is used to determine the invertibility of linear operator $L(\cdot,\cdot)$ appearing in the method. The invertibility of B can be assured by an additional condition of the form $||I - B|| < 1$ or in some other way. A possible choice for B is $B = B(q)$ or $B = F_1'(q)$.*
- *It follows from the definition of r_0 and r that $r_0 \geq r$.*

2.2. Semilocal Convergence Analysis

For the semilocal case, we also define some functions and parameters. Let $w_0 : [0, +\infty) \times [0, +\infty) \to [0, +\infty)$ be a continuous and nondecreasing function.

Assume that the equation

$$w_0(t, t) = 1, \tag{20}$$

has one smallest positive root that we denote by ρ_0. Let $w : [0, \rho_0) \times [0, \rho_0) \times [0, \rho_0) \to [0, +\infty)$ be a nondecreasing continuous function. Moreover, for $\eta, \bar{\eta} \geq 0$, define parameters C_1 and C_2 by

$$\begin{aligned} C_1 &= \frac{w(\bar{\eta}, \eta, 0)}{1 - w_0(0, \eta)}, \\ C_2 &= \frac{w(0, \frac{\eta}{1-C_1}, \eta)}{1 - w_0(\eta, \frac{\eta}{1-C_1})} \end{aligned}$$

and function $C : [0, \rho_0) \to [0, +\infty)$ by $C(t) = \frac{w(t,t,t)}{1-w_0(t,t)}$. Assume that the equation

$$(\frac{C_1 C_2}{1 - C(t)} + C_1 + 1)\eta - t = 0 \tag{21}$$

has one smallest positive root that we denote by ρ.

The semilocal convergence analysis of method (2) will be based on conditions (H):

- $(h1)$ There exists $x_{-1}, x_0 \in D$, and $B \in \mathfrak{L}(X, Y)$ such that $B^{-1} \in \mathfrak{L}(Y, X)$.
- $(h2)$ Condition (20) holds, and for each $x \in D$

  $$||B^{-1}(L(z, x) - B)|| \leq w_0(||z - x_0||, ||x - x_0||),$$

  where w_0 is defined previously and ρ_0 is given in (20).

  Set $D_1 = D \bigcap \bar{U}(x_0, \rho_0)$.
- $(h3)$ For $L(\cdot,\cdot) : D_1 \times D_1 \to \mathfrak{L}(X, Y)$, and each $x, y, z \in D_1$

  $$\begin{aligned} &||B^{-1}(F_1(y) - F_1(x) + F_2(y) - F_2(x) - L(z, x)(y - x))|| \\ &\quad \leq w(||z - x_0||, ||y - x_0||, ||x - x_0||)||y - x||, \end{aligned}$$

  where w is defined previously.
- $(h4)$ $\bar{U}(x_0, \rho) \subseteq D$ and condition (21) holds for ρ, where $||x_1 - x_0|| \leq \eta$ and $||x_{-1} - x_0|| \leq \bar{\eta}$.

Then, using the hypotheses (H), we obtain the estimates:

$$\begin{aligned}
\|x_2 - x_1\| &\leq \frac{w(\|x_{-1} - x_0\|, \|x_1 - x_0\|, \|x_0 - x_0\|)\|x_1 - x_0\|}{1 - w_0(\|x_0 - x_0\|, \|x_1 - x_0\|)} = C_1\|x_1 - x_0\|, \\
\|x_2 - x_0\| &\leq \|x_2 - x_1\| + \|x_1 - x_0\| \leq (1 + C_1)\|x_1 - x_0\| \\
&= \frac{1 - C_1^2}{1 - C_1}\|x_1 - x_0\| \\
&< \frac{\eta}{1 - C_1} < \rho, \\
\|x_3 - x_2\| &\leq \frac{w(\|x_0 - x_0\|, \|x_2 - x_0\|, \|x_1 - x_0\|)}{1 - w_0(\|x_1 - x_0\|, \|x_2 - x_0\|)}\|x_2 - x_1\| \\
&\leq \frac{w(0, \frac{\eta}{1-C_1}, \eta)}{1 - w_0(\eta, \frac{\eta}{1-C_1})}\|x_2 - x_1\| = C_2\|x_2 - x_1\|, \\
\|x_3 - x_0\| &\leq \|x_3 - x_2\| + \|x_2 - x_1\| + \|x_1 - x_0\| \\
&\leq C_2\|x_2 - x_1\| + C_1\|x_1 - x_0\| + \|x_1 - x_0\| \\
&\leq (C_2C_1 + C_1 + 1)\|x_1 - x_0\|, \\
\|x_4 - x_3\| &\leq \frac{w(\|x_1 - x_0\|, \|x_3 - x_0\|, \|x_2 - x_0\|)}{1 - w_0(\|x_2 - x_0\|, \|x_3 - x_0\|)}\|x_3 - x_2\| \\
&\leq C(\rho)\|x_3 - x_2\| \leq C(\rho)C_2\|x_2 - x_1\| \\
&\leq C(\rho)C_2C_1\|x_1 - x_0\|,
\end{aligned} \tag{22}$$

similarly for $i = 3, 4, \ldots$

$$\begin{aligned}
\|x_{i+1} - x_i\| &\leq C(\rho)\|x_i - x_{i-1}\| \leq C(\rho)^{i-2}\|x_3 - x_2\|, \\
\|x_{i+1} - x_0\| &\leq \|x_{i+1} - x_i\| + \ldots + \|x_4 - x_3\| + \|x_3 - x_0\| \\
&\leq C(\rho)\|x_i - x_{i-1}\| + \ldots + C(\rho)\|x_3 - x_2\| \\
&\quad + (C_2C_1 + C_1 + 1)\|x_1 - x_0\| \\
&\leq C(\rho)^{i-2}\|x_3 - x_2\| + \ldots + C(\rho)\|x_3 - x_2\| \\
&\quad + (C_2C_1 + C_1 + 1)\|x_1 - x_0\| \\
&\leq (\frac{1 - C(\rho)^{i-1}}{1 - C(\rho)}C_2C_1 + C_1 + 1)\|x_1 - x_0\| \\
&< (\frac{C_1C_2}{1 - C(\rho)} + C_1 + 1)\eta \leq \rho,
\end{aligned} \tag{23}$$

$$\begin{aligned}
\|x_{i+j} - x_i\| &\leq \|x_{i+j} - x_{i+j-1}\| + \|x_{i+j-1} - x_{i+j-2}\| + \ldots + \|x_{i+1} - x_i\| \\
&\leq (C(\rho)^{i+j-3} + \ldots + C(\rho)^{i-2})\|x_3 - x_2\| \\
&\leq C(\rho)^{i-2}\frac{1 - C(\rho)^{j-1}}{1 - C(\rho)}\|x_3 - x_2\| \\
&\leq C(\rho)^{i-2}\frac{1 - C(\rho)^{j-1}}{1 - C(\rho)}C_2C_1\|x_1 - x_0\| \\
&\leq C(\rho)^{i-2}\frac{1 - C(\rho)^{j-1}}{1 - C(\rho)}C_2C_1\eta.
\end{aligned} \tag{24}$$

It follows from (23) that $x_i \in U(x_0,\rho)$; and from (24) that sequence x_i is complete in a Banach space X. In particular, it converges to some $x^* \in \bar{U}(x_0,\rho)$. By letting $i \to +\infty$ in the estimate

$$\|B^{-1}(F_1(x_i)+F_2(x_i))\| = \|B^{-1}(F_1(x_i)+F_2(x_i)-F_1(x_{i-1})-F_2(x_{i-1})-L_{i-2,i-1}(x_i-x_{i-1}))\|$$

$$\leq \frac{w(\|x_{i-2}-x_0\|,\|x_i-x_0\|,\|x_{i-1}-x_0\|)\|x_i-x_{i-1}\|}{1-w_0(\|x_{i-1}-x_0\|,\|x_i-x_0\|)} \leq \frac{w(\rho,\rho,\rho)}{1-w_0(\rho,\rho)}\|x_i-x_{i-1}\|,$$

we obtain $F_1(x^*)+F_2(x^*)=0$. The uniqueness part is omitted as analogous to the one in the local convergence case.

Hence, we can present our semilocal convergence result associated to the method (2).

Theorem 2. *Assume that the conditions (H) hold. Then, sequence x_k, defined by the method (2) for $x_{-1}, x_0 \in D$, is well defined in $U(x_0,\rho)$; remains in $U(x_0,\rho)$; and converges to a solution $x^* \in \bar{U}(x_0,\rho)$ of Equation (1). On the other hand, the vector x^* is the only solution of Equation (1) in $\bar{U}(x_0,\rho)$.*

The same comments given in the previous remark hold.

3. Numerical Experiment

Consider the nondifferentiable system of equations

$$\begin{cases} 3x_1^2x_2+x_2^2-1+|x_1-1|^{3/2}=0, \\ x_1^4+x_1x_2^3-1+|x_2|^{3/2}=0. \end{cases} \tag{25}$$

We therefore have an operator $F:\mathbb{R}^2\to\mathbb{R}^2$ such that $F=F_1+F_2$, as in (1), with $F_1,F_2:\mathbb{R}^2\to\mathbb{R}^2$, $F_1=(F_1{}^1,F_1{}^2)$, $F_2=(F_2{}^1,F_2{}^2)$, being

$$F_1^1(x_1,x_2)=3x_1^2x_2+x_2^2-1 \text{ and } F_1^2(x_1,x_2)=x_1^4+x_1x_2^3-1,$$

$$F_2^1(x_1,x_2)=|x_1-1|^{3/2} \text{ and } F_2^2(x_1,x_2)=|x_2|^{3/2},$$

where the operator F_1 is continuously Fréchet-differentiable and F_2 is continuous but is a Fréchet nondifferentiable operator.

For $\mathbf{u}=(u_1,u_2^T)$, $\mathbf{v}=(v_1,v_2)^T\in\mathbb{R}^2$, we consider the divided difference of first order defined by $[\mathbf{u},\mathbf{v};F]=([\mathbf{u},\mathbf{v};F]_{ij})_{i,j=1}^2\in\mathcal{L}(\mathbb{R}^2,\mathbb{R}^2)$, where

$$[\mathbf{u},\mathbf{v};F]_{i1} = \begin{cases} \dfrac{F_i(u_1,u_2,\ldots,v_m)-F_i(u_1,v_2,\ldots,v_m)}{u_1-v_1}, & \text{if } u_2\neq v_2, \\ 0, \text{ if } u_1=v_1, \end{cases}$$

$$[\mathbf{u},\mathbf{v};F]_{i2} = \begin{cases} \dfrac{F_i(u_1,u_2,\ldots,v_m)-F_i(u_1,v_2,\ldots,v_m)}{u_2-v_2}, & \text{if } u_2\neq v_2, \\ 0, \text{ if } u_2=v_2, \end{cases}$$

for $i=1,2$.

The iterative processes given by (1) allow us to consider direct iterative processes, such as (5) and (6); as well as iterative processes that use the decomposition method, such as (7) and (8). In this experiment, for the nondifferentiable system (25), we check that the application of the iterative processes that use the decomposition method have better behavior than the direct methods.

To carry out this study, we will consider as an approximate solution of system (25):

$$\mathbf{x}^* = (0.9383410452297656, 0.3312445136375143),$$

the starting points $\mathbf{x_{-1}} = (5,5)$ and $\mathbf{x_0} = (1,0)$, and use a tolerance $\|\mathbf{x_{n+1}} - \mathbf{x_n}\| \leq 10^{-16}$. In these conditions, in Tables 1 and 2 we can see the results of the application of the direct iterative processes, the secant-type, and Kurchatov methods. Whereas in Tables 3 and 4 we can see the results of the application of the iterative processes that use the decomposition method, Newton-secant-type and Newton–Kuchatov methods. Observing the results obtained, it is evident that the best behavior of the iterative processes is given by (2) using the decomposition method.

Table 1. $\|\mathbf{x}^* - \mathbf{x_n}\|$ for secant-type methods (5) and different values of the parameter λ.

n	$\lambda = 0$	$\lambda = 0.5$	$\lambda = 0.99$
1	3.18484×10^{-1}	2.965×10^{-1}	4.54388×10^{-2}
2	5.21264×10^{-2}	4.13083×10^{-2}	3.74494×10^{-3}
3	3.66108×10^{-3}	2.35344×10^{-3}	2.94716×10^{-5}
4	2.59348×10^{-4}	7.2935×10^{-5}	3.69966×10^{-9}
5	1.30031×10^{-6}	1.24012×10^{-7}	2.10942×10^{-15}
6	4.42187×10^{-10}	6.07747×10^{-12}	
7	1.11022×10^{-15}	1.11022×10^{-16}	

Table 2. Kurchatov method (6).

n	$\|\mathbf{x}^* - \mathbf{x_n}\|$
1	2.99754×10^{-1}
2	1.07269×10^{-1}
3	4.20963×10^{-2}
4	8.37098×10^{-3}
5	2.78931×10^{-4}
6	9.0784×10^{-8}
7	2.9826×10^{-11}

Table 3. $\|\mathbf{x}^* - \mathbf{x_n}\|$ for Newton-secant-type methods (7) and different values of the parameter λ.

n	$\lambda = 0$	$\lambda = 0.5$	$\lambda = 0.99$
1	2.3538×10^{-1}	1.00278×10^{-1}	4.29554×10^{-2}
2	3.48717×10^{-1}	2.88094×10^{-2}	2.53626×10^{-3}
3	1.47537×10^{-1}	1.90518×10^{-3}	9.06208×10^{-6}
4	3.4371×10^{-2}	8.39107×10^{-6}	1.9925×10^{-10}
5	3.08399×10^{-3}	4.78016×10^{-9}	
6	4.63665×10^{-5}	7.38298×10^{-15}	
7	4.05776×10^{-8}		
8	2.96929×10^{-13}		

Table 4. Newton–Kurchatov method (8).

n	$\|\overline{x^*} - \overline{x_n}\|$
1	6.1659×10^{-1}
2	4.75269×10^{-3}
3	6.29174×10^{-5}
4	3.37027×10^{-9}

Remark 2. *In the above example, we have selected the initial guess in a region where the operator is not smooth. The methods can be applied to systems where the operator is not smooth at the solution.*

For instance, for the system:

$$\begin{cases} |x_1^2 - 1| + x_2 - 1 = 0, \\ x_1 + x_2^2 - 1 = 0, \end{cases} \tag{26}$$

the solution is $(1,1)$*. If we take as initial guess* $(0.5, 0.5)$*, the Steffensen method gives as errors* 3.12×10^{-2}, 5.48×10^{-4}, 1.92×10^{-7}, 2.15×10^{-14}, *we observe its second order.*

4. Boundary Value Problem: Discretization via the Multiple Shooting Method

We will use the multiple shooting method for the discretization of boundary problems of the type

$$y''(t) = f(t, y(t), y'(t)), \qquad y(a) = \alpha, \qquad y(b) = \beta. \tag{27}$$

Thus, we should find the solution of the following nonlinear system of equations $F(s) = 0$, where $F : \mathbb{R}^N \longrightarrow \mathbb{R}^N$ and

$$\begin{cases} F_1(s_0, s_1, \ldots, s_{N-1}) &= s_1 - y'(t_1; s_0) \\ F_2(s_0, s_1, \ldots, s_{N-1}) &= s_2 - y'(t_2; s_0, s_1) \\ & \vdots \\ F_{N-1}(s_0, s_1, \ldots, s_{N-1}) &= s_{N-1} - y'(t_{N-1}; s_0, s_1, \ldots, s_{N-2}) \\ F_N(s_0, s_1, \ldots, s_{N-1}) &= \beta - y(t_N; s_0, s_1, s_{N-2}, s_{N-1}). \end{cases}$$

for a discretization of $[a, b]$ with N subintervals,

$$TN\,j, \qquad T = b - a, \qquad j = 0, 1, \ldots, N.$$

We consider the secant-type method

$$\begin{cases} \text{Given } y_{-1},\ y_0 \in D, \\ z_n = \lambda_n y_n + (1 - \lambda_n) y_{n-1}, \quad \lambda_n \in [0, 1), \\ y_{n+1} = y_n - [z_n, y_n; F]^{-1} F(y_n), \quad n \geq 0, \end{cases} \tag{28}$$

where λ_n is such that $||z_n - x_n|| \leq Tol$ for a given tolerance, and Newton's method

$$\begin{cases} \text{Given } \ y_0 \in D, \\ y_{n+1} = y_n - F(y_n)^{-1} F(y_n), \quad n \geq 0. \end{cases} \tag{29}$$

We perform a numerical comparison between both methods. As we can see, in the multiple shooting method, the iterative schemes are used as black boxes.

For the initial slope $\vec{s}_0 = (s_0^0, s_1^0, \dots, s_{N-1}^0)$, we propose

$$\begin{cases} s_0^0 &= \dfrac{\beta-\alpha}{b-a} = \dfrac{y(t_N)-y(t_0)}{t_N-t_0}, \\ s_1^0 &= \dfrac{y(t_N)-y(t_1;s_0)}{t_N-t_1}, \\ s_2^0 &= \dfrac{y(t_N)-y(t_2;s_0,s_1)}{t_N-t_2}, \\ &\vdots \\ s_{N-1}^0 &= \dfrac{y(t_N)-y(t_{N-1};s_0,s_1,\dots,s_{N-2})}{t_N-t_{N-1}}. \end{cases}$$

We analyze this particular example ([21], p. 554):

$$\begin{aligned} y''(t) &= \tau\cdot\sinh(\tau\cdot y(t)), \\ y(0) &= 0, \quad y(1)=1. \end{aligned}$$

We take $\tau = 2.5$ and $N = 4$ subintervals.

This problem admits the solution:

$$y(t) = \frac{2}{\tau}\,\mathrm{arg\,sinh}\left(\frac{s}{2}\cdot\frac{\mathrm{sn}\,(\tau t, 1-s^2/4)}{\mathrm{cn}\,(\tau t, 1-s^2/4)}\right),$$

where

$$s = y'(0) = 0.3713363932677645$$

and $\mathrm{sn}(\cdot,\cdot)$ and $\mathrm{cn}(\cdot,\cdot)$ are the Jacobi elliptic functions.

Newton's method (29),

n	$\Vert F(\vec{s}_n)\Vert_\infty$	$\Vert y(t)-y_n\Vert_\infty$	$\Vert y'(t)-y'_n\Vert_\infty$
0	10^{0}	10^{-1}	10^{0}
1	10^{-1}	10^{-1}	10^{-1}
2	10^{-2}	10^{-2}	10^{-2}
3	10^{-4}	10^{-4}	10^{-4}
4	10^{-7}	10^{-7}	10^{-7}
5	10^{-15}	10^{-15}	10^{-14}

Secant-type method (28),

n	$\Vert F(\vec{s}_n)\Vert_\infty$	$\Vert y(t)-y_n\Vert_\infty$	$\Vert y'(t)-y'_n\Vert_\infty$	$\Vert F'(y_n)-[y_n,x_n;F]\Vert_\infty$
0	10^{0}	10^{-1}	10^{0}	10^{-6}
1	10^{-1}	10^{-1}	10^{-1}	10^{-6}
2	10^{-2}	10^{-2}	10^{-2}	10^{-7}
3	10^{-4}	10^{-4}	10^{-4}	10^{-6}
4	10^{-7}	10^{-7}	10^{-7}	10^{-6}
5	10^{-15}	10^{-15}	10^{-14}	10^{-6}

The methods using Jacobians obtain their order of convergence. However, in this example, the computation of the derivatives involves the approximation of a more complicated problem. For this reason, the methods free of derivatives are preferred, see [21]. Of course, we need to compute a good approximation to the Jacobian, this is the motivation of our parameters λ_n. For more similar examples and conclusions, we refer [22].

Remark 3. *In many cases, when we manipulate an image, some random noise appears. This noise makes the later steps of processing the image difficult and inaccurate.*

Let $f : \Omega \to \mathbb{R}$ be a noise signal or image.

Introducing the variable w:

$$w = \frac{\nabla u}{\sqrt{|\nabla u|^2}},$$

the Total-Variation model is equivalent to the nonlinear and nondifferentiable system:

$$\begin{aligned} -\nabla \cdot w + \lambda(u - f) &= 0, \\ w\sqrt{|\nabla u|^2} - \nabla u &= 0. \end{aligned}$$

This system should be discretized using finite differences and the associated nonlinear system of equations can be approximated by our family (see [23] for more details).

5. Conclusions

This paper was devoted to the analysis of a general family of two-point Newton-type methods for smooth and nonsmooth operators. We have considered two types of convergence—semilocal and local. The family includes a great number of methods. We have applied the schemes to several interesting problems, in particular to nonsmooth nonlinear systems, boundary problems, and image denoising models.

Author Contributions: All the authors have been contributed similarly and they are participated in all the work.

Acknowledgments: Research of the first and third authors supported in part by Programa de Apoyo a la investigación de la fundación Séneca-Agencia de Ciencia y Tecnología de la Región de Murcia 20928/PI/18 and by MTM2015-64382-P. Research of the fourth and fifth authors supported by Ministerio de Economía y Competitividad under grant MTM2014-52016-C2-1P.

References

1. Amat, S.; Busquier, S. A modified secant method for semismooth equations. *Appl. Math. Lett.* **2003**, *16*, 877–881. [CrossRef]
2. Amat, S.; Busquier, S. On a higher order secant method. *Appl. Math. Comput.* **2003**, *141*, 321–329. [CrossRef]
3. Amat, S.; Busquier, S. A two-step Steffensen's method under modified convergence conditions. *J. Math. Anal. Appl.* **2006**, *324*, 1084–1092. [CrossRef]
4. Argyros, I.K.; Magreñán, A.A. *Iterative Methods and Their Dynamics with Applications: A Contemporary Study*; CRC Press: Boca Raton, FL, USA, 2017.
5. Argyros, I.K.; Magreñán, A.A. *A Contemporary Study of Iterative Methods*; Academic Press: Cambridge, MA, USA, 2018.
6. Grau-Sánchez, M. Noguera, M.; Amat, S. On the approximation of derivatives using divided difference operators preserving the local convergence order of iterative methods. *J. Comput. Appl. Math.* **2013**, *237*, 363–372.
7. Amat, S.; Bermúdez, C.; Busquier, S.; Mestiri, D. A family of Halley-Chebyshev iterative schemes for non-Fréchet differentiable operators. *J. Comput. Appl. Math.* **2009**, *228*, 486-493. [CrossRef]

8. Bartoň, M. Solving polynomial systems using no-root elimination blending schemes. *Comput.-Aided Des.* **2011**, *43*, 1870–1878. [CrossRef]
9. Bartoň, M.; Elber, G.; Hanniel, I. Topologically guaranteed univariate solutions of underconstrained polynomial systems via no-loop and single-component tests. *Comput.-Aided Des.* **2011**, *43*, 1035–1044. [CrossRef]
10. Hanniel, I.; Elber, G. Subdivision termination criteria in subdivision multivariate solvers using dual hyperplanes representations. *Comput.-Aided Des.* **2007**, *39*, 369–378. [CrossRef]
11. Chen, J.; Shen, Z. Convergence analysis of the secant type methods. *Appl. Math. Comput.* **2007**, *188*, 514–524. [CrossRef]
12. Hongmin, R.; Qingiao, W. The convergence ball of the Secant method under Hölder continuous divided differences. *J. Comput. Appl. Math.* **2006**, *194*, 284–293. [CrossRef]
13. Kewei, L. Homocentric convergence ball of the Secant method. *Appl. Math. J. Chin. Univ. Ser. B* **2007**, *22*, 353–365.
14. Ren, H.; Argyros, I.K. Local convergence of efficient Secant-type methods for solving nonlinear equations. *Appl. Math. Comput.* **2012**, *218*, 7655–7664. [CrossRef]
15. Zincenko, A.I. *Some Approximate Methods of Solving Equations with Non-Differentiable Operators*; Dopovidi Akad Nauk: 1963; pp. 156–161.
16. Hernández, M.A.; Rubio, M.J. A uniparametric family of iterative processes for solving nondifferentiable equations. *J. Math. Anal. Appl.* **2002**, *275*, 821–834. [CrossRef]
17. Hernández, M.A.; Rubio, M.J.; Ezquerro, J.A. Secant-like methods for solving nonlinear integral equations of the Hammerstein type. *J. Comput. Appl. Math.* **2000**, *115*, 245–254. [CrossRef]
18. Kurchatov, V.A. On a method of linear interpolation for the solution of funcional equations. *Dolk. Akad. Nauk SSSR* **1971**, *198*, 524–526; translation in *Soviet Math. Dolk.* **1971**, *12*, 835–838. (In Russian)
19. Hernández, M.A.; Rubio, M.J. On a Newton-Kurchatov-type Iterative Process. *Numer. Funct. Anal.* **2016**, *37*, 65–79. [CrossRef]
20. Kantorovich, L.V.; Akilov, G.P. *Functional Analysis*; Pergamon Press: Oxford, UK, 1982.
21. Stoer, J.; Bulirsch, R. *Introduction to Numerical Analysis*, 2nd ed.; Springer: New York, NY, USA, 1993.
22. Alarcón, V.; Amat, S.; Busquier, S.; López, D.J. A Steffensen's type method in Banach spaces with applications on boundary value problems. *J. Comput. Appl. Math.* **2008**, *216*, 243–250. [CrossRef]
23. Amat, S.; Argyros, I.K.; Busquier, S.; Hernández-Verón, M.A.; Martínez, E. A unified convergence analysis for single step-type methods for non-smooth operators. 2019, submitted.

13

An Optimal Eighth-Order Family of Iterative Methods for Multiple Roots

Saima Akram *, Fiza Zafar and Nusrat Yasmin

Centre for Advanced Studies in Pure and Applied Mathematics, Bahauddin Zakariya University, Multan 60800, Pakistan
* Correspondence: saimaakram@bzu.edu.pk

Abstract: In this paper, we introduce a new family of efficient and optimal iterative methods for finding multiple roots of nonlinear equations with known multiplicity $(m \geq 1)$. We use the weight function approach involving one and two parameters to develop the new family. A comprehensive convergence analysis is studied to demonstrate the optimal eighth-order convergence of the suggested scheme. Finally, numerical and dynamical tests are presented, which validates the theoretical results formulated in this paper and illustrates that the suggested family is efficient among the domain of multiple root finding methods.

Keywords: nonlinear equations; optimal iterative methods; multiple roots; efficiency index

---

1. Introduction

The problem of solving nonlinear equation is recognized to be very old in history as many practical problems which arise are nonlinear in nature . Various one-point and multi-point methods are presented to solve nonlinear equations or systems of nonlinear equations [1–3]. The above-cited methods are designed for the simple root of nonlinear equations but the behavior of these methods is not similar when dealing with multiple roots of nonlinear equations. The well known Newton's method with quadratic convergence for simple roots of nonlinear equations decays to first order when dealing with multiple roots of nonlinear equations. These problems lead to minor troubles such as greater computational cost and severe troubles such as no convergence at all. The prior knowledge of multiplicity of roots make it simpler to manage these troubles. The strange behavior of the iterative methods while dealing with multiple roots has been well known since 19th century in the least when Schröder [4] developed a modification of classical Newton's method to conserve its second order of convergence for multiple roots. The nonlinear equations with multiple roots commonly arise from different topics such as complex variables, fractional diffusion or image processing, applications to economics and statistics (Lēvy distributions), etc. By knowing the practical nature of multiple root finders, various one-point and multi-point root solvers have been developed in recent past [5–18] but most of them are not optimal as defined by Kung and Traub [19], who stated that an optimal without memory method can achieve its convergence order at the most 2^n requiring $n+1$ evaluations of functions or derivatives. As stated by Ostrowski [1], if an iterative method possess order of convergence as O and total number of functional evaluations is n per iterative step, then the index defined by $E = O^{1/n}$ is recognized as efficiency index of an iterative method.

Sharma and Sharma [17] proposed the following optimal fourth-order multiple root finder with known multiplicity m as follows:

$$\begin{aligned} y_n &= x_n - \frac{2m}{m+2} \cdot \frac{f(x_n)}{f'(x_n)}, m > 1 \\ x_{n+1} &= x_n - \frac{m}{8}\Phi(x_n)\frac{f(x_n)}{f'(x_n)}, \end{aligned}$$

where $\Phi(x_n) = \Big\{(m^3 - 4m + 8) - (m+2)^2(\frac{m}{m+2})^m \frac{f'(x_n)}{f'(y_n)} \times 2(m-1)(m+2)(\frac{m}{m+2})^m \frac{f'(x_n)}{f'(y_n)}\Big\}$.

A two-step sixth-order non-optimal family for multiple roots presented by Geum et al. [9] is given by:

$$\begin{aligned} y_n &= x_n - m \cdot \frac{f(x_n)}{f'(x_n)}, m > 1, \\ x_{n+1} &= y_n - Q(r_n, s_n) \cdot \frac{f(y_n)}{f'(y_n)}, \end{aligned} \tag{1}$$

where, $r_n = \sqrt[m]{\frac{f(y_n)}{f(x_n)}}, s_n = \sqrt[m-1]{\frac{f'(y_n)}{f'(x_n)}}$ and $Q : \mathbb{C}^2 \to \mathbb{C}$ is holomorphic in a neighborhood of $(0,0)$. The following is a special case of their family:

$$\begin{aligned} y_n &= x_n - m \cdot \frac{f(x_n)}{f'(x_n)}, n \geq 0, m > 1, \\ x_{n+1} &= y_n - m\left[1 + 2(m-1)(r_n - s_n) - 4r_n s_n + s_n^2\right] \cdot \frac{f(y_n)}{f'(y_n)}. \end{aligned} \tag{2}$$

Another non-optimal family of three-point sixth-order methods for multiple roots by Geum et al. [10] is given as follows:

$$\begin{aligned} y_n &= x_n - m \cdot \frac{f(x_n)}{f'(x_n)}, m \geq 1, \\ w_n &= y_n - m \cdot G(p_n) \cdot \frac{f(x_n)}{f'(x_n)}, \\ x_{n+1} &= w_n - m \cdot K(p_n, v_n,) \cdot \frac{f(x_n)}{f'(x_n)}, \end{aligned} \tag{3}$$

where $p_n = \sqrt[m]{\frac{f(y_n)}{f(x_n)}}$ and $v_n = \sqrt[m]{\frac{f(w_n)}{f(x_n)}}$. The weight functions $G : \mathbb{C} \to \mathbb{C}$ is analytic in a neighborhood of 0 and $K : \mathbb{C}^2 \to \mathbb{C}$ is holomorphic in a neighborhood of $(0,0)$. The following is a special case of the family in Equation (3):

$$\begin{aligned} y_n &= x_n - m \cdot \frac{f(x_n)}{f'(x_n)}, m \geq 1, \\ w_n &= y_n - m \cdot \left[1 + p_n + 2p_n^2\right] \cdot \frac{f(x_n)}{f'(x_n)}, \\ x_{n+1} &= w_n - m \cdot \left[1 + p_n + 2p_n^2 + (1 + 2p_n)v_n\right] \cdot \frac{f(x_n)}{f'(x_n)}. \end{aligned} \tag{4}$$

The families in Equations (1) and (3) require four evaluations of function to produce convergence of order six having efficiency index $6^{\frac{1}{4}} = 1.5650$ and therefore are not optimal in the sense of the Kung–Traub conjecture [19].

Recently, Behl et al. [20] presented a multiple root finding family of iterative methods possessing convergence order eight given as:

$$\begin{aligned} y_n &= x_n - m\frac{f(x_n)}{f'(x_n)}, m \geq 1, \\ z_n &= y_n - u_n Q(h_n)\frac{f(x_n)}{f'(x_n)}, \\ x_{n+1} &= z_n - u_n t_n G(h_n, t_n)\frac{f(x_n)}{f'(x_n)}, \end{aligned} \tag{5}$$

where the functions $Q : \mathbb{C} \to \mathbb{C}$ and $G : \mathbb{C}^2 \to \mathbb{C}$ are restricted to be analytic functions in the regions nearby (0) and $(0, 0)$, respectively, with $u_n = \left(\frac{f(y_n)}{f(x_n)}\right)^{\frac{1}{m}}$, $h_n = \frac{u_n}{a_1 + a_2 u_n}$ and $t_n = \left(\frac{f(z_n)}{f(y_n)}\right)^{\frac{1}{m}}$, being a_1 and a_2 complex non-zero free parameters.

We take Case (27) for ($a_1 = 1, a_2 = 1, G_{02} = 0$) from the family of Behl et al. [20] and represent it by BM given by:

$$\begin{aligned} y_n &= x_n - m\frac{f(x_n)}{f'(x_n)}, \\ z_n &= y_n - \left(m + 2h_n m + \frac{1}{2}h_n^2(4m + 2m)\right)\frac{f(x_n)}{f'(x_n)}u_n \\ x_{n+1} &= z_n - \left(m + mt_n + 3mh_n^2 + mh_n(2 + 4t_n + h_n)\right)\frac{f(x_n)}{f'(x_n)}u_n t_n. \end{aligned} \tag{6}$$

Most recently, another optimal eighth-order scheme presented by Zafar et al. [21] is given as:

$$\begin{aligned} y_n &= x_n - m\frac{f(x_n)}{f'(x_n)}, m \geq 1, \\ z_n &= y_n - mu_n H(u_n)\frac{f(x_n)}{f'(x_n)}, \\ x_{n+1} &= z_n - u_n t_n(B_1 + B_2 u_n)P(t_n)G(w_n)\frac{f(x_n)}{f'(x_n)}, \end{aligned} \tag{7}$$

where B_1,$B_2 \in \mathbb{R}$ are suppose to be free parameters and weight functions $H : \mathbb{C} \to \mathbb{C}$, $P : \mathbb{C} \to \mathbb{C}$ and $G : \mathbb{C} \to \mathbb{C}$ are restricted to be analytic in the regions nearby 0 with $u_n = \left(\frac{f(y_n)}{f(x_n)}\right)^{\frac{1}{m}}$, $t_n = \left(\frac{f(z_n)}{f(y_n)}\right)^{\frac{1}{m}}$ and $w_n = \left(\frac{f(z_n)}{f(x_n)}\right)^{\frac{1}{m}}$.

From the eighth-order family of Zafar et al. [21], we consider the following special case denoted by ZM:

$$\begin{aligned} y_n &= x_n - m\frac{f(x_n)}{f'(x_n)}, \\ z_n &= y_n - mu_n\left(6u_n^3 - u_n^2 + 2u_n + 1\right)\frac{f(x_n)}{f'(x_n)}, \\ x_{n+1} &= z_n - mu_n t_n\,(1 + 2u_n)\,(1 + t_n)(1 + 2w_n)\frac{f(x_n)}{f'(x_n)}. \end{aligned} \tag{8}$$

The class of iterative methods referred as optimal is significant as compared to non-optimal methods due to their speed of convergence and efficiency index. Therefore, there was a need to develop optimal eighth-order schemes for finding multiple zeros ($m > 1$) and simple zeros ($m = 1$) due to their competitive efficiencies and order of convergence [1]; in addition, fewer iterations are needed to get desired accuracy as compared to iterative methods having order four and six given by Sharma and Geum [9,10,17], respectively. In this paper, our main concern is to find the optimal iterative methods for multiple root μ with known multiplicity $m \in \mathbb{N}$ of an adequately differentiable nonlinear function $f : I \subseteq \mathbb{R} \to \mathbb{R}$, where I represents an open interval. We develop an optimal eighth-order zero finder for multiple roots with known multiplicity $m \geq 1$. The beauty of the method lies in the fact that developed scheme is simple to implement with minimum possible number of functional evaluations. Four evaluations of the function are needed to obtain a family of convergence order eighth having efficiency index $8^{\frac{1}{4}} = 1.6817$.

The rest of the paper is organized as follows. In Section 2, we present the newly developed optimal iterative family of order eight for multiple roots of nonlinear equations. The discussion of analysis of convergence is also given in this section. In Section 3, some special cases of newly developed eighth-order schemes are presented. In Section 4, numerical results and comparison of the presented schemes with existing schemes of its domain is discussed. Concluding remarks are given in Section 5.

2. Development of the Scheme and Convergence Analysis

In this section, we suggest a new family of eighth-order method with known multiplicity $m \geq 1$ of the required multiple root as follows:

$$\begin{aligned} y_n &= x_n - m \cdot \frac{f(x_n)}{f'(x_n)}, n \geq 0, \\ z_n &= y_n - m \cdot t \cdot H(t) \cdot \frac{f(x_n)}{f'(x_n)}, \\ x_{n+1} &= z_n - m \cdot t \cdot L(s,u) \cdot \frac{f(x_n)}{f'(x_n)}, \\ \text{where } t &= \sqrt[m]{\frac{f(y_n)}{f(x_n)}}, s = \sqrt[m]{\frac{f(z_n)}{f(y_n)}}, u = \sqrt[m]{\frac{f(z_n)}{f(x_n)}}. \end{aligned} \tag{9}$$

where the function $H : \mathbb{C} \to \mathbb{C}$ is restricted to be analytic function in the regions nearby 0 and weight function $L : \mathbb{C}^2 \to \mathbb{C}$ is holomorphic in the regions nearby $(0,0)$ and t, s and u are *one-to-m* multiple-valued functions.

In the next theorem, it is demonstrated that the proposed scheme in Equation (9) achieves the optimal eighth order of convergence without increasing the number of functional evaluations.

Theorem 1. *Suppose $x = \mu$ (say) is a multiple root having multiplicity $m \geq 1$ of an analytic function $f : \mathbb{C} \to \mathbb{C}$ in the region enclosing a multiple zero μ of $f(x)$. Which implies that the family of iterative methods defined by Equation (9) has convergence of order eighth when the following conditions are fulfilled:*

$$H_0 = 1, H_1 = 2, H_2 = -2, H_3 = 36, L_{00} = 0, L_{10} = 1, L_{01} = 2, L_{11} = 4, L_{20} = 2. \tag{10}$$

Then, the proposed scheme in Equation (9) satisfies the following error equation:

$$\begin{aligned} e_{n+1} &= \frac{1}{24m^7}\{c_1(c_1^2(11+m) - 2mc_2)((677 + 108m + 7m^2)c_1^4 \\ &\quad -24m(9+m)c_1^2c_2 + 12m^2c_2^2 + 12m^2c_1c_3)e_n^8\} + O(e_n^9), \end{aligned} \tag{11}$$

where $e_n = x_n - \mu$ and $c_k = \frac{m!}{(m+k)!}\frac{f^{(m+k)}(\mu)}{f^{(m)}(\mu)}, k = 1, 2, 3, \cdots$.

Proof. Suppose $x = \mu$ is a multiple root of $f(x)$. We expand $f(x_n)$ and $f'(x_n)$ by Taylor's series expansion about $x = \mu$ using Mathematica (Computer based algebra software), to get

$$f(x_n) = \frac{f^{(m)}(\mu)}{m!}e_n^m\left(1 + c_1e_n + c_2e_n^2 + c_3e_n^3 + c_4e_n^4 + c_5e_n^5 + c_6e_n^6 + c_7e_n^7 + c_8e_n^8 + O(e_n^9)\right), \tag{12}$$

and

$$\begin{aligned} f'(x_n) &= \tfrac{f^{(m)}(\mu)}{m!}e_n^{m-1}m + c_1(m+1)e_n + c_2(m+2)e_n^2 + c_3(m+3)e_n^3 + c_4(m+4)e_n^4 \\ &\quad + c_5(m+5)e_n^5 + c_6(m+6)e_n^6 + c_7(m+7)e_n^7 + c_8(m+8)e_n^8 + O(e_n^9), \end{aligned}$$

respectively. By utilizing the above Equations (11) and (12) in the first substep of Equation (9), we obtain

$$y_n - \mu = \frac{c_1e_n^2}{m} + \frac{(2c_2m - c_1^2(m+1))e_n^3}{m^2} + \sum \lim_{k=0}^{4} G_k e_n^{k+4} + O(e_n^9), \tag{13}$$

where $G_k = G_k(m, c_1, c_2, \ldots, c_8)$ are expressed in terms of $m, c_1, c_2, c_3, \ldots, c_8$ and the two coefficients $G_0$and G_1 can be explicitly written as $G_0 = \frac{1}{m^3}\{3c_3m^2 + c_1^3(m+1)^2 - c_1c_2m(3m+4)\}$ and $G_1 = -\frac{1}{m^4}\{c_1^4(m+1)^3 - 2c_2c_1^2m(2m^2+5m+3) + 2c_3c_1m^2(2m+3) + 2m^2(c_2^2(m+2) - 2c_4m)\}$. By Taylor's expansion, we get

$$f(y_n) = f^{(m)}(\mu)e_n^{2m}\left[\frac{(\frac{c_1}{m})^m}{m!} + \frac{(2mc_2 - (m+1)c_1^2)(\frac{c_1}{m})^m e_n}{c_1 m!} + \sum_{k=0}^{6} \overline{G}_k e_n^{k+2} + O(e_n^9)\right]. \tag{14}$$

By using Equations (12) and (14), we get

$$u = \frac{c_1 e_n}{m} + \frac{(2mc_2 - (m+2)c_1^2)e_n^2}{m^2} + \psi_1 e_n^3 + \psi_2 e_n^4 + \psi_3 e_n^5 + O(e_n^6), \tag{15}$$

where $\psi_1 = \frac{1}{2m^3}[c_1^3(2m^2+7m+7) + 6c_3m^2 - 2c_1c_2m(3m+7)]$, $\psi_2 = -\frac{1}{6m^4}[c_1^4(6m^3+29m^2+51m+34) - 6c_2c_1^2m(4m^2+16m+17) + 12c_1c_3m^2(2m+5) + 12m^2(c_2^2(m+3) - 2c_4m)]$, $\psi_3 = \frac{1}{24m^5}[-24m^3(c_2c_3(5m+17) - 5c_5m) + 12c_3c_1^2m^2(10m^2+43m+49) + 12c_1m^2\{c_2^2(10m^2+47m+53) - 2c_4m(5m+13)\} - 4c_2c_1^3m(30m^3+163m^2+306m+209) + c_1^5(24m^4+146m^3+355m^2+418m+209)]$.

Taylor series of $H(t)$ about 0 is given by:

$$H(t) = H_0 + H_1 t + \frac{H_2}{2!}t^2 + \frac{H_3}{3!}t^3 + O(e_n^4) \tag{16}$$

where $H_j = H^j(0)$ for $0 \le j \le 3$. Inserting Equations (13)–(16) in the second substep of the scheme in Equation (9), we get

$$\begin{aligned} z_n &= \mu + \frac{-(1+H_0)c_1e_n^2}{m} - \frac{(1+H_1+m-H_0(3+m)c_1^2) + 2(-1+H_0)mc_2)e_n^3}{m^2} \\ &\quad + \frac{1}{2m^3}\Big[(2+10H_1 - H_2 + 4m + 4H_1m + 2m^2 - H_0(13+11m+2m^2))c_1^3 \\ &\quad +2m(-4-4H_1-3m+H_0(11+3m)c_1c_2 - 6(-1+H_0)m^2c_3)e_n^4\Big] + z_5e_n^5 \\ &\quad +z_6e_n^6 + z_7e_n^7 + O(e_n^8). \end{aligned}$$

By selecting $H_0 = 1$ and $H_1 = 2$, we obtain

$$z_n = \mu + \frac{(c_1^3(9-H_2+m) - 2mc_1c_2)}{2m^3}e_n^4 + z_5e_n^5 + z_6e_n^6 + z_7e_n^7 + O(e_n^8), \tag{17}$$

where $z_5 = -\frac{1}{6m^4}\{c_1^4(125+H_3+84m+7m^2-3H_2(7+3m) + 6m(-3H_2+4(7+m))c_1^2c_2 + 12c_2^2m^2 + 12c_2c_1m)$, $z_6 = \frac{1}{24m^5}\{1507 + 1850m + 677m^2 + 46m^3 + 4H_3(9+4m) - 6H2(59+53m+12m^2))c_1^5 - 4m(925 + 8H_3 + 594m + 53m^2 - 3H_2(53+21m)c_1^3c_2 + 12m^2(83 - 9H_2 + 13m)c_1^2c_3 - 168m^3c_2c_3 + 12m^2c_1(115 - 12H2 + 17m)c_2^2 - 6mc_4)$ and $z_7 = -\{12c_1^2c_3m^2(36\beta + 13m + 11) + (37 - 168c_2c_3m^3 + 4c_1^3c_2m(96\beta^2 + 252\beta + 53m^2 + 18(14\beta+5)m) + 12c_1m^2(c_2^2(48\beta + 17m + 19) - 6c_4m)\}$.

Again, we use the Taylor's expansion for Equation (17) to get:

$$\begin{aligned} f(z_n) &= f^{(m)}(\mu)e_n^{4m}\frac{2^{-m}\left(\frac{c_1^3(9-H_2+m)-2mc_1c_2}{m^3}\right)^m}{m!} - \frac{\left(2^{-m}\left(\frac{c_1^3(9-H_2+m)-2mc_1c_2}{m^3}\right)^{m-1}\rho_0\right)}{3(m^3m!)}e_n \\ &\quad + \sum \lim\nolimits_{j=0}^{7} \overline{H}_j e_n^{j+1} + O(e_n^9), \end{aligned}$$

where $\rho_0 = c_1^4(125 + H_3 + 84m + 7m^2 - 3H_2(7+3m))c_1^4 - 6m(-3H2 + 4(7+m))c_1^2c_2 + 12m^2c_2^2 + 12c_3c_1m^2)$. With the help of Equations (12) and (18), we have

$$s = \frac{c_1^2(9 - H_2 + m) - 2mc_2}{2m^2}e_n^2 + \rho_1 e_n^3 + \rho_2 e_n^4 + \rho_3 e_n^5 + O(e_n^6), \tag{18}$$

where

$$\rho_1 = -\frac{1}{6m^3}\{c_1^3(98 + H_3 + 4m^2 + 54m - 6H_2(3+m) - 12m(9 - H_2 + m)c_1c_2 + 12m^2c_3\},$$

$$\rho_2 = \frac{1}{24m^4}899 + 1002m + 313m^2 + 18m^3 + 4H_3(8+3m) - 6H_2(43 + 33m + 6m^2))c_1^4 -$$
$$12m(167 + 2H_3 + 87m + 6m^2 - H_2(33+10m)c_1^2c_2 + 24m^2(26 - 3H_2 + 3m)c_1c_3 +$$
$$12m^2(c_2^2(35 - 4H_2 + 3m) - 6mc_4)$$

and $\rho_3 = -\frac{1}{60m^5}[-4257 - 7270m - 4455m^2 - 101m^3 - 48m^4 - 10H_3(37 + 30m + 6m^2) + 30H_2(60 + 75m + 31m^2 + 4m^3)c_1^5 + 10m(1454 + 60H_3 + 1548m + 21H_3m + 454m^2 + 24m^3 - 18H_2(25 + 18m + 3m^2)c_1^3c_2 - 30m^2(234 + 3H_3 + 118m + 8m^2 - 2H2(24 + 7m)c_1^2c_3 - 60m^2c_1(141 + 2H_3 + 67m + 4m^2 - 2H_2(15 + 4m)c_2^2 + 2(-17 + 2H_2 - 2m)mc_4) - 120m^3(-25 + 3H_2 - 2m)c_2c_3 + 2mc_5\} + (\frac{1}{720m^6})((102047 + 180H_2^2 + 204435m + 187055m^2 + 81525m^3 + 14738m^4 + 600m^5 + 40H_3(389 + 498m + 214m^2 + 30m^3) - 45H_2(1223 + 2030m + 1353m^2 + 394m^3 + 40m^4)) - 30m(13629 + 22190m + 12915m^2 + 2746m^3 + 120m^4 + 16H_3(83 + 64m + 12m^2) - 6H_2(1015 + 1209m + 470m^2 + 56m^3)) + 120m^2(2063 + 2088m + 589m^2 + 30m^3 + H_3(88 + 30m) - 18H_2 + (36 + 25m + 4m^2)) + 80m^2(2323 + 2348m + 635m^2 + 30m^3 + 4H_3(28 + 9m) - 3H_2(259 + 173m + 26m^2)) - 2m(303 + 4H_3 + 149m + 10m^2 - 9H_2(7 + 2m)) - 720m^3((393 + 6H_3 + 178m + 10m^2 - H_2(87 + 22m))] + (-42 + 5H_2 - 5m)mc_5) + 20m^3((-473 - 8H_3 - 195m - 10m^2 + 12H_2(9 + 2m))c_2c_3 + 6m(65 - 8H2 + 5m)c_2 + 3m(71 - 9H2 + 5m)c_{10}mc_6$.

Since it is obvious from Equation (15) that u possess order e_n, the expansion of weight function $L_f(s,u)$ by Taylor's series is possible in the regions nearby origin given as follows:

$$L(s,u) = L_{00} + sL_{10} + uL_{01} + suL_{11} + \frac{s^2}{2!}L_{20} \tag{19}$$

where $L_{i,j} = \frac{1}{i!j!}\frac{\partial^{i+j}}{\partial s^j \partial u^j}L(s,u)\Big|_{(0,0)}$. By using Equations (12)–(19) in the proposed scheme in Equation (9), we have

$$e_{n+1} = M_2e_n^2 + M_3e_n^3 + M_4e_n^4 + M_5e_n^5 + M_6e_n^6 + M_7e_n^7 + O(e_n^8), \tag{20}$$

where the coefficients $M_i(2 \le i \le 7)$ depend generally on m and the parameters $L_{i,j}$. To obtain at least fifth-order convergence, we have to choose $L_{00} = 0, L_{10} = 1$ and get

$$e_{n+1} = \frac{((-2 + L_{01})c_1^2((-9 + H_2 - m)c_1^2 + 2mc_2)}{2m^4}e_n^5 + \bar{M}_6e_n^6 + \bar{M}_7e_n^7 + O(e_n^8).$$

where the coefficients $\bar{M}_i(6 \le i \le 7)$ depend generally on m and the parameters $L_{i,j}$. To obtain eighth-order convergence, we are restricted to choosing the values of parameters given by:

$$H_2 = -2, H_3 = 36, L_{00} = 0, L_{10} = 1, L_{01} = 2, L_{20} = 2, L_{11} = 4. \tag{21}$$

This leads us to the following error equation:

$$e_{n+1} = \frac{1}{24m^7}[c_1(c_1^2(11 + m) - 2mc_2)((677 + 108m + 7m^2)c_1^4 - 24m(9 + m)c_1^2c_2 + 12m^2c_2^2 + 12m^2c_1c_3)]e_n^8 + O(e_n^9). \tag{22}$$

The above error equation (Equation (22)) confirms that the presented scheme in Equation (9) achieves optimal order of convergence eight by utilizing only four functional evaluations (using $f(x_n), f'(x_n), f(y_n)$ and $f(z_n)$) per iteration. □

3. Special Cases of Weight Functions

From Theorem 1, several choices of weight functions can be obtained. We have considered the following:

Case 1: The polynomial form of the weight function satisfying the conditions in Equation (10) can be represented as:

$$\begin{aligned} H(t) &= 1+2t-t^2+6t^3 \\ L(s,u) &= s+2u+4su+s^2 \end{aligned} \tag{23}$$

The particular iterative method related to Equation (23) is given by:
SM-1:

$$\begin{aligned} y_n &= x_n - m\cdot\frac{f(x_n)}{f'(x_n)}, n\geq 0, \\ z_n &= y_n - m\cdot t\cdot(1+2t-t^2+6t^3)\frac{f(x_n)}{f'(x_n)}, \\ x_{n+1} &= z_n - m\cdot t\cdot(s+s^2+2u+4su)\cdot\frac{f(x_n)}{f'(x_n)} \\ \text{where } t &= \sqrt[m]{\frac{f(y_n)}{f(x_n)}}, s=\sqrt[m]{\frac{f(z_n)}{f(y_n)}}, u=\sqrt[m]{\frac{f(z_n)}{f(x_n)}} \end{aligned} \tag{24}$$

Case 2: The second suggested form of the weight functions in which $H(t)$ is constructed using rational weight function satisfying the conditions in Equation (10) is given by:

$$\begin{aligned} H(t) &= \frac{1+8t+11t^2}{1+6t} \\ L(s,u) &= s+2u+4su+s^2 \end{aligned} \tag{25}$$

The corresponding iterative method in Equation (25) can be presented as:
SM-2:

$$\begin{aligned} y_n &= x_n - m\cdot\frac{f(x_n)}{f'(x_n)}, n\geq 0, \\ z_n &= y_n - m\cdot t\cdot\left(\frac{1+8t+11t^2}{1+6t}\right)\frac{f(x_n)}{f'(x_n)}, \\ x_{n+1} &= z_n - m\cdot t\cdot(s+s^2+2u+4su)\cdot\frac{f(x_n)}{f'(x_n)} \\ \text{where } t &= \sqrt[m]{\frac{f(y_n)}{f(x_n)}}, s=\sqrt[m]{\frac{f(z_n)}{f(y_n)}}, u=\sqrt[m]{\frac{f(z_n)}{f(x_n)}} \end{aligned} \tag{26}$$

Case 3: The third suggested form of the weight function in which $H(t)$ is constructed using trigonometric weight satisfying the conditions in Equation (10) is given by:

$$\begin{aligned} H(t) &= \frac{5+18t}{5+8t-11t^2} \\ L(s,u) &= s+2u+4su+s^2 \end{aligned} \tag{27}$$

The corresponding iterative method obtained using Equation (27) is given by:
SM-3:

$$\begin{aligned} y_n &= x_n - m \cdot \frac{f(x_n)}{f'(x_n)}, n \geq 0, \\ z_n &= y_n - m \cdot t \cdot (\frac{5+18t}{5+8t-11t^2}) \frac{f(x_n)}{f'(x_n)}, \\ x_{n+1} &= z_n - m \cdot t \cdot (s + s^2 + 2u + 4su) \cdot \frac{f(x_n)}{f'(x_n)} \\ \text{where } t &= \sqrt[m]{\frac{f(y_n)}{f(x_n)}}, s = \sqrt[m]{\frac{f(z_n)}{f(y_n)}}, u = \sqrt[m]{\frac{f(z_n)}{f(x_n)}}. \end{aligned} \tag{28}$$

4. Numerical Tests

In this section, we show the performance of the presented iterative family in Equation (9) by carrying out some numerical tests and comparing the results with existing method for multiple roots. All numerical computations were performed in Maple 16 programming package using 1000 significant digits of precision. When μ was not exact, we preferred to take the accurate value which has larger number of significant digits rather than the assigned precision. The test functions along with their roots μ and multiplicity m are listed in Table 1 [22]. The proposed methods SM-1 (Equation (24)), SM-2 (Equation (26)) and SM-3 (Equation (28)) are compared with the methods of Geum et al. given in Equations (2) and (4) denoted by GKM-1 and GKM-2 and with method of Bhel given in Equation (6) denoted by BM and Zafar et al. method given in Equation (8) denoted by ZM. In Tables 1–8, the error in first three iterations with reference to the sought zeros ($|x_n - \mu|$) is considered for different methods. The notation $E(-i)$ can be considered as $E \times 10^{-i}$. The test function along with their initial estimates x_0 and computational order of convergence (COC) is also included in these tables, which is computed by the following expression [23]:

$$COC \approx \frac{\log|(x_{k+1}-\mu)/(x_k-\mu)|}{\log|(x_k-\mu)/(x_{k-1}-\mu)|}.$$

Table 1. Test functions.

Test Functions	Exact Root μ	Multiplicity m
$f_1(x) = (cos(\frac{\pi x}{2}) + x^2 - \pi)^5$	2.034724896...	5
$f_2(x) = (e^x + x - 20)^2$	2.842438953...	2
$f_3(x) = (\ln x + \sqrt{(x^4+1)} - 2)^9$	1.222813963...	9
$f_4(x) = (cosx - x)^3$	0.7390851332...	3
$f_5(x) = ((x-1)^3 - 1)^{50}$	2.0	50
$f_6(x) = (x^3 + 4x^2 - 10)^6$	1.365230013...	6
$f_7(x) = (8xe^{-x^2} - 2x - 3)^8$	−1.7903531791...	8

Table 2. Comparison of different methods for multiple roots.

$f_1(x), x_0 = 2.5$							
	GKM-1	GKM-2	SM-1	SM-2	SM-3	ZM	BM
$\|x_1 - \mu\|$	6.83(−4)	1.11(−3)	2.15(−4)	1.87(−4)	2.03(−4)	1.52(−4)	1.84(−4)
$\|x_2 - \mu\|$	3.42(−14)	2.53(−18)	2.37(−29)	3.53(−30)	1.25(−29)	9.69(−31)	2.89(−30)
$\|x_3 - \mu\|$	2.13(−55)	3.58(−106)	5.28(−229)	5.71(−236)	2.53(−231)	2.56(−240)	1.05(−236)
COC	4.00	6.00	8.00	8.00	8.00	8.00	8.00

Table 3. Comparison of different methods for multiple roots.

$f_2(x), x_0 = 3.0$							
	GKM-1	**GKM-2**	**SM-1**	**SM-2**	**SM-3**	**ZM**	**BM**
$\|x_1 - \mu\|$	1.18(−7)	5.27(−6)	2.33(−7)	1.21(−7)	1.90(−7)	1.40(−7)	1.16(−7)
$\|x_2 - \mu\|$	2.62(−37)	1.15(−32)	1.30(−53)	2.21(−56)	1.99(−54)	1.30(−55)	1.57(−56)
$\|x_3 - \mu\|$	3.07(−221)	1.25(−192)	1.19(−423)	2.67(−446)	2.87(−430)	7.37(−440)	1.73(−447)
COC	6.00	6.00	8.00	8.00	8.00	8.00	8.00

Table 4. Comparison of different methods for multiple roots.

$f_3(x), x_0 = 3.0$							
	GKM-1	**GKM-2**	**SM-1**	**SM-2**	**SM-3**	**ZM**	**BM**
$\|x_1 - \mu\|$	5.50(−1)	4.29(−2)	1.81(−2)	1.75(−2)	1.79(−2)	D *	D
$\|x_2 - \mu\|$	3.99(−7)	8.77(−10)	2.82(−15)	9.58(−16)	2.04(−15)	D	D
$\|x_3 - \mu\|$	1.13(−27)	7.51(−56)	2.06(−117)	8.21(−122)	6.49(−119)	D	D
COC	4.00	6.00	8.00	8.00	8.00	D	D

* D stands for divergence.

Table 5. Comparison of different methods for multiple roots.

$f_4(x), x_0 = 1.0$							
	GKM-1	**GKM-2**	**SM-1**	**SM-2**	**SM-3**	**ZM**	**BM**
$\|x_1 - \mu\|$	2.77(−4)	2.55(−5)	6.78(−8)	5.45(−8)	6.29(−8)	4.90(−8)	5.15(−8)
$\|x_2 - \mu\|$	3.28(−14)	6.83(−36)	7.95(−60)	8.55(−61)	3.83(−60)	4.06(−61)	4.91(−61)
$\|x_3 - \mu\|$	5.86(−49)	2.51(−213)	2.82(−475)	3.11(−483)	7.18(−478)	8.99(−486)	3.36(−485)
COC	3.50	6.00	8.00	8.00	8.00	7.99	7.99

Table 6. Comparison of different methods for multiple roots.

$f_5(x), x_0 = 2.1$							
	GKM-1	**GKM-2**	**SM-1**	**SM-2**	**SM-3**	**ZM**	**BM**
$\|x_1 - \mu\|$	7.68(−5)	1.12(−5)	7.58(−7)	4.85(−7)	6.52(−7)	4.77(−7)	4.65(−7)
$\|x_2 - \mu\|$	3.49(−17)	5.33(−29)	3.70(−47)	4.10(−49)	8.82(−48)	5.66(−49)	2.72(−49)
$\|x_3 - \mu\|$	1.46(−66)	6.11(−169)	1.19(−369)	1.06(−385)	9.93(−375)	2.22(−384)	3.79(−387)
COC	3.99	6.00	8.00	8.00	8.00	7.99	7.99

Table 7. Comparison of different methods for multiple roots.

$f_6(x), x_0 = 3.0$							
	GKM-1	**GKM-2**	**SM-1**	**SM-2**	**SM-3**	**ZM**	**BM**
$\|x_1 - \mu\|$	5.44(−2)	1.01(−1)	5.40(−2)	5.30(−2)	5.36(−2)	4.36(−2)	5.39(−2)
$\|x_2 - \mu\|$	7.40(−7)	5.37(−7)	1.10(−10)	4.72(−11)	8.60(−11)	1.36(−11)	4.92(−11)
$\|x_3 - \mu\|$	3.54(−26)	1.86(−38)	5.28(−80)	2.43(−83)	5.76(−81)	1.80(−87)	3.14(−83)
COC	3.97	5.96	8.00	7.98	7.97	7.97	7.97

Table 8. Comparison of different methods for multiple roots.

$f_7(x), x_0 = -1.2$							
	GKM-1	**GKM-2**	**SM-1**	**SM-2**	**SM-3**	**ZM**	**BM**
$\|x_1 - \mu\|$	2.65(−3)	2.15(−3)	4.38(−4)	4.24(−4)	4.32(−4)	3.41(−4)	4.26(−4)
$\|x_2 - \mu\|$	7.24(−12)	9.63(−17)	4.44(−27)	1.11(−27)	3.11(−27)	3.58(−28)	1.14(−27)
$\|x_3 - \mu\|$	4.05(−46)	7.81(−97)	4.97(−211)	2.55(−216)	2.28(−212)	5.27(−220)	3.06(−216)
COC	4.00	6.00	8.00	8.00	8.00	7.99	7.99

It is observed that the performance of the new method SM-2 is the same as BM for the function f_1 and better than ZM for the function f_2. The newly developed schemes SM-1, SM-2 and SM-3 are not only convergent but also their speed of convergence is better than GKM-1 and GKM-2 while ZM and BM show divergence for the function f_3. For functions f_4, f_5, f_6 and f_7, the newly developed schemes SM-1, SM-2 and SM-3 are comparable with ZM and BM. Hence, we conclude that the proposed family is comparable and robust among existing methods for multiple roots.

5. Dynamical Analysis

For the sake of stability comparison, we plot the dynamical planes corresponding to each scheme (SM-1, SM-2, SM-3, BM and ZM) for the nonlinear functions $f_1, f_2, f_3, f_4, f_5, f_6, f_7$ by using the procedure described in [24]. We draw a mesh of 400×400 points such that each point of the mesh is an initial-approximation of the required root of corresponding nonlinear function. The point is orange if the sequence of iteration method converges to the multiple root (with tolerance 10^{-3}) in fewer than 80 iterations and the point is black if the sequence does not converges to the multiple root. The multiple zero is represented by a white star in the figures. Figures 1–14 show that the basin of attraction drawn in orange is of the multiple zero only (i.e., a set of initial guesses converging to the multiple roots fills all the plotted regions of the complex plane). In general, convergence to other zeros or divergence can appear (referred to as strange stationary points). SM-1 has wider regions of convergence for f_1 as compared to ZM and BM in Figures 1 and 2; SM-1 and SM-3 have wider regions of convergence for f_2 as compared to ZM and BM in Figures 3 and 4. The convergence region of SM-2 for functionsf_3, f_4 and f_6 is comparable with ZM and BM, as shown in Figures 5–8, 11 and 12. For function f_5 in Figures 9 and 10, the convergence region of SM-3 is better than ZM and BM. For function f_7, SM-1 and SM-3 have better convergence regions than ZM and BM, as shown in Figures 13 and 14. Figures 1–14 show that the region in orange is comparable or bigger for the presented methods SM-1, SM-2 and SM-3 than the regions obtained by schemes BM and ZM, which confirms the fast convergence and stability of the proposed schemes.

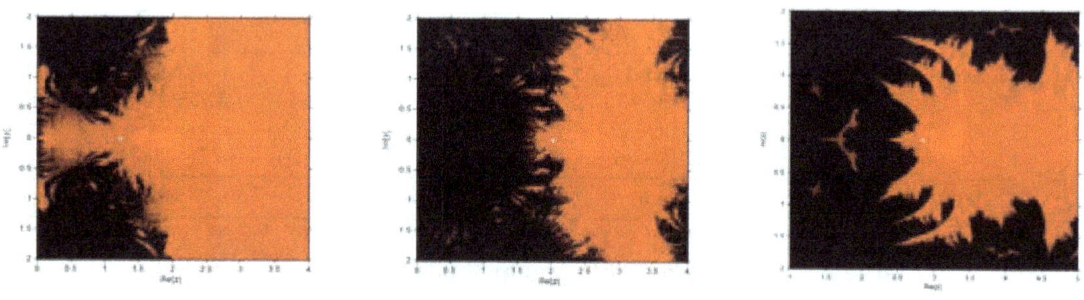

Figure 1. Basins of attraction of SM1 (**Left**), SM2 (**Middle**) and SM3 (**Right**) for $f_1(x)$.

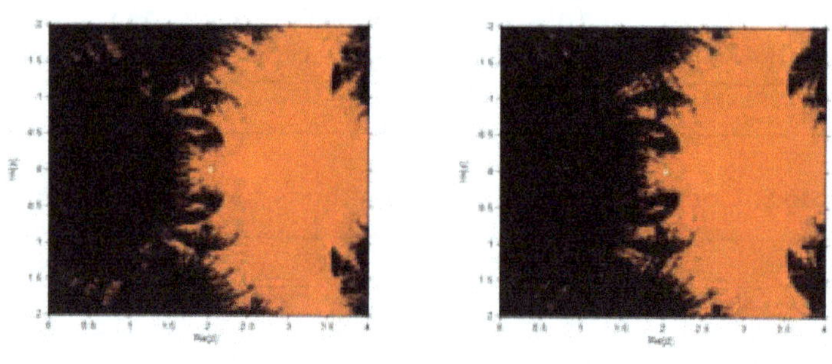

Figure 2. Basins of attraction of BM (**Left**) and ZM (**Right**) for $f_1(x)$.

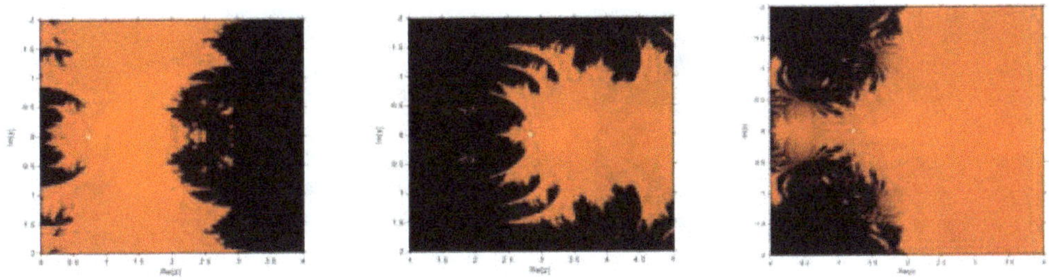

Figure 3. Basins of attraction of SM1 (**Left**), SM2 (**Middle**) and SM3 (**Right**) for $f_2(x)$.

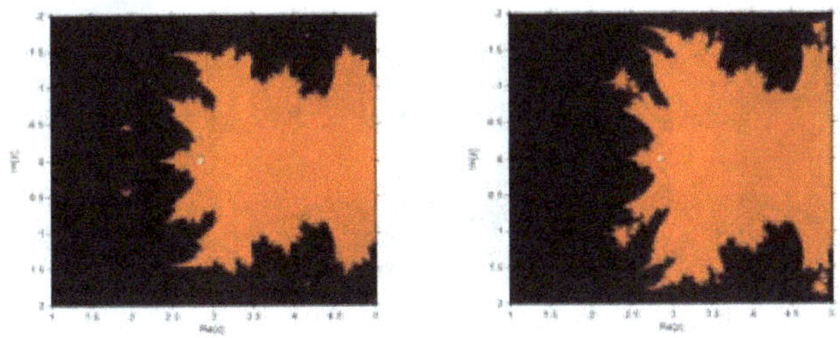

Figure 4. Basins of attraction of BM (**Left**) and ZM (**Right**) for $f_2(x)$.

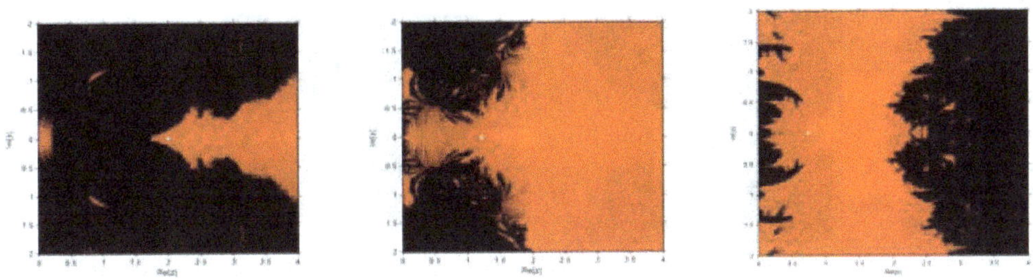

Figure 5. Basins of attraction of SM1 (**Left**), SM2 (**Middle**) and SM3 (**Right**) for $f_3(x)$.

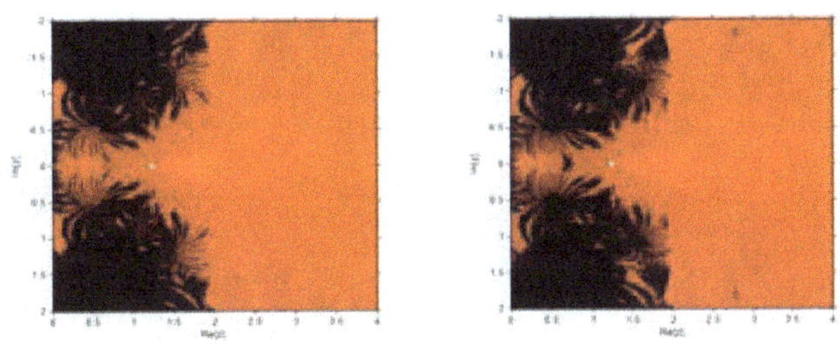

Figure 6. Basins of attraction of BM (**Left**) and ZM (**Right**) for $f_3(x)$.

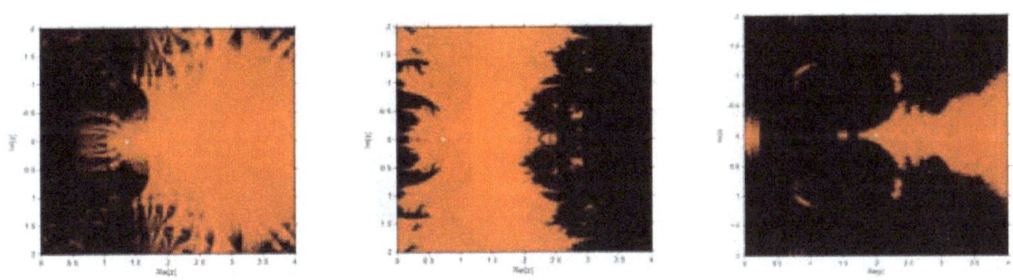

Figure 7. Basins of attraction of SM1 (**Left**), SM2 (**Middle**) and SM3 (**Right**) for $f_4(x)$.

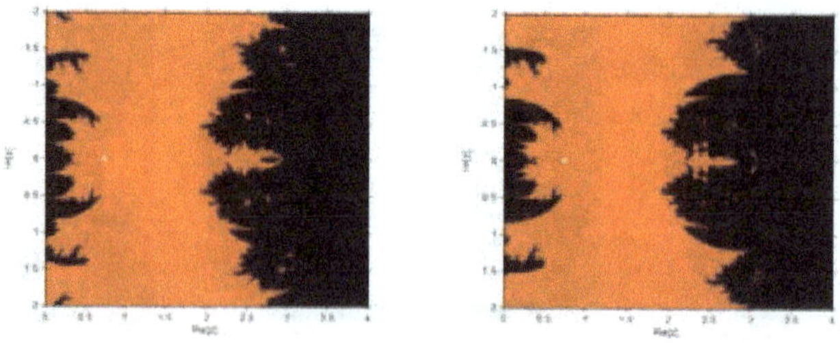

Figure 8. Basins of attraction of BM (**Left**) and ZM (**Right**) for $f_4(x)$.

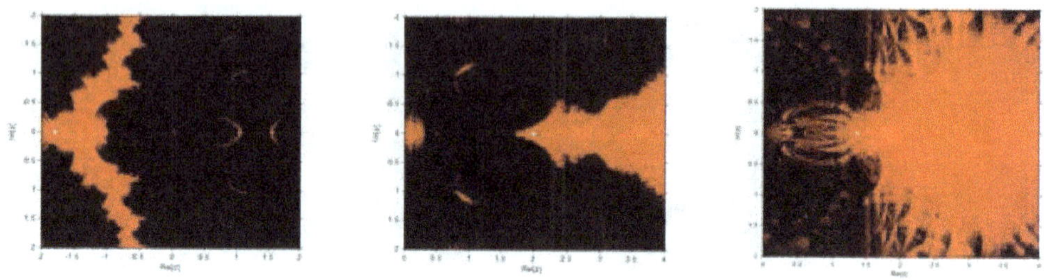

Figure 9. Basins of attraction of SM1 (**Left**), SM2 (**Middle**) and SM3 (**Right**) for $f_5(x)$.

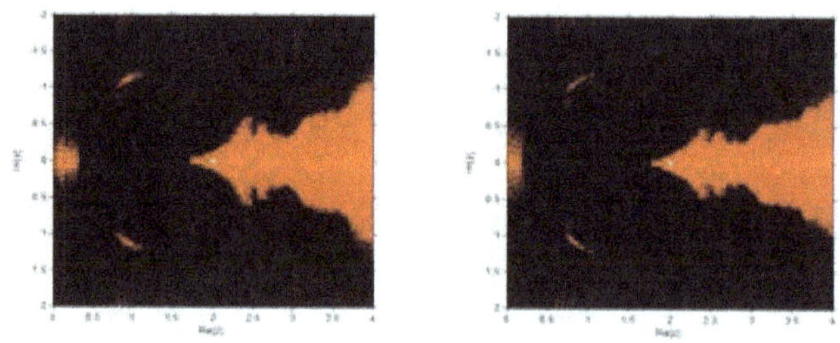

Figure 10. Basins of attraction of BM (**Left**) and ZM (**Right**) for $f_5(x)$.

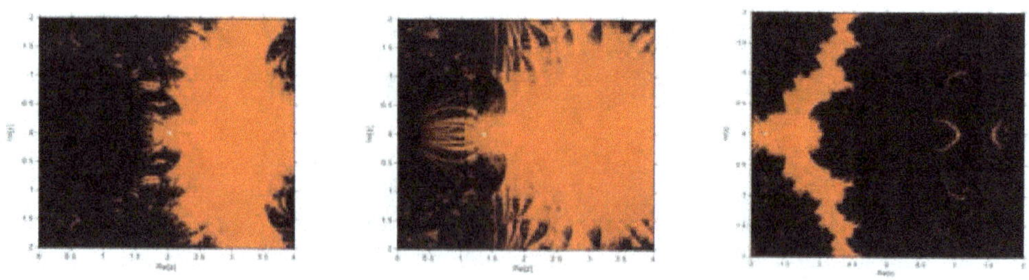

Figure 11. Basins of attraction of SM1 (**Left**), SM2 (**Middle**) and SM3 (**Right**) for $f_6(x)$.

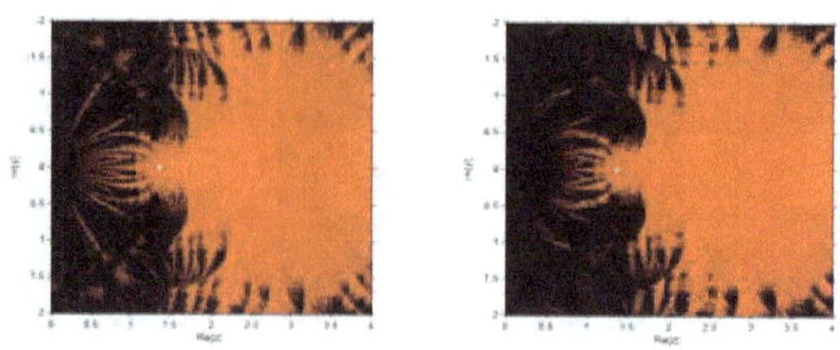

Figure 12. Basins of attraction of BM (**Left**) and ZM (**Right**) for $f_6(x)$.

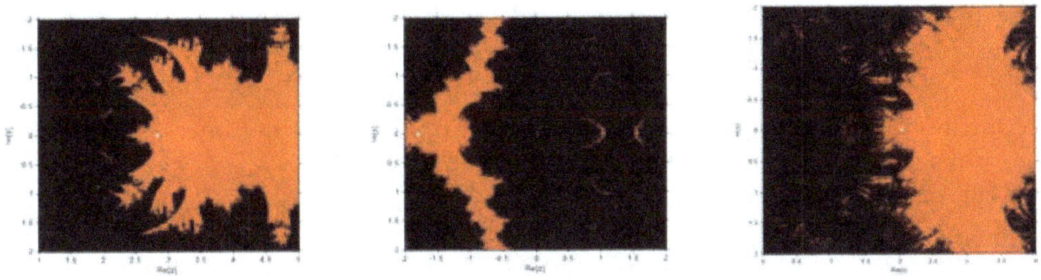

Figure 13. Basins of attraction of SM1 (**Left**), SM2 (**Middle**) and SM3 (**Right**) for $f_7(x)$.

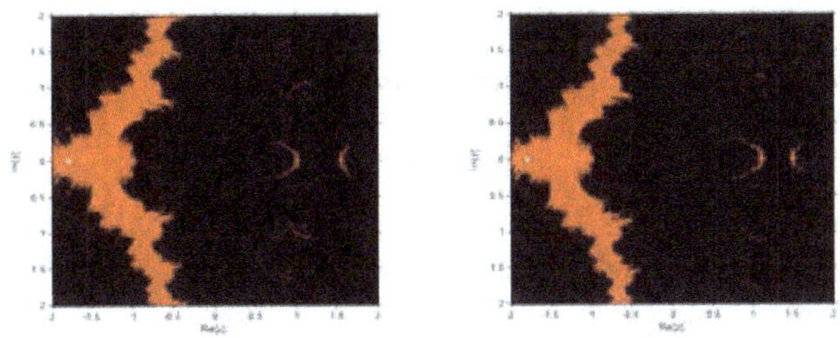

Figure 14. Basins of attraction of BM (**Left**) and ZM (**Right**) for $f_7(x)$.

6. Conclusions

In this paper, we present a new family of optimal eighth-order schemes to find multiple roots of nonlinear equations. An extensive convergence analysis is done, which verifies that the new family is optimal eighth-order convergent. The presented family required four functional evaluations to get optimal eighth-order convergence, having efficiency index $8^{\frac{1}{4}} = 1.6817$, which is higher than the efficiency index of the methods for multiple roots and of the families of Geum et al. [9,10]. Finally, numerical and dynamical tests confirmed the theoretical results and showed that the three members SM-1, SM-2 and SM-3 of the new family are better than existing methods for multiple roots. Hence, the proposed family is efficient among the domain of multiple root finding methods.

Author Contributions: Methodology, S.A.; writing—original draft preparation, S.A.; investigation, S.A.; writing—review and editing, F.Z. and N.Y.; and supervision, F.Z. and N.Y.

References

1. Ostrowski, A.M. *Solution of Equations and Systems of Equations*; Academic Press: New York, NY, USA, 1960.
2. Petkovic, M.S.; Neta, B.; Petkovic, L.D.; Dzunic, J. *Multipoint Methods for Solving Nonlinear Equations*; Academic Press: New York, NY, USA, 2013.
3. Traub, J.F. *Iterative Methods for the Solution of Equations*; Prentice-Hall: Englewood Cliffs, NJ, USA, 1964.
4. Schröder, E. Über unendlich viele algorithmen zur auflösung der gleichungen. *Math. Annal.* **1870**, 2, 317–365. [CrossRef]
5. Singh, A.; Jaiswal, P. An efficient family of optimal fourth-order iterative methods for finding multiple roots of nonlinear equations. *Proc. Natl. Acad. Sci. India Sec. A* **2015**, *85*, 439–450. [CrossRef]
6. Behl, R.; Cordero, A.; Motsa, S.S.; Torregrosa, J.R.; Kanwar, V. An optimal fourth-order family of methods for multiple roots and its dynamics. *Numer. Algorithms* **2016**, *71*, 775–796. [CrossRef]
7. Biazar, J.; Ghanbari, B. A new third-order family of nonlinear solvers for multiple roots. *Comput. Math. Appl.* **2010**, *59*, 3315–3319. [CrossRef]
8. Chun, C.; Bae, H.J.; Neta, B. New families of nonlinear third-order solvers for finding multiple roots. *Comput. Math. Appl.* **2009**, *57*, 1574–1582. [CrossRef]
9. Geum, Y.H.; Kim, Y.I.; Neta, B. A class of two-point sixth-order multiple-zero finders of modified double-Newton type and their dynamics. *Appl. Math. Comput.* **2015**, *270*, 387–400. [CrossRef]

10. Geum, Y.H.; Kim, Y.I.; Neta, B. A sixth-order family of three-point modified Newton-like multiple-root finders and the dynamics behind their extraneous fixed points. *Appl. Math. Comput.* **2016**, *283*, 120–140. [CrossRef]
11. Hueso, J.L.; Martinez, E.; Treuel, C. Determination of multiple roots of nonlinear equation and applications. *J. Math. Chem.* **2015**, *53*, 880–892. [CrossRef]
12. Lee, S.; Choe, H. On fourth-order iterative methods for multiple roots of nonlinear equation with high efficiency. *J. Comput. Anal. Appl.* **2015**, *18*, 109–120.
13. Lin, R.I.; Ren, H.M.; Šmarda, Z.; Wu, Q.B.; Khan, Y.; Hu, J.L. New families of third-order iterative methods for finding multiple roots. *J. Appl. Math.* **2014**, *2014*. [CrossRef]
14. Li, S.; Cheng, L.; Neta, B. Some fourth-order nonlinear solvers with closed formulae for multiple roots. *Comput. Math. Appl.* **2010**, *59*, 126–135. [CrossRef]
15. Ahmad, N.; Singh, V.P. Some New Three step Iterative methods for solving nonlinear equation using Steffensen's and Halley method. *Br. J. Math. Comp. Sci.* **2016**, *19*, 1–9. [CrossRef] [PubMed]
16. Neta, B. Extension of Murakami's high-order non-linear solver to multiple roots. *Int. J. Comput. Math.* **2010**, *87*, 1023–1031. [CrossRef]
17. Sharma, J.R.; Sharma, R. Modified Jarratt method for computing multiple roots. *Appl. Math. Comput.* **2010**, *217*, 878–881. [CrossRef]
18. Zhou, X.; Chen, X.; Song, Y. Constructing higher-order methods for obtaining the multiple roots of nonlinear equations. *J. Comput. Appl. Math.* **2011**, *235*, 4199–4206. [CrossRef]
19. Kung, H.T.; Traub, J.F. Optimal order of one-point and multipoint iteration. *J. Assoc. Comput. Mach.* **1974**, *21*, 643–651. [CrossRef]
20. Behl, R.; Cordero, A.; Motsa, S.S.; Torregrosa, J.R. An eighth-order family of optimal multiple root finders and its dynamics. *Numer. Algorithms* **2018**, *77*, 1249–1272. [CrossRef]
21. Zafar, F.; Cordero, A.; Quratulain, R.; Torregrosa, J.R. Optimal iterative methods for finding multiple roots of nonlinear equations using free parameters. *J. Math. Chem.* **2017**, *56*, 1884–1091. [CrossRef]
22. Neta, B.; Chun, C.; Scott, M. On the development of iterative methods for multiple roots. *Appl. Math. Comput.* **2013**, *224*, 358–361. [CrossRef]
23. Weerakoon, S.; Fernando, T.G.I. A variant of Newton's method with accelerated third-order convergence. *Appl. Math. Lett.* **2000**, *13*, 87–93. [CrossRef]
24. Chicharro, F.I.; Cordero, A.; Torregrosa, J.R. Drawing dynamical and pa- rameters planes of iterative families and methods. *Sci. World J.* **2013**, *2013*. [CrossRef] [PubMed]

14

An Improved Curvature Circle Algorithm for Orthogonal Projection onto a Planar Algebraic Curve

Zhinan Wu [1,†] **and Xiaowu Li** [2,*,†]

1 School of Mathematics and Computer Science, Yichun University, Yichun 336000, China; zhi_nan_7@163.com
2 College of Data Science and Information Engineering, Guizhou Minzu University, Guiyang 550025, China
* Correspondence: lixiaowu002@126.com.
† These authors contributed equally to this work.

Abstract: Point orthogonal projection onto planar algebraic curve plays an important role in computer graphics, computer aided design, computer aided geometric design and other fields. For the case where the test point $\mathbf{p}$ is very far from the planar algebraic curve, we propose an improved curvature circle algorithm to find the footpoint. Concretely, the first step is to repeatedly iterate algorithm (the Newton's steepest gradient descent method) until the iterated point could fall on the planar algebraic curve. Then seek footpoint by using the algorithm (computing footpoint $\mathbf{q}$) where the core technology is the curvature circle method. And the next step is to orthogonally project the footpoint $\mathbf{q}$ onto the planar algebraic curve by using the algorithm (the hybrid tangent vertical foot algorithm). Repeatedly run the algorithm (computing footpoint $\mathbf{q}$) and the algorithm (the hybrid tangent vertical foot algorithm) until the distance between the current footpoint and the previous footpoint is near 0. Furthermore, we propose Second Remedial Algorithm based on Comprehensive Algorithm B. In particular, its robustness is greatly improved than that of Comprehensive Algorithm B and it achieves our expected result. Numerical examples demonstrate that Second Remedial Algorithm could converge accurately and efficiently no matter how far the test point is from the plane algebraic curve and where the initial iteration point is.

Keywords: point projection; intersection; planar algebraic curve; Newton's iterative method; the improved curvature circle algorithm

1. Introduction

Reconstructing curve/surface is an important work in the field of computer aided geometric design, especially in geometric modeling and processing where it is crucial to fit curve/surface in high accuracy and reduce the error of representation curve/surface. The representation of the four curve types are the explicit-type, implicit-type, parametric-type and subdivision-type. Because implicit representation has unique advantage in the process of computer aided geometric design, it has wide and far-reaching applications. From scattered and unorganized three-dimensional data, Bajaj et al. [1] reconstructed surface and functions on surfaces. They [2,3] have constructed the algebraic B-spline surfaces with least-squares fitting feature using tensor product technique. Schulz et al. [4] constructed an enveloping algebraic surface using gradually approximate algebraization method. Kanatani et al. [5] applied the algebraic curve to construct geometric ellipse fitting using unified strict maximum likelihood estimation method. Mullen et al. [6] reconstructed robust and accurate algebraic surface functions to sign the unsigned from scattered and unorganized three-dimensional data point sets. Upreti et al. [7] used a technique to sign algebraic level sets on NURBS surface and algebraic Boolean level sets on NURBS surfaces. Rouhani et al. [8] applied the algebraic function for polynomial representation system. And L.G. Zagorchev et al. [9] applied the algebraic function for general algebraic surface.

Up to now, there are three main types of methods to solve the problem of point orthogonal projection onto planar algebraic curve: local method, global method and compromise method between these two methods. Here are three typical approaches.

According to the most basic geometric characteristic, orthogonal projection of test point $\mathbf{p}$ onto the planar algebraic curve is actually the point $\mathbf{x}$ on the curve such that cross product of vectors $\overrightarrow{\mathbf{xp}}$ and $\nabla f(\mathbf{x})$ is 0.

$$\nabla f(\mathbf{x}) \times (\mathbf{p} - \mathbf{x}) = 0. \tag{1}$$

Equation (1) can be transformed into Newton's iteration formula (3). Furthermore, Sullivan et al. [10] adopted a hybrid method with Lagrange multiplier and Newton's iterative method to compute the closest point on the planar algebraic curve for each test point. Some orthogonal projection problems can be transformed into solving system of nonlinear equations. The common characteristic of methods [10,11] is that they converge locally and fast, while methods [10,11] are dependent on the initial points.

The first global method of solving system of nonlinear equations is the Homotopy continuous method [12,13]. They constructed Homotopy continuous formula.

$$H(\mathbf{x}, t) = (1 - t)P(\mathbf{x}) + tQ(\mathbf{x}), \quad t \in [0, 1] \tag{2}$$

where t is a parameter of continuous transformation from 0 to 1, $P(\mathbf{x}) = 0$ is the original system of nonlinear equations to be solved, $Q(\mathbf{x})$ is the objective solution of system of nonlinear equations $P(\mathbf{x}) = 0$. All isolated solutions of system of nonlinear equations $P(\mathbf{x}) = 0$ can be computed by the numerical continuous Homotopy methods [12,13]. So the Homotopy methods [12,13] are global convergence. The Homotopy methods' robustness is proved by [14] and their high time-consuming property is verified in [15]. Of course, the Homotopy methods [12,13] are ideal in theory, but it is difficult to find or construct the objective system of nonlinear equations $Q(\mathbf{x}) = 0$ in practical engineering applications.

The second global resultant methods convert system of nonlinear equations into the expression of the resultants and then solve the resultants [16–19]. According to classical elimination theory, system of two nonlinear equations with two variables can be turned into a resultant polynomial with one variable, which is equivalent to the two simultaneous equations. The Sylvester's resultant and Cayley's statement of Bézout's method are the most famous resultant methods [16–19]. Because the resultant methods [16–19] can solve all roots if the degree of the planar algebraic curve is less than 4, they are good global methods. However, if the degree of the planar algebraic curve is more than quintic, it becomes harder and harder with increasing degree to solve two-polynomial system with the resultant methods.

The third global method is the adoption of the Bézier clipping technique [20–22]. In the first step, solving the nonlinear system of Equation (1) is transformed into solving all roots of Bernstein-Bézier representation with convex hull property. In the second step, if the parts of the domains do not include the solution, we clip the parts of the domains by using convex hull box with Bernstein-Bézier form such that the discarded parts of the region has no solution and all the solutions are in the retained parts of the region. In the third step, the de Casteljau subdivision rule is used to segment the remaining part of the curve obtained by elimination in step 2. Repeat steps 2 and 3 until we can find all the solutions to Equation (1). The advantage of this method is that all solutions of Equation (1) can be found. But this global clipping method has one difficulty: sometimes Equation (1) is difficult or even impossible to convert into Bernstein-Bézier form. For example, specific Equation (1) $\nabla f(\mathbf{x}) \times (\mathbf{p} - \mathbf{x}) = -36p_1y^{17} + 36xy^{17} + 6p_2x^5 - 6x^5y - 4p_1x + 4p_2y + 4x^2 - 4y^2$ is impossible to convert into Bernstein-Bézier form where $f(\mathbf{x}) = x^6 + 4xy + 2y^{18} - 1 = 0$.

The compromise method is between local and global methods. Consisting of the geometric property with computing the nearest point is proposed by Hartmann [23,24] named as the first compromise method. Repeatedly run the Newton's steepest gradient descent method (3) until the iterative point falls on the planar algebraic curve, where the initial iterative point is unrestricted.

$$\mathbf{x}_{n+1} = \mathbf{x}_n - (f(\mathbf{x}_n) / \langle \nabla f(\mathbf{x}_n), \nabla f(\mathbf{x}_n) \rangle) \nabla f(\mathbf{x}_n). \tag{3}$$

$$\mathbf{q} = \mathbf{p} - (\langle \mathbf{p} - \mathbf{y}_n, \nabla f(\mathbf{y}_n) \rangle / \langle \nabla f(\mathbf{y}_n), \nabla f(\mathbf{y}_n) \rangle) \nabla f(\mathbf{y}_n). \tag{4}$$

Running the iterative formula (4) one time, the method [23,24] can obtain the vertical foot point $\mathbf{q}$ where the iterative point $\mathbf{y}_n$ of the formula (4) is the final iterative point obtained by formula (3). Continuously iterate the above two steps until the vertical foot point $\mathbf{q}$ is on the planar algebraic curve $f(\mathbf{x})$. Unluckily, progressive geometric tangent approximation iteration method with computing vertical foot point $\mathbf{q}$ fails for some planar algebraic curves $f(\mathbf{x})$.

The second compromise method is developed by Nicholas [25] who adopted the osculating circle technique to realize orthogonal projection onto the planar algebraic curve. Calculate the corresponding curvature at one point on the planar algebraic curve, and then the radius and center of the curvature circle. The line segment formed by the test point and the center of the curvature circle intersects the curvature circle at footpoint $\mathbf{q}$. Approximately take the footpoint $\mathbf{q}$ as a point on the planar algebraic curve. For the new point on the planar algebraic curve, repeat the above procedure to get a new footpoint and corresponding new approximate point on the planar algebraic curve. Repeat the above behavior until the footpoint $\mathbf{q}$ is the orthogonal projection point $\mathbf{p}_\Gamma$. Because the planar algebraic curve does not have parametric control like parametric curve, taking the footpoint as an approximate point on the planar algebraic curve will bring about large errors. So it makes the operation of the whole algorithm unstable.

The third compromise method is the circle shrinking technique [26]. Repeatedly run the iterative formula (3) such that the final iterative point $\mathbf{p}_c$ falls on the planar algebraic curve as far as possible, where the selection of initial iterative point is arbitrary. The next iterative point on the planar algebraic curve is obtained through a series of combined operations of circle and the planar algebraic curve, where the center and radius of the circle are test point $\mathbf{p}$ and $\|\mathbf{p} - \mathbf{p}_c\|$, respectively. A series of combined operations include the two most important steps: Find a point $\mathbf{p}^+$ on the circle by means of the mean value theorem; Seek the intersection of the line segment $\overline{\mathbf{pp}^+}$ and the circle where we call this intersection as the current intersection point $\mathbf{p}_c$. Repeatedly run this series of combined operations until the distance between the current point $\mathbf{p}_c$ and the previous point $\mathbf{p}_c$ is 0. The circle shrinking technique [26] takes a lot of time to seek point $\mathbf{p}^+$ each time. The algorithm has one difficulty: if the degree of the planar algebraic curve is higher than 5, the intersection point $\mathbf{p}_c$ of line segment $\overline{\mathbf{pp}^+}$ and the planar algebraic curve cannot be solved directly by formula or the iterative methods to find the intersection $\mathbf{p}_c$ will lead to instability.

The four compromise method is a circle double-and-bisect algorithm [27]. The circle doubling algorithm begins with a very small circle where the center is the test point $\mathbf{p}$ and the radius is very small r_1. Keep the same center of the circle, take the radius r_2 twice of r_1 to draw a new circle. If there is no intersection between the new circle and the curve, draw a new circle with radius twice of r_2. Continuously repeat the above process until new circle can intersect with the planar algebraic curve and the former circle does not. Naturally, the former circle and the current circle are called interior circle and exterior circle, respectively. Moreover, the bisecting technology implements the rest of the process. Continue to draw a new circle with new radius $r = (r_1 + r_2)/2$. If the new current circle whose radius is r intersects with the curve, substitute r for r_2, else for r_1. Repeatedly run the above progress until the difference between the two radii is approximate zero($|r_1 - r_2| < \varepsilon$). But this method is very difficult to judge whether the exterior circle intersects the planar algebraic curve or not [27].

The fifth compromise method is the integrated hybrid second order algorithm [28]. It includes two sub-algorithms: the hybrid second order algorithm and the initial iterative value estimation algorithm. They mainly exploint three ideas: (1) the tangent orthogonal vertical foot method coupled with calibration method; (2) Newton's steepest gradient descent iterative method to impel the iteration point to be on the planar implicit curve; (3) Newton's iterative method to speed up the whole iteration process. Before running the hybrid second order algorithm, the initial iterative value estimation algorithm is used to force the initial iterative value of the formula (17) of the hybrid second order algorithm and the orthogonal projection point $\mathbf{p}_\Gamma$ as close as possible. After a lot of tests, if the distance between the test point $\mathbf{p}$ and the curve is not very far, the advantages of this algorithm are obvious in term of robustness and efficiency. But when the test point is very far from the curve, the integrated hybrid second order algorithm is invalid.

2. The Improved Curvature Circle Algorithm

In Reference [28], when the test point $\mathbf{p}$ is not particularly far away, the integrated hybrid algorithm can have ideal result. But if the test point $\mathbf{p}$ is very far from the curve, the algorithm is invalid where the test point $\mathbf{p}$ can not be robustly and effectively orthogonally projected onto the planar algebraic curve. In order to overcome this difficulty, we propose an improved curvature circle algorithm to ensure robustness and effective convergence with the test point $\mathbf{p}$ being arbitrarily far away. No matter how far the test point $\mathbf{p}$ is from the planar algebraic curve, if the initial iteration point $\mathbf{x}_0$ is very close to the orthogonal projection point of the test point $\mathbf{p}$, the preconceived algorithm can converge well. So we attempt to construct an algorithm to find an initial iterative point very close to the orthogonal projection point $\mathbf{p}_\Gamma$ of the test point $\mathbf{p}$. The general idea is the following. Repeatedly iterate the formula (3) by utilizing the Newton's steepest gradient descent method until the iteration point fall on the planar algebraic curve as far as possible, written as $\mathbf{p}_c$. This time, the distance between the iteration point $\mathbf{p}_c$ and the orthogonal projection point $\mathbf{p}_\Gamma$ is much smaller than that between the original iteration point $\mathbf{x}_0$ and the orthogonal projection point $\mathbf{p}_\Gamma$. The iteration point $\mathbf{p}_c$ is closer to the orthogonal projection point $\mathbf{p}_\Gamma$. In order to further promote the iteration point $\mathbf{p}_c$ and the orthogonal projection point $\mathbf{p}_\Gamma$ to be closer, we introduce a key step with curvature circle algorithm. Draw a curvature circle through point $\mathbf{p}_c$ on the planar algebraic curve with the radius R determined by the curvature k and the center $\mathbf{m}$ being a normal direction point of point $\mathbf{p}_c$ on the planar algebraic curve. Line segment $\overline{\mathbf{mp}}$ determined by the test point $\mathbf{p}$ and the center $\mathbf{m}$ intersects curvature circle at point $\mathbf{q}$. We take the intersection point $\mathbf{q}$ as the next iteration point for the iteration point $\mathbf{p}_c$. Of course, the distance between the intersection point $\mathbf{q}$ and the orthogonal projection point $\mathbf{p}_\Gamma$ is much smaller than the previous one. We use the intersection point $\mathbf{q}$ as the new test point, and run the hybrid algorithm again where the initial iterative point at this moment can be set as $\mathbf{q} - (0.1, 0.1)$. Repeatedly iterate until the iteration point falls on the planar algebraic curve $f(\mathbf{x})$, written as $\mathbf{p}_c$. We repeat the last two key steps in this procedure until the iteration point $\mathbf{p}_c$ and the orthogonal projection point $\mathbf{p}_\Gamma$ overlap (See Figure 1).

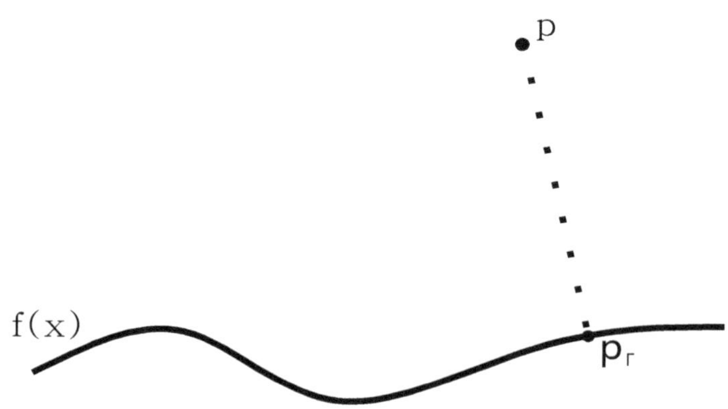

Figure 1. Test point $\mathbf{p}$ orthogonal projection onto planar algebraic curve $f(\mathbf{x})$.

Let's elaborate on the general idea. Let $\mathbf{p}$ be a test point on the plane. There is an planar algebraic curve Γ on the plane.

$$f(x, y) = 0. \tag{5}$$

The plane algebraic curve (5) can be simply written as

$$f(\mathbf{x}) = 0, \tag{6}$$

where $\mathbf{x} = (x, y)$. The goal of this paper is to find a point $\mathbf{p}_\Gamma$ on the planar algebraic curve $f(\mathbf{x})$ to satisfy the basic relationship

$$\|\mathbf{p} - \mathbf{p}_\Gamma\| = \min_{\mathbf{x}\in\Gamma} \|\mathbf{p} - \mathbf{x}\|. \tag{7}$$

The above problem can be written as

$$\begin{cases} f(\mathbf{p}_\Gamma) = 0, \\ \nabla f(\mathbf{p}_\Gamma) \times (\mathbf{p} - \mathbf{p}_\Gamma) = 0, \\ \|\mathbf{p} - \mathbf{p}_\Gamma\| = \min_{\mathbf{x}\in\Gamma} \|\mathbf{p} - \mathbf{x}\|, \end{cases} \tag{8}$$

where $\nabla f = \left[\frac{\partial f}{\partial x}, \frac{\partial f}{\partial y}\right]$ is Hamiltonian operator and symbol $\times$ is cross product. We take s as the arc length parameter of the planar algebraic curve $f(\mathbf{x})$ and $\mathbf{t} = \left[\frac{dx}{ds}, \frac{dy}{ds}\right]$ is unit tangent vector along the planar algebraic curve $f(\mathbf{x})$. Take derivative of Equation (6) with respect to arc length parameter s and combine with unit tangent vector condition $\|\mathbf{t}\| = 1$, we obtain the following simultaneous system of nonlinear equations,

$$\begin{cases} \langle \mathbf{t}, \nabla f \rangle = 0, \\ \|\mathbf{t}\| = 1. \end{cases} \tag{9}$$

It is easy to get the solution of Equation (9).

$$\mathbf{t} = \left[-\frac{\partial f}{\partial y}, \frac{\partial f}{\partial x}\right] / \|\nabla f\|. \tag{10}$$

Repeatedly iterate Equation (3) called as the Newton's steepest gradient descent method until until the iterative termination criteria $|f(\mathbf{x}_{n+1})| < \varepsilon$, where the initial iterative point is $\mathbf{x}_0 = \mathbf{p} - (0.1, 0.1)$ and refer to the iterative point $\mathbf{x}_{n+1}$ as $\mathbf{p}_c$. The first advantage of the Newton's steepest gradient descent method (3) is to make the iteration point fall on the planar algebraic curve $f(\mathbf{x})$ as far as possible. Its second advantage of the Newton's steepest gradient descent method (3) is that the iteration point fallen on the planar algebraic curve is relatively close to the orthogonal projection point $\mathbf{p}_\Gamma$, and it brings great convenience to implementation of the subsequent sub-algorithms. The Newton's steepest gradient descent method (Algorithm 1) can be specifically described as (See Figure 2).

Algorithm 1: The Newton's steepest gradient descent method.

Input: The test point $\mathbf{p}$ and the planar algebraic curve $f(\mathbf{x}) = 0$
Output: The iterative point $\mathbf{p}_c$ fallen on planar algebraic curve $f(\mathbf{x}) = 0$
Description:
Step 1:

$\mathbf{x}_{n+1} = \mathbf{p} - (0.1, 0.1)$;
Do {
  $\mathbf{x}_n = \mathbf{x}_{n+1}$;
  Update $\mathbf{x}_{n+1}$ according to the iterative Equation (3);
}while ($|f(\mathbf{x}_{n+1})| > \varepsilon$&& $\|\mathbf{x}_{n+1} - \mathbf{x}_n\| > \varepsilon$);

Step 2:

$\mathbf{p}_c = \mathbf{x}_{n+1}$;
Return $\mathbf{p}_c$;

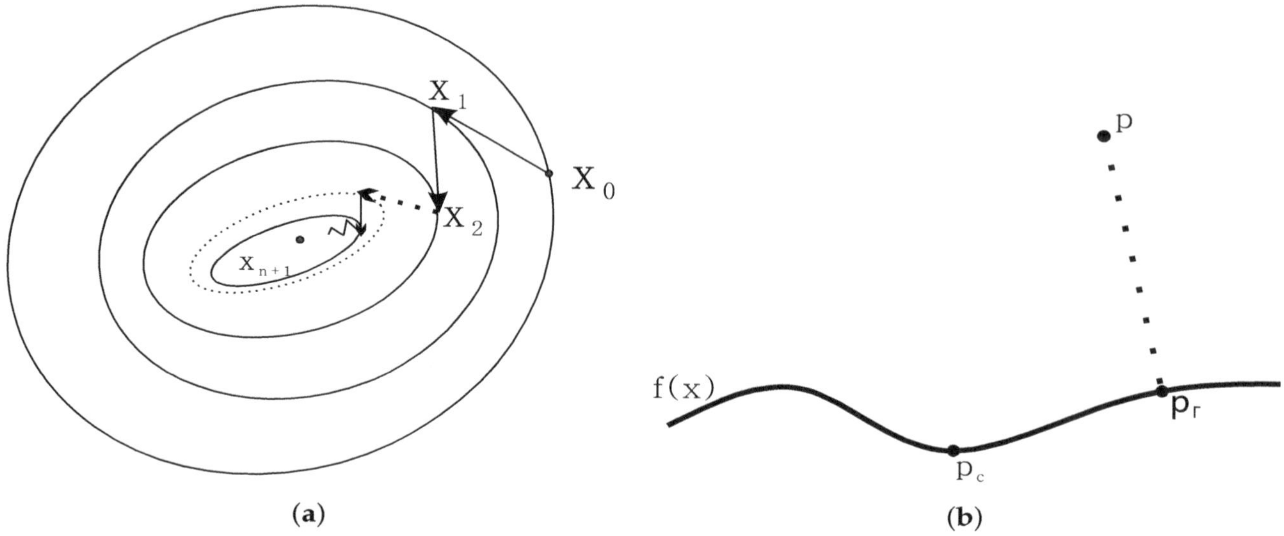

Figure 2. The entire graphic demonstration of Algorithm 1. (**a**) The whole iterative process of the Newton's steepest gradient descent method; (**b**) The last step of the iterative point $\mathbf{p}_c$ fallen on the planar algebraic curve $f(\mathbf{x})$ through the Newton's steepest gradient descent method.

In this case, if the iterative point $\mathbf{p}_c$ fallen on the planar algebraic curve $f(\mathbf{x})$ is taken as the initial iterative point of the hybrid algorithm, convergence or divergence may occur where divergence can not improve the algorithm. As for divergence, it can not achieve the purpose of improving the algorithm. From another point of view, the distance between iteration point $\mathbf{p}_c$ and orthogonal projection point $\mathbf{p}_\Gamma$ of the test point $\mathbf{p}$ should be closer. It lays a good foundation for the implementation of subsequent sub-algorithms. In order to get the iteration point and the orthogonal projection point $\mathbf{p}_\Gamma$ closer, we adopt curvature circle way to promote the iteration point and the orthogonal projection point $\mathbf{p}_\Gamma$ being closer. Because the iterative point is on the planar algebraic curve, the curvature k at the iterative point $\mathbf{p}_c$ fallen on the planar algebraic curve $f(\mathbf{x})$ is defined as [29],

$$k = k(x,y) = \frac{\left[-f_y, f_x\right] \mathbf{G} \left[-f_y, f_x\right]^T}{\|\nabla f\|^3}, \tag{11}$$

where $\mathbf{G} = \begin{pmatrix} f_{xx} & f_{xy} \\ f_{yx} & f_{yy} \end{pmatrix}$. The radius R and the center $\mathbf{m}$ of the curvature circle $\bigcirc$ directed by the curvature k are

$$R = |1/k|, \tag{12}$$

and

$$\mathbf{m} = \mathbf{p}_c + \frac{\overrightarrow{\mathbf{n}}}{k}, \tag{13}$$

where the unit normal vector $\overrightarrow{\mathbf{n}}$ is $\overrightarrow{\mathbf{n}} = \frac{\nabla f}{\|\nabla f\|}$. The line segment $\overline{\mathbf{mp}}$ determined by the test point $\mathbf{p}$ and the center $\mathbf{m}$ of the curvature circle $\bigcirc$ intersects the curvature circle $\bigcirc$ at point $\mathbf{q}$ which is named as footpoint $\mathbf{q}$. From elementary geometric knowledge, the parametric equation of the line segment $\overline{\mathbf{mp}}$ can be expressed as

$$\mathbf{x} = \mathbf{p} + (\mathbf{m} - \mathbf{p})w, \tag{14}$$

where parametric $0 \le w \le 1$ is undetermined. In addition, the equation of the curvature circle $\bigcirc$ can be written as

$$\|\mathbf{m} - \mathbf{x}\| = R. \tag{15}$$

By solving Equation (14) and Equation (15) together, the analytic expression of the intersection $\mathbf{q}$ is obtained

$$\mathbf{q} = \mathbf{p} + (\mathbf{m} - \mathbf{p})w, \tag{16}$$

where the undetermined parameter w is accurately identified as $w = 1 - \frac{R}{\|\mathbf{m} - \mathbf{p}\|}$. The computation of the footpoint $\mathbf{q}$ can be realized through Algorithm 2 (See Figure 3).

Algorithm 2: Computing footpoint $\mathbf{q}$ via the curvature circle ◯ and the line segment $\overline{\mathbf{mp}}$.

Input: The test point $\mathbf{p}$, the planar algebraic curve $f(\mathbf{x}) = 0$ and the iterative point $\mathbf{p}_c$ on the planar algebraic curve $f(\mathbf{x}) = 0$.

Output: The footpoint $\mathbf{q}$.

Description:

Step 1:

Compute the curvature k of the iterative point $\mathbf{p}_c$ fallen on the planar algebraic curve $f(\mathbf{x}) = 0$ by the curvature calculation formula (11).

Step 2:

Calculate the radius R and the center $\mathbf{m}$ of the curvature circle ◯ through the formulas (12) and (13), respectively.

Step 3:

Compute the footpoint $\mathbf{q}$ by the formula (16).

Return $\mathbf{q}$;

Remark 1. *The important formula for computing the curvature k is the formula (11). If the denominator of the curvature k with the formula (11) is 0, the whole iteration process will degenerate. In order to solve this special degeneration, we adopt a small perturbation of the curvature k of the formula (11) in programming implementation of Algorithm 2. Namely, the denominator of the curvature k with the formula (11) could be incremented by a small positive constant ε, the denominator of the curvature k is the denominator of the curvature k +ε, and Algorithm 2 continues to calculate the center and the radius of the curvature circle corresponding to the curvature after disturbance. Of course, in all subsequent formulas or iterative formulas, we also do the same denominators perturbation treatment for the case of the zero denominators of the formulas or the iterative formulas.*

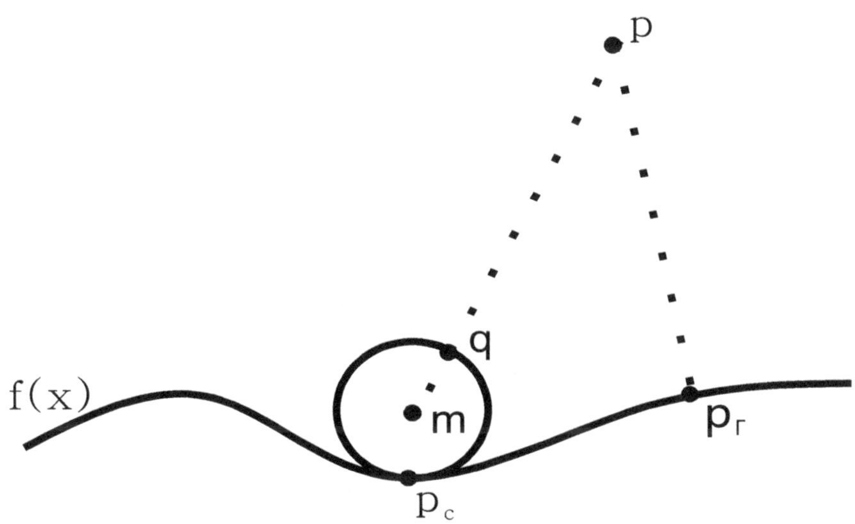

Figure 3. Graphic demonstration for Algorithm 2.

Under this circumstance, if the footpoint point $\mathbf{q}$ at this moment is taken as the initial iteration point of the hybrid algorithm, the convergence probability of the hybrid algorithm is much greater than that of using the point $\mathbf{p}_c$ in Algorithm 1 as the initial iterative point of the hybrid algorithm. The reason is that the distance $\|\mathbf{q}-\mathbf{p}_\Gamma\|$ is smaller than the distance $\|\mathbf{p}_c-\mathbf{p}_\Gamma\|$. But divergence may happen in this case. In order to further guarantee the robustness,we orthogonally project the footpoint $\mathbf{q}$ onto the planar algebraic curve $f(\mathbf{x})$ by using the hybrid algorithm, instead of directly using the footpoint $\mathbf{q}$ as the initial iterative point. At this time we still call the orthogonal projection point of the footpoint $\mathbf{q}$ as the point $\mathbf{p}_c$ which is just fallen on the planar algebraic curve $f(\mathbf{x})$. Because at this time the footpoint $\mathbf{q}$ is close to the planar algebraic curve $f(\mathbf{x})$, the algorithm can ensure complete convergence. The distance between the iterative point $\mathbf{p}_c$ and the orthogonal projection point $\mathbf{p}_\Gamma$ of the test point $\mathbf{p}$ becomes smaller again. The core iterative formula (17) of the hybrid algorithm is as follows (See [28]).

$$\begin{cases} \mathbf{y}_n = \mathbf{x}_n - \left(f(\mathbf{x}_n)/\left\langle \nabla f(\mathbf{x}_n), \nabla f(\mathbf{x}_n)\right\rangle\right) \nabla f(\mathbf{x}_n), \\ \mathbf{z}_n = \mathbf{y}_n - \left(\mathbf{F}(\mathbf{y}_n)/\left\langle \nabla \mathbf{F}(\mathbf{y}_n), \nabla \mathbf{F}(\mathbf{y}_n)\right\rangle\right) \nabla \mathbf{F}(\mathbf{y}_n), \\ \mathbf{Q} = \mathbf{q} - \left(\left\langle (\mathbf{q}-\mathbf{z}_n), \nabla f(\mathbf{z}_n)\right\rangle / \left\langle \nabla f(\mathbf{z}_n), \nabla f(\mathbf{z}_n)\right\rangle\right) \nabla f(\mathbf{z}_n), \\ \mathbf{u}_n = \mathbf{z}_n + sign(\langle \mathbf{q}-\mathbf{z}_n, \mathbf{t}_0\rangle)\mathbf{t}_0 \Delta s, \\ \mathbf{v}_n = \mathbf{u}_n - \left(\mathbf{F}(\mathbf{u}_n)/\left\langle \nabla \mathbf{F}(\mathbf{u}_n), \nabla \mathbf{F}(\mathbf{u}_n)\right\rangle\right) \nabla \mathbf{F}(\mathbf{u}_n), \\ \mathbf{x}_{n+1} = \mathbf{v}_n + [-\Delta e, 0]\left[\nabla f^T, (\Delta \mathbf{v}_n)^T\right]^{-1} \quad (if \left|\left[\nabla f^T, (\Delta \mathbf{v}_n)^T\right]\right| = 0, \mathbf{x}_{n+1} = \mathbf{v}_n), \end{cases} \tag{17}$$

where $\mathbf{F}_0(\mathbf{x}) = [(\mathbf{q}-\mathbf{x}) \times \nabla f(\mathbf{x})] = 0$, $\mathbf{F}(\mathbf{x}) = \dfrac{\mathbf{F}_0(\mathbf{x})}{\sqrt{\langle \nabla f(\mathbf{x}), \nabla f(\mathbf{x})\rangle}}$, $\mathbf{t}_0 = \dfrac{[-f_y, f_x]}{\|\Delta f\|}$, $\Delta s = \|\mathbf{Q}-\mathbf{z}_n\|$, $f(\mathbf{v}_n) = \Delta e$, $\Delta \mathbf{v}_n = -(\mathbf{F}(\mathbf{u}_n)/\langle \nabla \mathbf{F}(\mathbf{u}_n), \nabla \mathbf{F}(\mathbf{u}_n)\rangle) \nabla \mathbf{F}(\mathbf{u}_n)$.

The iterative formula (17) mainly contains four techniques. The core technology is the tangent foot vertical method with the third step and the fourth step of the iterative formula (17). Draw a tangent line $\mathbf{L}$ from a point on a plane algebraic curve $f(\mathbf{x})$. Through the footpoint $\mathbf{q}$ (The footpoint $\mathbf{q}$ at this time is as the test point of iterative formula (17)), make a vertical line of the tangent $\mathbf{L}$ and get its corresponding vertical foot point $\mathbf{Q}$, which is equivalent to the third step in the formula (17). From the fourth step of the iterative formula (17), we get the next iteration point of particular importance for the initial iteration point. When the next iteration point is not very close to the planar algebraic curve $f(\mathbf{x})$, we adopt the second important technique with the iteration point correction method, equivalent to the sixth step of the iterative formula (17). The iteration point is to move to the plane algebra curve as close as possible such that the distance between the correction point of the iteration point and the plane algebra curve $f(\mathbf{x})$ is as close as possible. These two techniques are pure geometric techniques. When the distance between the test point and the planar algebraic curve is very close, the effect of convergence is obvious. Of course, when the distance between the test point and the planar algebraic curve is relatively long, sometimes there will be non-convergence. In order to improve the robustness of convergence, we add the Newton's steepest gradient descent method before the first technique with the third step and the fourth step of the iterative formula (17). Its first aim is to bring the initial iteration point closer to the planar algebraic curve $f(\mathbf{x})$. Its second aim is to promote the accuracy of subsequent iterations. In order to accelerate the whole iteration process of the iterative formula (17), we once again incorporate the fourth technology of Newton's iterative method which is closely related to the footpoint $\mathbf{q}$. This technique not only accelerates the convergence rate of the whole iteration process but also improves the iteration robustness. Furthermore, the accuracy of the whole iteration process can be improved by the fourth technique. So we add Newton's iterative method after the first step with the second technique, and then add it again before the last step with the third technique. Based on the above explanation and illustration, we get the following the hybrid tangent vertical foot algorithm (Algorithm 3).

Algorithm 3: The hybrid tangent vertical foot algorithm (See Figure 4).

Input: The footpoint $\mathbf{q}$ and the planar algebraic curve $f(\mathbf{x}) = 0$.
Output: The point $\mathbf{p}_c$ fallen on the planar algebraic curve $f(\mathbf{x}) = 0$.
Description:
Step 1:
$\mathbf{x}_{n+1} = \mathbf{q} - (0.1, 0.1)$;
Do {
$\mathbf{x}_n = \mathbf{x}_{n+1}$;
Execute $\mathbf{x}_{n+1}$ according to the iterative Equation (17);
}while ($\|\mathbf{x}_{n+1} - \mathbf{x}_n\|^2 > \varepsilon \&\& |f(\mathbf{x}_{n+1})| > \varepsilon$)
Step 2:
$\mathbf{p}_c = \mathbf{x}_{n+1}$;
Return $\mathbf{p}_c$;

With the description of the above three algorithms, we propose a comprehensive and complete algorithm (Algorithm 4) closely related to Algorithm 2 (See Figure 4).

Algorithm 4: The first improved curvature circle algorithm (Comprehensive Algorithm A).

Input: Test point $\mathbf{p}$ and the planar algebraic curve $f(\mathbf{x})$.
Output: Orthogonal projection point $\mathbf{p}_\Gamma$ of the test point $\mathbf{p}$ which orthogonally projects the test point $\mathbf{p}$ onto a planar algebraic curve $f(\mathbf{x})$.
Description:
Step 1: Starting from the neighbor point of the test point $\mathbf{p}$, calculate the point $\mathbf{p}_c$ fallen on the $f(\mathbf{x})$ via Algorithm 1.
Do{
Step 2: Compute the footpoint $\mathbf{q}$ via Algorithm 2.
Step 3: Project footpoint $\mathbf{q}$ onto the planar algebraic curve $f(\mathbf{x})$ via Algorithm 3, then get the new iterative point $\mathbf{p}_c$ fallen on the $f(\mathbf{x})$.
}while (distance (the current $\mathbf{p}_c$, the previous $\mathbf{p}_c$)$> \epsilon$).
$\mathbf{p}_\Gamma = \mathbf{p}_c$;
Return $\mathbf{p}_\Gamma$;

Through a series of rigorous deductions, Comprehensive Algorithm A is the important algorithm of our paper. No matter how far the test point $\mathbf{p}$ is from the planar algebraic curve $f(\mathbf{x})$, test point $\mathbf{p}$ could very robustly orthogonally projects onto the planar algebraic curve $f(\mathbf{x})$. This has achieved our desired result. After a lot of testing and observation, when the point on the curve is close to the orthogonal projection point, we find that Comprehensive Algorithm A presents two characteristics: (1) difference between the first distance and the second distance decreases slower and slower, where the first distance and the second distance are the one between the previous iterative point $\mathbf{p}_c$ on the planar algebraic curve and the orthogonal projection point $\mathbf{p}_\Gamma$, and the one between the current iterative point $\mathbf{p}_c$ on the planar algebraic curve and the orthogonal projection point $\mathbf{p}_\Gamma$, respectively; (2) the rate goes even slower at which the absolute value of the inner product gradually approaches zero. These two characteristics are what we don't want to obtain because they are contrary to the efficiency of computer systems. On the premise of ensuring robustness, we try our best to improve and excavate a certain degree of efficiency for the problem of point orthogonal projection onto planar algebraic curve. We have an ingenious discovery. After each running of Algorithm 3, we run the Newton's iterative method associated with the original test point $\mathbf{p}$, which can improve the convergence and ensure the orthogonality. Namely, that is to add this step after the last step of the formula (17). Thus the iterative formula (18) is obtained.

$$\begin{cases} \mathbf{y}_n = \mathbf{x}_n - \left(f(\mathbf{x}_n) / \langle \nabla f(\mathbf{x}_n), \nabla f(\mathbf{x}_n) \rangle\right) \nabla f(\mathbf{x}_n), \\ \mathbf{z}_n = \mathbf{y}_n - \left(\mathbf{F}(\mathbf{y}_n) / \langle \nabla \mathbf{F}(\mathbf{y}_n), \nabla \mathbf{F}(\mathbf{y}_n) \rangle\right) \nabla \mathbf{F}(\mathbf{y}_n), \\ \mathbf{Q} = \mathbf{q} - \left(\langle (\mathbf{q} - \mathbf{z}_n), \nabla f(\mathbf{z}_n) \rangle \, / \, \langle \nabla f(\mathbf{z}_n), \nabla f(\mathbf{z}_n) \rangle\right) \nabla f(\mathbf{z}_n), \\ \mathbf{u}_n = \mathbf{z}_n + sign(\langle \mathbf{q} - \mathbf{z}_n, \mathbf{t}_0 \rangle) \mathbf{t}_0 \Delta s, \\ \mathbf{v}_n = \mathbf{u}_n - \left(\mathbf{F}(\mathbf{u}_n) / \langle \nabla \mathbf{F}(\mathbf{u}_n), \nabla \mathbf{F}(\mathbf{u}_n) \rangle\right) \nabla \mathbf{F}(\mathbf{u}_n), \\ \mathbf{w}_n = \mathbf{v}_n + [-\Delta e, 0] \left[\nabla f^T, (\Delta \mathbf{v}_n)^T\right]^{-1} \left(if \left| \left[\nabla f^T, (\Delta \mathbf{v}_n)^T\right] \right| = 0, then \ \mathbf{w}_n = \mathbf{v}_n\right), \\ \mathbf{x}_{n+1} = \mathbf{w}_n - \left(\mathbf{G}(\mathbf{w}_n) / \langle \nabla \mathbf{G}(\mathbf{w}_n), \nabla \mathbf{G}(\mathbf{w}_n) \rangle\right) \nabla \mathbf{G}(\mathbf{w}_n), \end{cases} \tag{18}$$

where $\mathbf{F}_0(\mathbf{x}) = [(\mathbf{q} - \mathbf{x}) \times \nabla f(\mathbf{x})] = 0$, $\mathbf{F}(\mathbf{x}) = \dfrac{\mathbf{F}_0(\mathbf{x})}{\sqrt{\langle \nabla f(\mathbf{x}), \nabla f(\mathbf{x}) \rangle}}$, $\mathbf{t}_0 = \dfrac{[-f_y, f_x]}{\|\Delta f\|}$, $\Delta s = \|\mathbf{Q} - \mathbf{z}_n\|$, $f(\mathbf{v}_n) = \Delta e$, $\Delta \mathbf{v}_n = -(\mathbf{F}(\mathbf{u}_n) / \langle \nabla \mathbf{F}(\mathbf{u}_n), \nabla \mathbf{F}(\mathbf{u}_n) \rangle) \nabla \mathbf{F}(\mathbf{u}_n)$, $\mathbf{G}_0(\mathbf{x}) = [(\mathbf{p} - \mathbf{x}) \times \nabla f(\mathbf{x})] = 0$, $\mathbf{G}(\mathbf{x}) = \dfrac{\mathbf{G}_0(\mathbf{x})}{\sqrt{\langle \nabla f(\mathbf{x}), \nabla f(\mathbf{x}) \rangle}}$. Because the iterative formula (17) of Algorithm 3 naturally becomes the iterative formula (18), so Algorithm 3 naturally becomes the following Algorithm 5.

Algorithm 5: The hybrid tangent vertical foot algorithm.

Input: The footpoint $\mathbf{q}$ and the planar algebraic curve $f(\mathbf{x}) = 0$.
Output: The point $\mathbf{p}_c$ fallen on planar algebraic curve $f(\mathbf{x}) = 0$.
Description:
Step 1:

$\mathbf{x}_{n+1} = \mathbf{q} - (0.1, 0.1)$;
Do {
  $\mathbf{x}_n = \mathbf{x}_{n+1}$;
  Execute $\mathbf{x}_{n+1}$ according to the iterative Equation (18);
}while ($\|\mathbf{x}_{n+1} - \mathbf{x}_n\|^2 > \varepsilon \&\& |f(\mathbf{x}_{n+1})| > \varepsilon$)

Step 2:

$\mathbf{p}_c = \mathbf{x}_{n+1}$;
Return $\mathbf{p}_c$;

Now let's replace Algorithm 3 of Comprehensive Algorithms A with Algorithm 5. We get the following Comprehensive Algorithm B (Algorithm 6).

Algorithm 6: The second improved curvature circle algorithm (Comprehensive Algorithm B).

Input: Test point $\mathbf{p}$ and the planar algebraic curve $f(\mathbf{x})$.
Output: Orthogonal projection point $\mathbf{p}_\Gamma$ of the test point $\mathbf{p}$ which orthogonally projects the test point $\mathbf{p}$ onto the planar algebraic curve $f(\mathbf{x})$.
Description:
Step 1: Starting from the neighbor point of the test point $\mathbf{p}$, calculate the point $\mathbf{p}_c$ fallen on the $f(\mathbf{x})$ via Algorithm 1.
Do{
  Step 2: Compute the footpoint $\mathbf{q}$ via Algorithm 2.
  Step 3: Project footpoint $\mathbf{q}$ onto the planar algebraic curve $f(\mathbf{x})$ via Algorithm 5, then get new point $\mathbf{p}_c$ fallen on the $f(\mathbf{x})$.
}while(distance(the current $\mathbf{p}_c$, the previous $\mathbf{p}_c$)$> \epsilon$).
$\mathbf{p}_\Gamma = \mathbf{p}_c$;
Return $\mathbf{p}_\Gamma$;

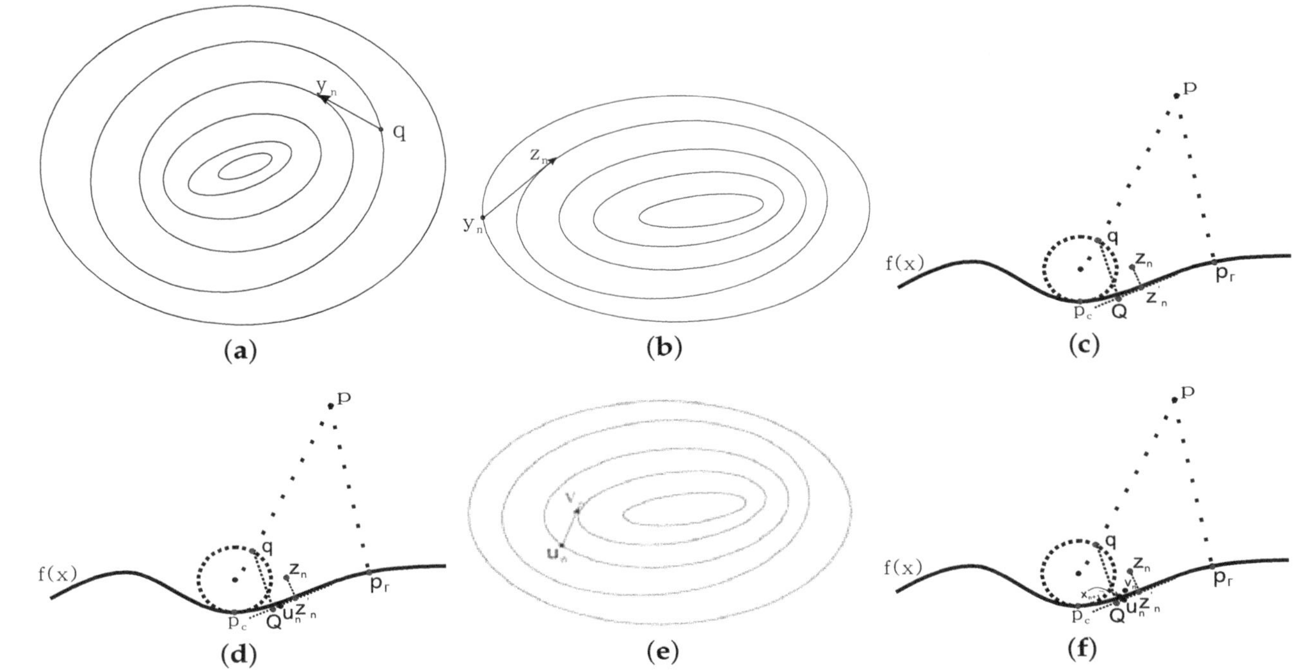

Figure 4. Graphic interpretation of the whole iteration process in Algorithm 3. (**a**) Newton's steepest gradient descent method in the first step; (**b**) The Newton's iteration method related to the test point in the second step; (**c**) The vertical foot point **Q** being the footpoint **q** orthogonal projection onto tangent line induced by the iterative point $\mathbf{z}_n$ on the planar algebraic curve in the third step; (**d**) Calculating line incremental iterative value in the fourth step; (**e**) Once again running the Newton's iteration method related to the test point in the fifth step; (**f**) Correcting the previous iteration value to improve the robustness of iteration in the last step.

Comprehensive Algorithm A and Comprehensive Algorithm B share common advantage: the robustness of the two algorithms is substantially improved than that of the existing algorithms because our algorithms are not subject to any restrictions on test points and initial iteration points. By comparison, Comprehensive Algorithm B has four advantages over Comprehensive Algorithm A. (1) The last step of the iterative formula (18) in Comprehensive Algorithm B can make corrections continuously; (2) The last step of the iterative formula (18) in Comprehensive Algorithm B accelerates the whole Comprehensive Algorithm B; (3) The last step of the iterative formula (18) in Comprehensive Algorithm B accelerates the inner product of two vectors to 0, where the first vector refers to the vector connecting the test point $\mathbf{p}$ and the iteration point $\mathbf{z}_{n+1}$ of Comprehensive Algorithm B and the second vector $\left[-\frac{\partial f}{\partial y}, \frac{\partial f}{\partial x}\right]_{|\mathbf{x}=\mathbf{x}_{n+1}}$ is the tangent vector derived from the iteration point $\mathbf{x}_{n+1}$ on the planar algebraic curve, respectively; (4) Comprehensive Algorithm B overcomes two shortcomings of Comprehensive Algorithm A.

Of course, when the test point is not too far from the plane algebra curve, Comprehensive Algorithm is also convergent for any initial iterative point. However, Comprehensive Algorithm A takes more time than directly using the hybrid second order algorithm. In practical applications such as computer graphics, it's hard to know if the test point $\mathbf{p}$ is close to or far from a planar algebraic curve. Because the main reason is that the degree and the type of the planar algebraic curve restrict the relative distance between the test point $\mathbf{p}$ and the planar algebraic curve. In order to optimize time efficiency, we take advantage of Comprehensive Algorithm A and the hybrid second order algorithm such that no matter where the test point $\mathbf{p}$ is located, it can be orthogonally projected onto the planar algebraic curve efficiently and robustly. First, the hybrid second order algorithm is iterated. If it does not converge after 100 iterations, it will be changed to Comprehensive Algorithm A to iterate until the iteration point reaches the orthogonal projection point $\mathbf{p}_\Gamma$. Specific algorithm implementation is the following Comprehensive Integrated Algorithm A (Algorithm 7).

Algorithm 7: The first comprehensive integrated improved curvature circle algorithm (Comprehensive Integrated Algorithm A).

Input: Test point $\mathbf{p}$ and the planar algebraic curve $f(\mathbf{x})$.
Output: Orthogonal projection point $\mathbf{p}_\Gamma$ of the test point $\mathbf{p}$.
Description:
Step 1:

```
x_{n+1} = p − (0.1, 0.1);
for(i = 0; i < N; i++) {
    x_n = x_{n+1};
    x_{n+1}=Hybrid second order algorithm(f, p, x_n);
    if(||x_{n+1} − x_n|| < ε)    break;
}
```

Step 2:

```
if(i ≥ N&&d ≥ 1e − 15) {
    x_n = x_{n+1};
    x_{n+1}=Comprehensive Algorithm A(f, p, x_n);
}
p_Γ = x_{n+1};
Return p_Γ;
```

Number N is an empirical value of the iterative times where the value N is specified as 5 or 6.

Similar to Comprehensive Algorithm A, by replacing Algorithm 3 with Algorithm 5, the following Comprehensive Integrated Algorithm B (Algorithm 8) can be obtained naturally.

Algorithm 8: The second comprehensive integrated improved curvature circle algorithm (Comprehensive Integrated Algorithm B).

Input: The test point $\mathbf{p}$ and the planar algebraic curve $f(\mathbf{x})$.
Output: Orthogonal projection point $\mathbf{p}_\Gamma$ of the test point $\mathbf{p}$.
Description:
Step 1:

```
x_{n+1} = p − (0.1, 0.1);
for(i = 0; i < N; i++) {
    x_n = x_{n+1};
    x_{n+1}=Hybrid second order algorithm(f, p, x_n);
    if(||x_{n+1} − x_n|| < ε)    break;
}
```

Step 2:

```
if(i ≥ N&&d ≥ 1e − 15) {
    x_n = x_{n+1};
    x_{n+1}=Comprehensive Algorithm B(f, p, x_n);
}
p_Γ = x_{n+1};
Return p_Γ;
```

Number N is an empirical value of the iterative times where the value N is specified as 5 or 6.

To sum up, we have presented four synthesis algorithms altogether. After analysis and judgment, Comprehensive Algorithm B and Comprehensive Integrated Algorithm B are the most robust and efficient. On the problem of orthogonal projection of point onto planar algebraic curve, if the distance between the test point and the planar algebraic curve is close, we recommend the hybrid second order algorithm, if the distance between the test point and the planar algebraic curve is not close, we recommend Comprehensive Algorithm B. Of course, if the distance between the test point and the planar algebraic curve cannot be known to be very far or close, Comprehensive Integrated Algorithm B is the best choice.

Remark 2. *In sum, Comprehensive Algorithm B has strong superiority over existing algorithms [10–28]. If the distance between the test point and the planar algebraic curve is very far away, the test point can be ideally orthogonally projected onto the planar algebraic curve. But when there are singular points* $\frac{\partial f}{\partial x}\cdot\frac{\partial f}{\partial x}+\frac{\partial f}{\partial y}\cdot\frac{\partial f}{\partial y}=0$ *in the planar algebraic curve, this case will seriously hinder the correct execution and implementation of Comprehensive Algorithm B. In order to solve the problem in the case of singularities in the planar algebraic curves, we propose a remedy to Comprehensive Algorithm B (Algorithm 9). The specific description is as follows (See Figure 5).*

Algorithm 9: The first remedial algorithm of Comprehensive Algorithm B.

Input: Test point $\mathbf{p}$ and the planar algebraic curve $f(\mathbf{x})$.

Output: Orthogonal projection point $\mathbf{p}_\Gamma$ of the test point $\mathbf{p}$.

Description:

Step 1.

Starting from the neighbor point of the test point $\mathbf{p}$, calculate the iterative point $\mathbf{p}_c$ fallen on the planar algebraic curve $f(\mathbf{x})$ via Algorithm 1.

Step 2.

Judge whether to use curvature circle method or tangent method in the next step.

Step 3.

Find the left endpoint $\mathbf{L}_0$ on the other side of $f(\mathbf{x})$ relative to the test point $\mathbf{p}$. According to the result of **step** 2, if use curvature circle method, then the left endpoint $\mathbf{L}_0$ is equal to the intersection point $\mathbf{q}$ which is computed by the curvature circle method with the formula (16). If not, then the left endpoint $\mathbf{L}_0$ is equal to the vertical foot $\mathbf{Q}$ which is computed by the tangent method with the third step of the formula (17).

Step 4.

Calculate the intersection point $\mathbf{p}_c$ of the line segment $\overline{\mathbf{L}_0\mathbf{p}}$ connecting the current left endpoint $\mathbf{L}_0$ and the test point $\mathbf{p}$ and the planar algebraic curve $f(\mathbf{x})$ by the hybrid method of combining Newton's iterative method and binary search method. The intersection point $\mathbf{p}_c$ is called as the current iterative point $\mathbf{p}_c$;

Step 5.

Repeat **Step** 2,**Step** 3 and **Step** 4 until the distance between the current iterative point $\mathbf{p}_c$ and the previous iterative point $\mathbf{p}_c$ is near zero;

Step 6.

$\mathbf{p}_\Gamma = \mathbf{p}_c$;
Return $\mathbf{p}_\Gamma$;

Now let's describe the hybrid method of combining Newton's iterative method and binary search method in detail. The parameter equation of the line segment $\overline{\mathbf{L}_0\mathbf{p}}$ can be expressed as

$$\begin{cases} x = L_1 + (p_1 - L_1)w, \\ y = L_2 + (p_2 - L_2)w, \end{cases} \tag{19}$$

where $\mathbf{L}_0 = (L_1, L_2)$, $\mathbf{p} = (p_1, p_2)$, and $0 \leq w \leq 1$ is a parameter of Equation (19). Substitute Equation (19) into Equation (6) of the planar algebraic curve to get a equation on the parameter w,

$$K(w) = f(x, y) = 0, \tag{20}$$

where the x and y of Equation (20) are completely determined by the x and y of Equation (19). So the most basic Newton's iterative formula corresponding to Equation (20) is not difficult to write as,

$$w_{n+1} = w_n - \frac{K(w_n)}{DK(w_n)}, \tag{21}$$

where $DK(w)$ is the first derivative of $K(w)$ about the parameter w. Now we start to iterate the Newton's iterative formula (21) with the initial iterative value $w_0 = 0.0$. Based on the actual situation, the intersection of the line segment $\overline{\mathbf{L}_0\mathbf{p}}$ and the planar algebraic curve is much closer to the left endpoint $\mathbf{L}_0$ and much farther from the original test point $\mathbf{p}$, therefore, the initial interval of the binary search method can be specified as $[a, b] = [0.0, 0.5]$. The detailed description of the hybrid method of combining Newton's iterative method and binary search method is as following Algorithm 10.

Algorithm 10: The hybrid method of combining Newton's iterative method and binary search method.

Input: The planar algebraic curve $f(\mathbf{x})$, the original test point $\mathbf{p} = (p_1, p_2)$, the iterative point $\mathbf{p}_c$ via Algorithm 1.

Output: The intersection $\mathbf{p}_c$ between the line segment $\overline{\mathbf{L}_0\mathbf{p}}$ and the planar algebraic curve $f(\mathbf{x})$.

Description:

Step 1:

The initial interval of the binary search method $[a, b] = [0.0, 0.5]$, the initial iterative value $w = 0.0$;

Step 2:

$w = w - K(w) / DK(w)$;
kmin=min($K(a)$,$K(b)$);
kmax=max($K(a)$,$K(b)$);
if ($K(w) <$ kmin or $K(w) >$ kmax)
  $w = (a + b) / 2$;
sa=sign($K(a)$);
sw=sign($K(w)$);
if($sa == sw$)
  $a = w$;
else
  $b = w$;

Step 3:

Repeatedly iterate **Step 2** until $|a - b| < \varepsilon$;

Step 4:

$\mathbf{p}_c = \mathbf{L}_0 + (\mathbf{p} - \mathbf{L}_0)w$;
Return $\mathbf{p}_c$;

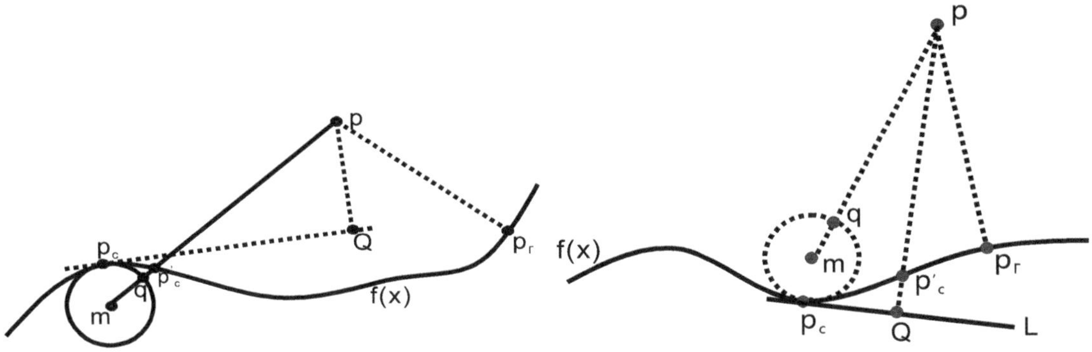

(**a**) (**b**)

Figure 5. Graphical interpretation for the first remedial algorithm of Comprehensive Algorithm B. (**a**) The intersection point $\mathbf{q}$ between the line segment $\overline{\mathbf{mp}}$ and the curvature circle and the test point $\mathbf{p}$ on the opposite side of the planar algebraic curve $f(\mathbf{x})$; (**b**)The vertical foot $\mathbf{Q}$ of the tangential line $\mathbf{L}$ and the test point $\mathbf{p}$ on the opposite side of the planar algebraic curve $f(\mathbf{x})$.

The robustness of the first remedial algorithm of Comprehensive Algorithm B is much better than that of Comprehensive Algorithm B while the first remedial algorithm of Comprehensive Algorithm B takes much more time than Comprehensive Algorithm B. The hybrid method of combining Newton's iterative method and binary search method is a hybrid method which binary search method ensures global convergence and the Newton's iterative method plays an accelerating role. In order to ensure robustness and improve efficiency, we have fully excavated Comprehensive Algorithm B. We have developed Second Remedial Algorithm (Algorithm 11). The specific description is as follows (See Figure 6).

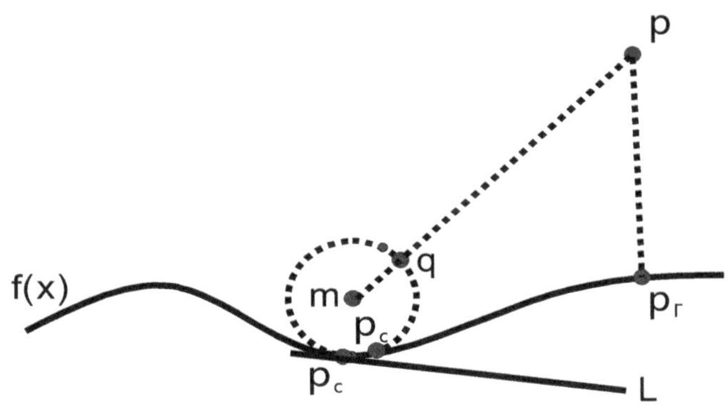

Figure 6. Graphic demonstration for Second Remedial Algorithm.

Algorithm 11: Second Remedial Algorithm.

Input: Test point $\mathbf{p}$ and the planar algebraic curve $f(\mathbf{x})$.
Output: Orthogonal projection point $\mathbf{p}_\Gamma$ of the test point $\mathbf{p}$ which orthogonally projects the test point $\mathbf{p}$ onto the planar algebraic curve $f(\mathbf{x})$
Description:
Step 1: Starting from a certain percentage of the test point $\mathbf{p}$, calculate the point $\mathbf{p}_c$ fallen on the $f(\mathbf{x})$ via Algorithm 1.
Do{
  Step 2: Compute the footpoint $\mathbf{q}$ via Algorithm 2.
  Step 3: Starting from the footpoint $\mathbf{q}$, compute the iterative point $\mathbf{p}_c$ fallen on the $f(\mathrm{x})$ via Algorithm 1.
}while(distance(the current $\mathbf{p}_c$, the previous $\mathbf{p}_c$)$> \epsilon$).
Step 4: Compute the orthogonal projection point $\mathbf{p}_\Gamma$ of the test point $\mathbf{p}$ via Algorithm 12.
  Return $\mathbf{p}_\Gamma$;

Algorithm 12: The hybrid Newton-type iterative algorithm.

Input: The current iterative point $\mathbf{p}_c$ fallen on the planar algebraic curve $f(\mathbf{x})$ and the planar algebraic curve $f(\mathbf{x})$.

Output: Orthogonal projection point $\mathbf{p}_\Gamma$ of the test point $\mathbf{p}$ which orthogonally projects the test point $\mathbf{p}$ onto the planar algebraic curve $f(\mathbf{x})$.

Description:

Step 1:

$\mathbf{x}_{n+1} = \mathbf{p}_c$;

Do {

$\mathbf{x}_n = \mathbf{x}_{n+1}$;

Compute $\mathbf{x}_{n+1}$ according to the iterative formula (22);

}while ($\|\mathbf{x}_{n+1} - \mathbf{x}_n\|^2 > \varepsilon \&\& |f(\mathbf{x}_{n+1})| > \varepsilon$)

Step 2:

$\mathbf{p}_\Gamma = \mathbf{x}_{n+1}$;

Return $\mathbf{p}_\Gamma$;

The expression of the iterative formula (22) is as follow,

$$\begin{cases} \mathbf{y}_n = \mathbf{x}_n - \left(f(\mathbf{x}_n) / \langle \nabla f(\mathbf{x}_n), \nabla f(\mathbf{x}_n) \rangle\right) \nabla f(\mathbf{x}_n), \\ \mathbf{z}_n = \mathbf{y}_n - \left(\mathbf{F}(\mathbf{y}_n) / \langle \nabla \mathbf{F}(\mathbf{y}_n), \nabla \mathbf{F}(\mathbf{y}_n) \rangle\right) \nabla \mathbf{F}(\mathbf{y}_n), \\ \mathbf{x}_{n+1} = \mathbf{z}_n + [-\Delta e, 0] \left[\nabla f^T, (\Delta \mathbf{z}_n)^T\right]^{-1} (if\ \left|[\nabla f^T, (\Delta \mathbf{z}_n)^T]\right| = 0, then\ \mathbf{x}_{n+1} = \mathbf{z}_n), \end{cases} \tag{22}$$

where $\mathbf{F}_0(\mathbf{x}) = [(\mathbf{p} - \mathbf{x}) \times \nabla f(\mathbf{x})] = 0$, $\mathbf{F}(\mathbf{x}) = \dfrac{\mathbf{F}_0(\mathbf{x})}{\sqrt{\langle \nabla f(\mathbf{x}), \nabla f(\mathbf{x}) \rangle}}$, $f(\mathbf{z}_n) = \Delta e$, $\Delta \mathbf{z}_n = -(\mathbf{F}(\mathbf{y}_n) / \langle \nabla \mathbf{F}(\mathbf{y}_n), \nabla \mathbf{F}(\mathbf{y}_n) \rangle) \nabla \mathbf{F}(\mathbf{y}_n)$.

Remark 3. *In this remark, we present the geometric interpretation of Second Remedial Algorithm. The purpose of the first step is to make the iterative point $\boldsymbol{p}_c$ fall on the planar algebraic curve as much as possible through Newton's steepest gradient descent method of Algorithm 1, where the coordinates of the initial iterative point take proportional value of that of the test point $\boldsymbol{p}$ to ensure that Algorithm 1 converges successfully. Otherwise, the distance between the initial iterative point and the planar algebraic curve is very large, which easily leads to the divergence of Algorithm 1. The purpose of* **Do ... While** *cycle body in Second Remedial Algorithm is to continuously and gradually move the iterative point $\boldsymbol{p}_c$ to fall on the planar algebraic curve to the orthogonal projection point $\boldsymbol{p}_\Gamma$. The second step in* **Do... While** *cycle body in Second Remedial Algorithm has two characteristics. Since the footpoint $\boldsymbol{q}$ is the unique intersection point of the curvature circle and the straight line segment $\overline{\boldsymbol{mp}}$ connecting the centre $\boldsymbol{m}$ of the curvature circle and the test point $\boldsymbol{q}$, the footpoint $\boldsymbol{q}$ is always closely related to the iterative point $\boldsymbol{p}_c$ fallen on the planar algebraic curve and the test point $\boldsymbol{p}$. The first characteristic is that the footpoint $\boldsymbol{q}$ can guarantee the global convergence of the whole algorithm (Second Remedial Algorithm). The second characteristic is that the distance between the footpoint $\boldsymbol{q}$ and the planar algebraic curve is much smaller than the distance between the test point $\boldsymbol{p}$ and the planar algebraic curve. So the third step with Algorithm 1 in* **Do ... While** *cycle body can very robustly iterate the footpoint $\boldsymbol{q}$ onto the planar algebraic curve. The core thought of* **Do ... While** *cycle body in Second Remedial Algorithm is to keep the iterative point $\boldsymbol{p}_c$ to fall on the planar algebraic curve and to move towards the orthogonal projection point $\boldsymbol{p}_\Gamma$ such that the distance $\|\boldsymbol{p}_c - \boldsymbol{p}_\Gamma\|$ between the iterative point $\boldsymbol{p}_c$ and the orthogonal projection point $\boldsymbol{p}_\Gamma$ becomes smaller and smaller. As the distance $\|\boldsymbol{p}_c - \boldsymbol{p}_\Gamma\|$ gets smaller and smaller, we have found that there is a defect in* **Do ... While** *cycle body in Second Remedial Algorithm. The decreasing speed of the distance $\|\boldsymbol{p}_c - \boldsymbol{p}_\Gamma\|$ is getting slower and slower. Especially the second formula of the formula (8) is very difficult to be satisfied. Namely, it is difficult to orthogonalize the vector $\overrightarrow{\boldsymbol{pp}_c}$ and the vector $\nabla f(\boldsymbol{p}_c)$. In order to overcome the difficulty, we add Algorithm 12 behind* **Do ... While** *cycle body in Second Remedial Algorithm.*

Algorithm 12 includes three components: (1) The Newton's steepest gradient descent method in the first step; (2) The Newton's iterative method closely associated with the test point $\boldsymbol{p}$ *in the second step; (3) Correcting method in the third step. Algorithm 12 has four advantages and important roles: (1) Algorithm 12 plays a role for accelerating the whole algorithm (Second Remedial Algorithm); (2) The first step plays a role for making the iteration point fall on the planar algebraic curve as far as possible; (3) The second step plays a role for accelerating orthogonalization between the vector* $\overrightarrow{\boldsymbol{p}\boldsymbol{p}_c}$ *and the vector* $\nabla f(\boldsymbol{p}_c)$*; (4) The third step plays a dual role for the accelerating orthogonalization and the promoting the iterative point to fall on the planar algebraic curve. The numerical tests of Second Remedial Algorithm achieve our expected results. No matter how far the test point* $\boldsymbol{p}$ *is from the planar algebraic curve* $f(\boldsymbol{x})$*, Second Remedial Algorithm can converge very robustly and efficiently. Second Remedial Algorithm is the best one in our paper. Of course, the robustness and the efficiency of Second Remedial Algorithm are better than that of the existing algorithms. We are very happy about that.*

Remark 4. *In order to further improve the efficiency of the test point* $\mathbf{p}$ *orthogonal projecting onto plane algebraic curve* $f(\boldsymbol{x})$*, we construct a Comprehensive Integrated Algorithm C which includes two parts: the hybrid second order algorithm in [28] and Second Remedial Algorithm. Firstly run the hybrid second order algorithm in [28]. If the hybrid second order algorithm converges, then it means that Comprehensive Integrated Algorithm C is finished. Otherwise, Second Remedial Algorithm runs until it converges successfully. That is the end of Comprehensive Integrated Algorithm C. The specific description of Comprehensive Integrated Algorithm C is similar to that of Comprehensive Integrated Algorithm B. Here, we are not giving a detailed description of Comprehensive Integrated Algorithm C. When the distance between the test point* $\mathbf{p}$ *and the planar algebraic curve* $f(\boldsymbol{x})$ *is not far, the hybrid second order algorithm in [28] has very high robustness and efficiency. When the distance between the test point* $\mathbf{p}$ *and the planar algebraic curve* $f(\boldsymbol{x})$ *is particularly far, the hybrid second order algorithm does not converge and fails, while Second Remedial Algorithm converges particularly robustly and successfully. To sum up, Comprehensive Integrated Algorithm C absorbs the advantages of two sub-algorithms and overcomes their respective shortcomings such that the robustness and the efficiency of Comprehensive Integrated Algorithm C are maximized.*

3. Convergence Analysis

This section proves that several Comprehensive Algorithms do not depend on the initial iteration points.

Theorem 1. *Comprehensive Algorithm A is independent of the initial iterative point.*

Proof. Firstly, we state the whole operation process of Comprehensive Algorithm A. Comprehensive Algorithm A contains three sub-algorithms (Algorithms 1–3). The role of Algorithm 1 is to repeatedly iterate the iterative formula (3) through Newton's steepest gradient descent method such that the final iteration point $\mathbf{x}_{n+1}$ could fall on the planar algebraic curve where the final iteration point $\mathbf{x}_{n+1}$ is denoted as $\mathbf{p}_c$. The function of Algorithm 2 is to seek the footpoint $\mathbf{q}$. The curvature circle $\bigcirc$ determined by the point $\mathbf{p}_c$ is obtained from the iterative point $\mathbf{p}_c$ on the planar algebraic curve $f(\mathbf{x})$ of Algorithm 1, where the curvature k, the radius R and the center $\mathbf{m}$ are determined by formulas (11)–(13), respectively. The intersection of the line segment $\overline{\mathbf{mp}}$ connecting the center $\mathbf{m}$ and the test point $\mathbf{p}$ and the curvature circle $\bigcirc$ is foot point $\mathbf{q}$. The footpoint $\mathbf{q}$ could be orthogonally projected onto the planar algebraic curve $f(\mathbf{x})$ by repeated iteration of Algorithm 3 where at this moment the test point is not the original test point $\mathbf{p}$ but the footpoint point $\mathbf{q}$ solved by Algorithm 2. Repeatedly run Algorithm 2 and Algorithm 3 bound together until the distance between the current footpoint $\mathbf{q}$ and the previous footpoint $\mathbf{q}$ is near zero.

Secondly, the Comprehensive Algorithm A is independent of the initial iterative point. No matter how far the original test point $\mathbf{p}$ is from the planar algebraic curve $f(\mathbf{x})$, no matter where the initial iterative point $\mathbf{x}_0$ is located, Algorithms 1 can ensure that the final iterative point $\mathbf{x}_{n+1}$ or $\mathbf{p}_c$ of the initial iterative point can fall on the planar algebraic curve $f(\mathbf{x})$. It is obvious that the distance $\|\mathbf{p}_c - \mathbf{p}_\Gamma\|$

between the iteration point $\mathbf{p}_c$ and the orthogonal projection point $\mathbf{p}_\Gamma$ is much smaller than the distance $\|\mathbf{p} - \mathbf{p}_\Gamma\|$ between the orthogonal projection point $\mathbf{p}_\Gamma$ and the original test point $\mathbf{p}$. From the iterative point $\mathbf{p}_c$ fallen on the planar algebraic curve $f(\mathbf{x})$, we can calculate the corresponding curvature k and its center $\mathbf{m}$ and radius R. The intersection point $\mathbf{q}$ of the curvature circle $\bigcirc$ and the line segment $\overline{\mathbf{mp}}$ connecting the original test point $\mathbf{p}$ and the center $\mathbf{m}$ of the curvature circle $\bigcirc$ is just the footpoint $\mathbf{q}$. That is to say, the footpoint $\mathbf{q}$ is directly generated by the curvature circle $\bigcirc$ and the line segment $\overline{\mathbf{mp}}$, and the curvature circle $\bigcirc$ is controlled by the iterative point $\mathbf{p}_c$ fallen on the planar algebraic curve $f(\mathbf{x})$. So the footpoint $\mathbf{q}$ is directly controlled by the original test point $\mathbf{p}$ and the iterative point $\mathbf{p}_c$, while the current footpoint $\mathbf{q}$ is between the orthogonal projection point $\mathbf{p}_\Gamma$ and the current iterative point $\mathbf{p}_c$. It also shows that Algorithm 2 plays a decisive role in the convergence robustness of Comprehensive Algorithm. In addition, we can also know that the distance between the footpoint point $\mathbf{q}$ and the planar algebraic curve $f(\mathbf{x})$ is much smaller than the distance between the original test point $\mathbf{p}$ and the planar algebra curve $f(\mathbf{x})$. At this point, we keep running Algorithm 3 with the footpoint point $\mathbf{q}$ as the current test point until the current test point can be orthogonally projected onto the plane algebraic curve $f(\mathbf{x})$ with guaranteed convergence of Algorithm 3. And now we can call the orthogonal projection point of the footpoint point $\mathbf{q}$ as also the iterative point $\mathbf{p}_c$ fallen on the planar algebraic curve $f(\mathbf{x})$. The first reason is the distance between the current iterative point $\mathbf{p}_c$ and the orthogonal projection point $\mathbf{p}_\Gamma$ of the original test point point $\mathbf{p}$ is smaller than the one between the previous iterative point $\mathbf{p}_c$ and the orthogonal projection point $\mathbf{p}_\Gamma$ of the original test point $\mathbf{p}$. The second reason is that it establishes a solid foundation for the convergence robustness of the subsequent sub-algorithms implementation. Then according to the requirements of Comprehensive Algorithm A, the second step and third step of Comprehensive Algorithm A are executed once per cycle, the distance $\|\mathbf{p}_c - \mathbf{p}_\Gamma\|$ between the current iterative point $\mathbf{p}_c$ on the planar algebraic curve and the orthogonal projection point $\mathbf{p}_\Gamma$ of the original test point $\mathbf{p}$ of the execution result is smaller than that between the previous iterative point $\mathbf{p}_c$ on the planar algebraic curve $f(\mathbf{x})$ and the orthogonal projection point $\mathbf{p}_\Gamma$ of the original test point $\mathbf{p}$. The distance $\|\mathbf{p}_c - \mathbf{p}_\Gamma\|$ between the current iterative point $\mathbf{p}_c$ and the orthogonal projection point $\mathbf{p}_\Gamma$ of the original test point $\mathbf{p}$ is becoming smaller. So repeatedly iterate the second step and the third step of Comprehensive Algorithm A until the distance $\|\mathbf{p}_c - \mathbf{p}_\Gamma\|$ between the current iterative point $\mathbf{p}_c$ and the orthogonal projection point $\mathbf{p}_\Gamma$ of the original test point $\mathbf{p}$ is becoming smaller and smaller. Ultimately, the distance $\|\mathbf{p}_c - \mathbf{p}_\Gamma\|$ between the current iterative point $\mathbf{p}_c$ and the orthogonal projection point $\mathbf{p}_\Gamma$ of the original test point $\mathbf{p}$ is becoming zero. It also demonstrates that Comprehensive Algorithm A is completely convergent. This further proves that Comprehensive Algorithm A can completely converge no matter how far away the original test point $\mathbf{p}$ is from the planar algebraic curve and no matter where the initial iterative point $\mathbf{x}_0$ of Comprehensive Algorithm A is on the plane. This means Comprehensive Algorithm A is independent of the initial iterative point. □

Theorem 2. *Comprehensive Algorithm B is independent of the initial iterative point.*

Proof. In the last step of the iterative formula (18) in Algorithm 5, Newton's iteration method, which is closely related to the original test point $\mathbf{p}$, is added. In this way, the iterative formula (17) is transformed into the iterative formula (18) in Algorithm 5. Algorithm 5 has several advantages over Algorithm 3. It can speed up the iteration, improve its accuracy and promote the orthogonalization of the tangent vector derived from the iteration point on the planar algebraic curve and the tangent vector connecting the test point and the iterative point. Replace Algorithm 3 of Comprehensive Algorithm A with Algorithm 5 to get Comprehensive Algorithm B. Since Comprehensive Algorithm A is independent of the initial iterative point, so Comprehensive Algorithm B is naturally independent of the initial iterative point. □

Theorem 3. *Comprehensive Integrated Algorithm A is independent of the initial iterative point.*

Proof. Comprehensive Integrated Algorithm A consists of two parts: the hybrid second order algorithm and Comprehensive Algorithm A. Whether the test point is very far or very close to the planar algebraic curve, the hybrid second order algorithm is executed several times. If this algorithm converges, then it represents that the execution of Comprehensive integrated Algorithm A is over. So Comprehensive Integrated Algorithm A is independent of the initial iterative point. If the hybrid second order algorithm does not converge, then run Comprehensive Algorithm A of the second step of Comprehensive Integrated Algorithm A. Because whether the test point is very far from or very close to the planar algebraic curve, we know from Theorem 1 that Comprehensive Algorithm A is independent of the initial iterative point. To sum up, Comprehensive Integrated Algorithm A is independent of the initial iterative point. □

In a similar way to the proof of Theorem 3, we can state the following result.

Theorem 4. *Comprehensive Integrated Algorithm B is independent of the initial iterative point.*

Theorem 5. *The first remedial algorithm of Comprehensive Algorithm B is independent of the initial iterative point.*

Proof. From the Figure 5, for any initial iterative point, the final iterative point $\mathbf{p}_c$ of Algorithm 1 in the first step of the first remedial algorithm of Comprehensive Algorithm B can ensure that it falls on the planar algebraic curve $f(\mathbf{x})$. The left endpoint $\mathbf{L}_0$ is the only one that can be determined through third step of the first remedial algorithm of Comprehensive Algorithm B. Graphic display shows that the left endpoint $\mathbf{L}_0$ and the original test point $\mathbf{p}$ are on both sides of the planar algebraic curve. Namely, there is only one intersection point (also written as $\mathbf{p}_c$) between the line segment $\overline{\mathbf{L}_0\mathbf{p}}$ and the planar algebraic curve $f(\mathbf{x})$. Because the hybrid method of combining Newton's iterative method and binary search method is global convergence method, the intersection $\mathbf{p}_c$ of the line segment $\overline{\mathbf{L}_0\mathbf{p}}$ and the planar algebraic curve $f(\mathbf{x})$ can be accurately and uniquely solved by this method. Then repeatedly iterate and run Step 2, Step 3 and Step 4, the distance $\|\mathbf{p}_c - \mathbf{p}_\Gamma\|$ between the current intersection point $\mathbf{p}_c$ and the orthogonal projection point $\mathbf{p}_\Gamma$ of the original test point $\mathbf{p}$ continues to shrink to zero. So we have this conclusion that the first remedial algorithm of Comprehensive Algorithm B is independent of the initial iterative point. □

Theorem 6. *Second Remedial Algorithm is independent of the initial iterative point.*

Proof. In Remark 3, we give a detailed description of the geometric interpretation of Second Remedial Algorithm. In this proof, we only elaborate on the most important geometric significance of Second Remedial Algorithm. The first step of Second Remedial Algorithm is to let the initial iteration point fall on the planar algebraic curve as much as possible through Newton's steepest gradient descent method of Algorithm 1. Moreover, there is few restriction on the selection of the initial iterative point. The purpose of **Do** ... **While** cycle body in Second Remedial Algorithm is to continuously and gradually move the iterative point $\mathbf{p}_c$ to fall on the planar algebraic curve to the orthogonal projection point $\mathbf{p}_\Gamma$. The second step in **Do**... **While** cycle body in Second Remedial Algorithm has two characteristics. Since the footpoint $\mathbf{q}$ is the unique intersection point of the curvature circle and the straight line segment $\overline{\mathbf{mp}}$ connecting the centre $\mathbf{m}$ of the curvature circle and the test point $\mathbf{p}$, the footpoint $\mathbf{q}$ is always closely related to the iterative point $\mathbf{p}_c$ fallen on the planar algebraic curve and the test point $\mathbf{p}$. The first characteristic is that the footpoint $\mathbf{q}$ can guarantee the global convergence of the whole algorithm (Second Remedial Algorithm). The second characteristic is that the distance between the footpoint $\mathbf{q}$ and the planar algebraic curve is much smaller than the distance between the test point $\mathbf{p}$ and the planar algebraic curve. So the third step with Algorithm 1 in **Do** ... **While** cycle body can very robustly iterate the footpoint $\mathbf{q}$ onto the planar algebraic curve. The core thought of **Do** ... **While** cycle body in Second Remedial Algorithm is to keep the iterative point $\mathbf{p}_c$ fallen on the planar algebraic curve moving towards the orthogonal projection point $\mathbf{p}_\Gamma$ such that the distance $\|\mathbf{p}_c - \mathbf{p}_\Gamma\|$ between the iterative point $\mathbf{p}_c$ and the orthogonal projection point $\mathbf{p}_\Gamma$ becomes smaller and smaller. If the distance $\|\mathbf{p}_c - \mathbf{p}_\Gamma\|$ gets smaller and smaller, we have found that the decreasing speed of the distance $\|\mathbf{p}_c - \mathbf{p}_\Gamma\|$ is getting slower and slower. Especially the second formula of the formula (8) is very difficult to be satisfied. Algorithm 12 behind the loop body has four advantages and important roles: (1) Algorithm 12 plays a role for accelerating the whole algorithm (Second Remedial Algorithm); (2) The first step plays a role for making the iteration point fall on the planar algebraic curve as far as possible; (3) The second step plays a role for accelerating orthogonalization between the vector $\overrightarrow{\mathbf{p}\mathbf{p}_c}$ and the vector $\nabla f(\mathbf{p}_c)$; (4) The third step plays a dual role for the accelerating orthogonalization and the promoting the iterative point to fall on the planar algebraic curve. No matter how far the test point is from the planar algebraic curve, Second Remedial Algorithm converges very robustly and efficiently. By adding this step, the efficiency and the robustness for Algorithm 12 of Second Remedial Algorithm is further improved. Then the robustness and the efficiency of Second Remedial Algorithm is also further improved. So Second Remedial Algorithm is independent of the initial iterative point. In addition, in a similar way to the proof of Theorem 3, it is not difficult to know that Comprehensive Integrated Algorithm C is also independent of the initial iterative point. □

4. Numerical Comparison Results

We now present some examples to illustrate the efficiency and the comparison of the newly developed method of Comprehensive Algorithm B and Second Remedial Algorithm to show the two algorithms' high robustness and efficiency for very remote test points. We have three examples to represent closed planar algebraic curve, two sub-closed planar algebraic curves, two branches but not closed planar algebra curves and a single branch not closed the planar algebra curve, respectively. All computations were done using VC++6.0. We used $\varepsilon = 10^{-16}$. The following stopping criteria is used for Comprehensive Algorithm B and Second Remedial Algorithm . In Tables 1–3, the four symbols $\mathbf{p}$, $\mathbf{p}_\Gamma$, $|f(\mathbf{p}_\Gamma)|$ and $|\langle \mathbf{V}_1, \mathbf{V}_2\rangle|$ are the original test point, the orthogonal projection point of the original test point, the deviation degree of the orthogonal projection point on the planar algebraic curve and the absolute value of the inner product of two vectors $\mathbf{V}_1$ and $\mathbf{V}_2$, respectively, where $\mathbf{V}_1$ is $\overrightarrow{\mathbf{p}\mathbf{p}_\Gamma}$ and $\mathbf{V}_2$ is the tangent vector $\left[-\frac{\partial f}{\partial y}, \frac{\partial f}{\partial x}\right]$ of the orthogonal projection point $\mathbf{p}_\Gamma$ on the planar algebraic curve $f(\mathbf{x})$. Thanks to the suggestions by the reviewers, the fourth quadrant result values of the three tables are implemented by Second Remedial Algorithm in Maple 18 environment.

Example 1 (Reference to [28]). *Suppose a planar algebraic curve $f(x,y) = x^6 + 2x^5y - 2x^3y^2 + x^4 - y^3 + 2y^8 - 4 = 0$ (See Figure 7). In each of the four quadrants, randomly select four distant test points. We calculate the corresponding orthogonal projection point for each test point via computation by Comprehensive Algorithm B and Second Remedial Algorithm. The specific results are shown in Table 1).*

Example 2. *Suppose a planar algebraic curve $f(x,y) = x^{10} + 6xy + 2y^{18} - 2 = 0$(See Figure 8). In each of the four quadrants, randomly select four distant test points. We calculate the corresponding orthogonal projection point for each test point via computation by Comprehensive Algorithm B and Second Remedial Algorithm. The specific results are shown in Table 2.*

Example 3. *Suppose a planar algebraic curve $f(x,y) = x^{10} + 6xy + 2y^{16} + 2 = 0$(See Figure 9). In each of the four quadrants, randomly select four distant test points. We calculate the corresponding orthogonal projection point for each test point via computation by Comprehensive Algorithm B and Second Remedial Algorithm. The specific results are shown in Table 3.*

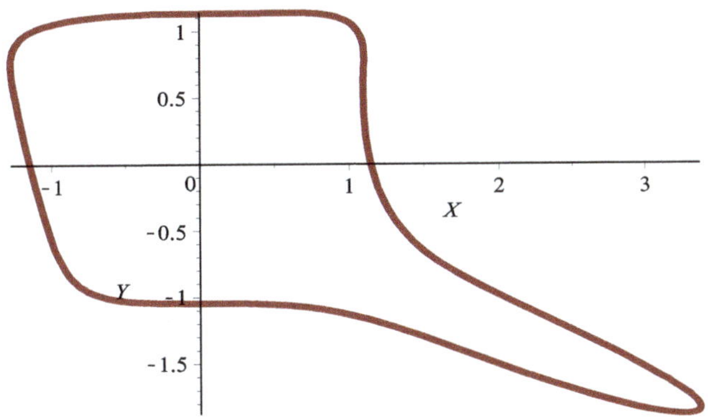

Figure 7. Graphical representation of the planar algebraic curve for Example 1.

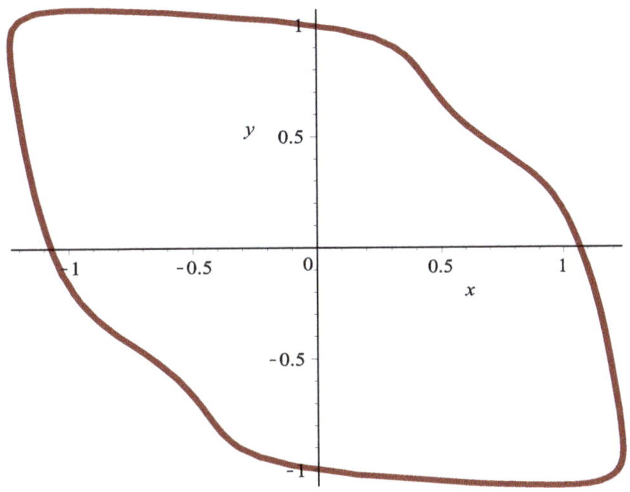

Figure 8. Graphical representation of the planar algebraic curve for Example 2.

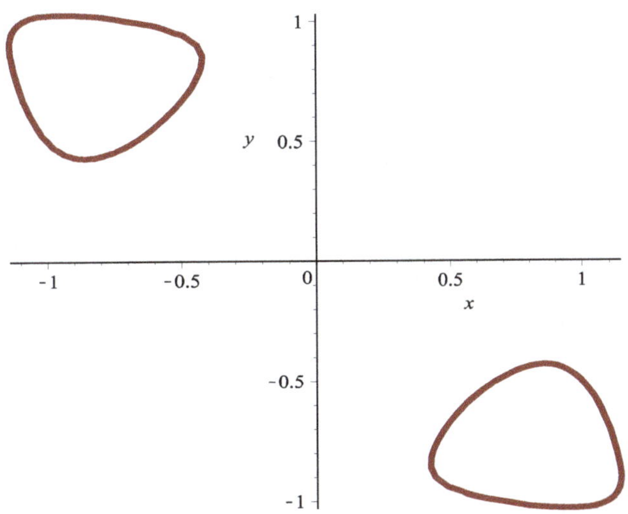

Figure 9. Graphical representation of the planar algebraic curve for Example 3.

Table 1. Test results of Comprehensive Algorithm B for Example 1.

$\mathbf{p}$	(1325, 7447)	(779, 325)	(990, 1375)	(0.59, 1377)	(−0.5, 8623)	(−16, 598)	(−3, 231)	(−21, 247)
$\mathbf{p}_\Gamma$	(3.1087, −1.8666)	(3.3647, −1.8184)	(3.3695, −1.8380)	(0.60, 1.140)	(−0.01582, 1.13368)	(−0.3353, 1.13074)	(−0.2268, 1.1328)	(−0.6071, 1.1166)
$\lvert f(\mathbf{p}_\Gamma)\rvert$	$3.9 \cdot 10^{-10}$	$7.9 \cdot 10^{-11}$	$1.4 \cdot 10^{-11}$	$2.7 \cdot 10^{-12}$	$2.3 \cdot 10^{-12}$	$4.0 \cdot 10^{-12}$	$1.9 \cdot 10^{-12}$	$2.3 \cdot 10^{-12}$
$\lvert\langle \mathbf{V}_1, \mathbf{V}_2\rangle\rvert$	$2.3 \cdot 10^{-10}$	$4.4 \cdot 10^{-11}$	$2.9 \cdot 10^{-09}$	$4.8 \cdot 10^{-12}$	$2.1 \cdot 10^{-13}$	$1.5 \cdot 10^{-12}$	$1.4 \cdot 10^{-11}$	$9.6 \cdot 10^{-12}$
$\mathbf{p}$	(−42, −127)	(−5, −38)	(−9,−579)	(−537, −11)	(78, −123)	(168, −12)	(537, −31)	(91, −221)
$\mathbf{p}_\Gamma$	(0.6633, -0.9941)	(0.4951, −1.0292)	(−0.2183, −1.0445)	(−1.2752, 0.7396)	(3.3519, −1.8685)	(3.3694, −1.8417)	(3.3694, −1.8414)	(3.3415, −1.8736)
$\lvert f(\mathbf{p}_\Gamma)\rvert$	$7.4 \cdot 10^{-12}$	$1.4 \cdot 10^{-12}$	$1.7 \cdot 10^{-12}$	$6.1 \cdot 10^{-12}$	$4.9 \cdot 10^{-12}$	$1.8 \cdot 10^{-13}$	$5.4 \cdot 10^{-13}$	$8.6 \cdot 10^{-12}$
$\lvert\langle \mathbf{V}_1, \mathbf{V}_2\rangle\rvert$	$1.7 \cdot 10^{-13}$	$1.2 \cdot 10^{-12}$	$1.1 \cdot 10^{-12}$	$1.7 \cdot 10^{-12}$	$2.3 \cdot 10^{-13}$	$5.7 \cdot 10^{-14}$	$2.8 \cdot 10^{-13}$	$4.5 \cdot 10^{-12}$

Table 2. Test results of Comprehensive Algorithm B for Example 2.

$\mathbf{p}$	(565, 945)	(979, 325)	(375, 405)	(1959, 1377)	(−9356, 8623)	(−816, 798)	(−3987, 1231)	(−4821, 647)
$\mathbf{p}_\Gamma$	(0.2978, −0.9108)	(1.0773, −0.0164)	(0.6094, 0.5450)	(0.4780, 0.6961)	(−1.2055, 1.0209)	(−1.2035, 1.0230)	(−1.2288, 0.9794)	(−1.2341, 0.9554)
$\lvert f(\mathbf{p}_\Gamma)\rvert$	$1.1 \cdot 10^{-14}$	$6.0 \cdot 10^{-13}$	$8.0 \cdot 10^{-18}$	$9.2 \cdot 10^{-15}$	$5.8 \cdot 10^{-15}$	$1.1 \cdot 10^{-14}$	$2.5 \cdot 10^{-15}$	$1.1 \cdot 10^{-14}$
$\lvert\langle \mathbf{V}_1, \mathbf{V}_2\rangle\rvert$	0	$1.8 \cdot 10^{-12}$	$2.3 \cdot 10^{-13}$	$4.5 \cdot 10^{-12}$	$6.9 \cdot 10^{-10}$	$2.1 \cdot 10^{-11}$	$1.7 \cdot 10^{-10}$	$1.1 \cdot 10^{-10}$
$\mathbf{p}$	(−7942, −275)	(−598, −98)	(−3709, −1979)	(−2937, −1391)	(9708, −323)	(2608, −1912)	(7347, −931)	(5091, −1921)
$\mathbf{p}_\Gamma$	(−1.2367, 0.8930)	(−1.1847, 0.4851)	(−1.0035, −0.1601)	(−1.0225, 0.1224)	(1.1427, −0.9100)	(1.1224, −0.9787)	(1.1414, −0.9273)	(1.1344,−0.9563)
$\lvert f(\mathbf{p}_\Gamma)\rvert$	$2.3 \cdot 10^{-15}$	$1.3 \cdot 10^{-11}$	$9.5 \cdot 10^{-15}$	$1.4 \cdot 10^{-13}$	$4.7 \cdot 10^{-15}$	$3.2 \cdot 10^{-15}$	$4.4 \cdot 10^{-15}$	$1.7 \cdot 10^{-17}$
$\lvert\langle \mathbf{V}_1, \mathbf{V}_2\rangle\rvert$	$4.0 \cdot 10^{-11}$	$3.6 \cdot 10^{-12}$	$8.7 \cdot 10^{-11}$	$7.2 \cdot 10^{-12}$	$5.8 \cdot 10^{-11}$	$8.7 \cdot 10^{-11}$	$7.2 \cdot 10^{-12}$	$2.1 \cdot 10^{-10}$

Table 3. Test results of Comprehensive Algorithm B for Example 3.

$\mathbf{p}$	(1387, 645)	(1879, 395)	(3075, 205)	(1956, 777)	(−9256, 4603)	(−836, 1798)	(−5987, 1031)	(−4181, 1247)
$\mathbf{p}_\Gamma$	(−1.0629, 0.6023)	(0.4232, −0.8650)	(0.4214, −0.8515)	(0.4276, −0.8795)	(−1.1305, 0.9655)	(−1.0823, 1.0103)	(−0.4232, 0.8220)	(−1.1369, 0.9490)
$\lvert f(\mathbf{p}_\Gamma)\rvert$	$1.7 \cdot 10^{-13}$	$1.3 \cdot 10^{-16}$	$8.1 \cdot 10^{-17}$	$1.0 \cdot 10^{-16}$	$5.3 \cdot 10^{-17}$	$1.8 \cdot 10^{-15}$	$6.1 \cdot 10^{-17}$	$1.0 \cdot 10^{-17}$
$\lvert\langle \mathbf{V}_1, \mathbf{V}_2\rangle\rvert$	$3.6 \cdot 10^{-12}$	$3.2 \cdot 10^{-12}$	$6.3 \cdot 10^{-13}$	$6.3 \cdot 10^{-12}$	$1.7 \cdot 10^{-10}$	$4.3 \cdot 10^{-11}$	$1.8 \cdot 10^{-12}$	$6.5 \cdot 10^{-11}$
$\mathbf{p}$	(−7342, −1275)	(−5098, −918)	(−3217, −2079)	(−2337, −1251)	(9508, −375)	(6608, −712)	(2347, −931)	(1491, −1321)
$\mathbf{p}_\Gamma$	(−0.4226, 0.8617)	(−0.4227, 0.8623)	(−1.0274, 0.5370)	(−1.0471, 0.5706)	(1.2367, −0.9267)	(1.2353, −0.9460)	(1.2255, −0.9888)	(1.2068, −1.0194)
$\lvert f(\mathbf{p}_\Gamma)\rvert$	$5.7 \cdot 10^{-17}$	$1.3 \cdot 10^{-16}$	$4.4 \cdot 10^{-14}$	$1.1 \cdot 10^{-13}$	$2.2 \cdot 10^{-15}$	$2.8 \cdot 10^{-15}$	$1.8 \cdot 10^{-15}$	$1.1 \cdot 10^{-15}$
$\lvert\langle \mathbf{V}_1, \mathbf{V}_2\rangle\rvert$	$1.0 \cdot 10^{-11}$	$5.5 \cdot 10^{-12}$	$7.3 \cdot 10^{-12}$	$9.1 \cdot 10^{-12}$	$2.2 \cdot 10^{-11}$	$3.6 \cdot 10^{-12}$	$5.4 \cdot 10^{-11}$	$2.2 \cdot 10^{-11}$

Remark 5. *Besides all test points of the three examples mentioned above are tested by Comprehensive Algorithm B, we have tested them again with the Second Remedial Algorithm. All the test results are consistent with those of Comprehensive Algorithm B and convergent. In addition, in the region* $[-3000, 3000] \times [-3000, 3000]$ *of each example, we randomly select a large number of test points, the probability of non-convergence is particularly low by Second Remedial Algorithm. Further, we use Second Remedial Algorithm other examples with test points in a very large area, and the probability of non-convergence is also very low. Second Remedial Algorithm is verified to be the best one again in our paper. Of course, the robustness and the efficiency of Second Remedial Algorithm is better than that of the existing algorithms.*

5. Conclusions and Future Work

In this paper, we have constructed a Comprehensive Algorithm which is an improved curvature circle algorithm for orthogonal projecting onto planar algebraic curve. Based on an integrated hybrid second-order algorithm [28], the Comprehensive Algorithm (the improved curvature circle algorithm) has also incorporated the curvature circle technique and Newton's gradient steepest descent method such that it can converge robustly and efficiently no matter how far the test point is from the planar algebraic curve and no matter where the initial iterative point is located. Furthermore, we propose Second Remedial Algorithm based on Comprehensive Algorithm B. In particular, its robustness and efficiency is greatly improved than that of Comprehensive Algorithm B and it achieves our expected result. The numerical examples show that the improved curvature circle algorithm is superior to the existing ones. In future work, we try to refine the idea of Comprehensive Algorithm and Second Remedy Algorithm. And the idea is applied to point orthogonal projecting onto spatial algebraic curve and algebraic surface.

Author Contributions: The contribution of all the authors is the same. All of the authors team up to develop the current draft. Z.W. is responsible for investigating, providing resources and methodology, the original draft, writing, reviewing, validation, editing and supervision of this work. X.L. is responsible for software, algorithm, program implementation, formal analysis, visualization, writing, reviewing, editing and supervision of this work.

Acknowledgments: We take the opportunity to thank the anonymous reviewers for their thoughtful and meaningful comments.

References

1. Bajaj, C.L.; Bernardini, F.; Xu, G. Reconstructing surfaces and functions on surfaces from unorganized three-dimensional data. *Algorithmica* **1997**, *19*, 243–261. [CrossRef]
2. Jüttler, B.; Felis, A. Least-squares fitting of algebraic spline surfaces. *Adv. Comput. Math.* **2002**, *17*, 135–152.
3. Song, X.; Jüttler, B. Modeling and 3D object reconstruction by implicitly defined surfaces with sharp features. *Comput. Graph.* **2009**, *33*, 321–330. [CrossRef]
4. Schulz, T.; Jüttler, B. Envelope computation in the plane by approximate implicitization. *Appl. Algebra Eng. Commun. Comput.* **2011**, *22*, 265–288. [CrossRef]
5. Kanatani, K.; Sugaya, Y. Unified computation of strict maximum likelihood for geometric fitting. *J. Math. Imaging Vis.* **2010**, *38*, 1–13. [CrossRef]
6. Mullen, P.; De Goes, F.; Desbrun, M.; Cohen-Steiner, D.; Alliez, P. Signing the unsigned: Robust surface reconstruction from raw pointsets. *Comput. Graph. Forum* **2010**, *29*, 1733–1741. [CrossRef]
7. Upreti, K.; Subbarayan, G. Signed algebraic level sets on NURBS surfaces and implicit Boolean signed algebraic level sets on NURBS surfaces and implicit Boolean. *Comput. Aided Des.* **2017**, *82*,112–126. [CrossRef]
8. Rouhani, M., Sappa, A.D. Implicit polynomial representation through a fast fitting error estimation. *IEEE Trans. Image Process.* **2012**, *21*, 2089–2098. [CrossRef]
9. Zagorchev, L.G.; Goshtasby, A.A. A curvature-adaptive implicit surface reconstruction for irregularly spaced points. *IEEE Trans. Vis. Comput. Graph.* **2012**, *18*, 1460–1473. [CrossRef]
10. William, H.P.; Brian, P.F.; Teukolsky, S.A.; William, T.V. *Numerical Recipes in C: The Art of Scientific Computing*, 2nd ed.; Cambridge University Press: Cambridge, UK, 1992.

11. Steve, S.; Sandford, L.; Ponce, J. Using geometric distance fits for 3-D object modeling and recognition. *IEEE Trans. Pattern Anal. Mach. Intell.* **1994**, *16*, 1183–1196.
12. Morgan, A.P. Polynomial continuation and its relationship to the symbolic reduction of polynomial systems. In *Symbolic and Numerical Computation for Artificial Intelligence*; Academic Press: Cambridge, MA, USA, 1992; pp. 23–45.
13. Layne, T.W.; Billups, S.C.; Morgan, A.P. Algorithm 652: HOMPACK: A suite of codes for globally convergent homotopy algorithms. *ACM Trans. Math. Softw.* **1987**, *13*, 281–310.
14. Berthold, K.P.H. Relative orientation revisited. *J. Opt. Soc. Am. A* **1991**, *8*, 1630–1638.
15. Dinesh, M.; Krishnan, S. Solving algebraic systems using matrix computations. *ACM Sigsam Bull.* **1996**, *30*, 4–21.
16. Chionh, E.-W. Base Points, Resultants, and the Implicit Representation of Rational Surfaces. Ph.D. Thesis, University of Waterloo, Waterloo, ON, Canada, 1990.
17. De Montaudouin, Y.; Tiller, W. The Cayley method in computer aided geometric design. *Comput. Aided Geom. Des.* **1984**, *1*, 309–326. [CrossRef]
18. Albert, A.A. *Modern Higher Algebra*; D.C. Heath and Company: New York, NY, USA, 1933.
19. Thomas, W.; David, S.; Anderson, C.; Goldman, R.N. Implicit representation of parametric curves and surfaces. *Comput. Vis. Graph. Image Proc.* **1984**, *28*, 72–84.
20. Nishita, T.; Sederberg, T.W.; Kakimoto, M. Ray tracing trimmed rational surface patches. *ACM SIGGRAPH Comput. Graph.* **1990**, *24*, 337–345. [CrossRef]
21. Elber, G.; Kim, M.-S. Geometric Constraint Solver Using Multivariate Rational Spline Functions. In Proceedings of the 6th ACM Symposium on Solid Modeling and Applications, Ann Arbor, MI, USA, 4–8 June 2001; pp. 1–10.
22. Sherbrooke, E.C.; Patrikalakis, N.M. Computation of the solutions of nonlinear polynomial systems. *Comput. Aided Geom. Des.* **1993**, *10*, 379–405. [CrossRef]
23. Hartmann, E. The normal form of a planar curve and its application to curve design. In *Mathematical Methods for Curves and Surfaces II*; Vanderbilt University Press: Nashville, TN, USA, 1997; pp. 237–244.
24. Hartmann, E. On the curvature of curves and surfaces defined by normal forms. *Comput. Aided Geom. Des.* **1999**, *16*, 355–376. [CrossRef]
25. Nicholas, J.R. Implicit polynomials, orthogonal distance regression, and the closest point on a curve. *IEEE Trans. Pattern Anal. Mach. Intell.* **2000**, *22*, 191–199.
26. Martin, A.; Bert, J. Robust computation of foot points on implicitly defined curves. In *Mathematical Methods for Curves and Surfaces: Troms?* Nashboro Press: Brentwood, TN, USA, 2004; pp. 1–10.
27. Hu, M.; Zhou, Y.; Li, X. Robust and accurate computation of geometric distance for Lipschitz continuous implicit curves. *Vis. Comput.* **2017**, *33*, 937–947. [CrossRef]
28. Li, X.; Pan, F.; Cheng, T.; Wu, Z.; Liang, J.; Hou, L. Integrated hybrid second order algorithm for orthogonal projection onto a planar implicit curve. *Symmetry* **2018**, *10*, 164. [CrossRef]
29. Goldman R. Curvature formulas for implicit curves and surfaces. *Comput. Aided Geom. Des.* **2005**, *22*, 632–658. [CrossRef]

15

Advances in the Semilocal Convergence of Newton's Method with Real-World Applications

Ioannis K. Argyros [1], Á. Alberto Magreñán [2,*], Lara Orcos [3] and Íñigo Sarría [4]

[1] Department of Mathematics Sciences Lawton, Cameron University, Lawton, OK 73505, USA; iargyros@cameron.edu

[2] Departamento de Matemáticas y Computación, Universidad de La Rioja, 26006 Logroño, Spain

[3] Departamento de Matemática Aplicada, Universidad Politècnica de València, 46022 València, Spain; laorpal@doctor.upv.es

[4] Escuela Superior de Ingeniería y Tecnología, Universidad Internacional de La Rioja, 26006 Logroño; Spain; inigo.sarria@unir.net

[*] Correspondence: angel-alberto.magrenan@unirioja.es

Abstract: The aim of this paper is to present a new semi-local convergence analysis for Newton's method in a Banach space setting. The novelty of this paper is that by using more precise Lipschitz constants than in earlier studies and our new idea of restricted convergence domains, we extend the applicability of Newton's method as follows: The convergence domain is extended; the error estimates are tighter and the information on the location of the solution is at least as precise as before. These advantages are obtained using the same information as before, since new Lipschitz constant are tighter and special cases of the ones used before. Numerical examples and applications are used to test favorable the theoretical results to earlier ones.

Keywords: Banach space; Newton's method; semi-local convergence; Kantorovich hypothesis

1. Introduction

In this study we are concerned with the problem of approximating a locally unique solution z^* of equation

$$G(x) = 0, \tag{1}$$

where G is a Fréchet-differentiable operator defined on a nonempty, open convex subset D of a Banach space E_1 with values in a Banach space E_2.

Many problems in Computational disciplines such us Applied Mathematics, Optimization, Mathematical Biology, Chemistry, Economics, Medicine, Physics, Engineering and other disciplines can be solved by means of finding the solutions of equations in a form like Equation (1) using Mathematical Modelling [1–7]. The solutions of this kind of equations are rarely found in closed form. That is why most solutions of these equations are given using iterative methods. A very important problem in the study of iterative procedures is the convergence region. In general this convergence region is small. Therefore, it is important to enlarge the convergence region without additional hypotheses.

The study of convergence of iterative algorithms is usually centered into two categories: Semi-local and local convergence analysis. The semi-local convergence is based on the information around an initial point, to obtain conditions ensuring the convergence of theses algorithms while the local convergence is based on the information around a solution to find estimates of the computed radii of the convergence balls.

Newton's method defined for all $n = 0, 1, 2, \ldots$ by

$$z_{n+1} = z_n - G'(z_n)^{-1}G(z_n), \tag{2}$$

is undoubtedly the most popular method for generating a sequence $\{z_n\}$ approximating z^*, where z_0 is an initial point. There is a plethora of convergence results for Newton's method [1–4,6,8–14]. We shall increase the convergence region by finding a more precise domain where the iterates $\{z_n\}$ lie leading to smaller Lipschitz constants which in turn lead to a tighter convergence analysis for Newton's method than before. This technique can apply to improve the convergence domain of other iterative methods in an analogous way.

Let us consider the conditions:

There exist $z_0 \in \Omega$ and $\eta \geq 0$ such that

$$G'(z_0)^{-1} \in \mathbb{L}(E_2, E_1) \text{ and } \|G'(z_0)^{-1}G(z_0)\| \leq \eta;$$

There exists $T \geq 0$ such that the Lipschitz condition

$$\|G'(z_0)^{-1}(G'(x) - G'(y))\| \leq T\|x - y\|$$

holds for all $x, y \in \Omega$.

Then, the sufficient convergence condition for Newton's method is given by the famous for its simplicity and clarity Kantorovich sufficient convergence criterion for Newton's method

$$h_K = 2T\eta \leq 1. \tag{3}$$

Let us consider a motivational and academic example to show that this condition is not satisfied. Choose $E_1 = E_2 = \mathbb{R}$, $z_0 = 1$, $p \in [0, 0.5)$, $D = S(z_0, 1-p)$ and define function G on D by

$$G(x) = z^3 - p.$$

Then, we have $T = 2(2-p)$. Then, the Kantorovich condition is not satisfied, since $h_K > 1$ for all $p \in (0, 0.5)$. We set $I_K = \emptyset$ to be the set of point satisfying Equation (3). Hence, there is no guarantee that Newton's sequence starting at z_0 converges to $z^* = \sqrt[3]{p}$.

The rest of the paper is structured as follows: In Section 2 we present the semi-local convergence analysis of Newton's method Equation (2). The numerical examples and applications are presented in Section 3 and the concluding Section 4.

2. Semi-Local Convergence Analysis

We need an auxiliary result on majorizing sequences for Newton's method.

Lemma 1. *Let $H > 0$, $K > 0$, $L > 0$, $L_0 > 0$ and $\eta > 0$ be parameters. Suppose that:*

$$h_4 = L_4\eta \leq 1, \tag{4}$$

where

$$L_4^{-1} = \begin{cases} \dfrac{1}{L_0 + H}, & \text{if } b = LK + 2\delta L_0(K - 2H) = 0 \\ 2\dfrac{-\delta(L_0 + H) + \sqrt{\delta^2(L_0 + H)^2 + \delta(LK + 2\delta L_0(K - 2H))}}{LK + 2\delta L_0(K - 2H)}, & \text{if } b > 0 \\ -2\dfrac{\delta(L_0 + H) + \sqrt{\delta^2(L_0 + H)^2 + \delta(LK + 2\delta L_0(K - 2H))}}{LK + 2\delta L_0(K - 2H)}, & \text{if } b < 0 \end{cases}$$

and

$$\delta = \frac{2L}{L + \sqrt{L^2 + 8L_0L}}.$$

holds. Then, scalar sequence $\{t_n\}$ *given by*

$$\begin{aligned} t_0 &= 0, \quad t_1 = \eta, \quad t_2 = t_1 + \tfrac{K\,(t_1 - t_0)^2}{2\,(1 - H\,t_1)}, \\ t_{n+2} &= t_{n+1} + \frac{L\,(t_{n+1} - t_n)^2}{2\,(1 - L_0\,t_{n+1})} \quad \textit{for all} \quad n = 1, 2, \cdots, \end{aligned} \tag{5}$$

is well defined, increasing, bounded from above by

$$t^{**} = \eta + \left(1 + \frac{\delta_0}{1 - \delta}\right) \frac{K\,\eta^2}{2\,(1 - H\,\eta)} \tag{6}$$

and converges to its unique least upper bound t^* *which satisfies*

$$t_2 \le t^* \le t^{**}, \tag{7}$$

where $\delta_0 = \dfrac{L(t_2 - t_1)}{2(1 - L_0 t_2)}$*. Moreover, the following estimates hold:*

$$0 < t_{n+2} - t_{n+1} \le \delta_0\,\delta^{n-1}\,\frac{K\,\eta^2}{2\,(1 - H\,\eta)} \quad \textit{for all} \quad n = 1, 2, \cdots \tag{8}$$

and

$$t^* - t_n \le \frac{\delta_0\,(t_2 - t_1)}{1 - \delta}\,\delta^{n-2} \quad \textit{for all} \quad n = 2, 3, \cdots. \tag{9}$$

Proof. By induction, we show that

$$0 < \frac{L\,(t_{k+1} - t_k)}{2\,(1 - L_0\,t_{k+1})} \le \delta \tag{10}$$

holds for all $k = 1, 2, \cdots$. Estimate Equation (10) is true for $k = 1$ by Equation (4). Then, we have by Equation (5)

$$\begin{aligned} 0 < t_3 - t_2 \le \delta_0\,(t_2 - t_1) \quad &\Longrightarrow \quad t_3 \le t_2 + \delta_0\,(t_2 - t_1) \\ &\Longrightarrow \quad t_3 \le t_2 + (1 + \delta_0)\,(t_2 - t_1) - (t_2 - t_1) \\ &\Longrightarrow \quad t_3 \le t_1 + \tfrac{1 - \delta_0^2}{1 - \delta_0}\,(t_2 - t_1) < t^{**} \end{aligned}$$

and for $m = 2, 3, \cdots$

$$\begin{aligned} t_{m+2} &\le t_{m+1} + \delta_0\,\delta^{m-1}\,(t_2 - t_1) \\ &\le t_m + \delta_0\,\delta^{m-2}\,(t_2 - t_1) + \delta_0\,\delta^{m-1}\,(t_2 - t_1) \\ &\le t_1 + (1 + \delta_0\,(1 + \delta + \cdots + \delta^{m-1}))\,(t_2 - t_1) \\ &= t_1 + (1 + \delta_0\,\tfrac{1 - \delta^m}{1 - \delta})\,(t_2 - t_1) \le t^{**}. \end{aligned}$$

Assume that Equation (10) holds for all natural integers $n \leq m$. Then, we get by Equations (5) and (10) that

$$0 < t_{m+2} - t_{m+1} \leq \delta_0 \, \delta^{m-1} \, (t_2 - t_1) \leq \delta^m \, (t_2 - t_1)$$

and

$$t_{m+2} \leq t_1 + (1 + \delta_0 \, \frac{1 - \delta^m}{1 - \delta}) \, (t_2 - t_1) \leq t_1 + \frac{1 - \delta^{m+1}}{1 - \delta} \, (t_2 - t_1) < t^{**}.$$

Evidently estimate Equation (10) is true, if m is replaced by $m + 1$ provided that

$$\frac{L}{2} \, (t_{m+2} - t_{m+1}) \leq \delta \, (1 - L_0 \, t_{m+2})$$

or

$$\frac{L}{2} \, (t_{m+2} - t_{m+1}) + \delta \, L_0 \, t_{m+2} - \delta \leq 0$$

or

$$\frac{L}{2} \, \delta^m \, (t_2 - t_1) + \delta \, L_0 \left(t_1 + \frac{1 - \delta^{m+1}}{1 - \delta} \, (t_2 - t_1) \right) - \delta \leq 0. \tag{11}$$

Estimate Equation (11) motivates us to define recurrent functions $\{\psi_k\}$ on $[0, 1)$ by

$$\psi_m(s) = \frac{L}{2} \, (t_2 - t_1) \, t^{m+1} + s \, L_0 \, (1 + s + t^2 + \cdots + t^m) \, (t_2 - t_1) - (1 - L_0 \, t_1) \, s.$$

We need a relationship between two consecutive functions ψ_k. We get that

$$\begin{aligned} \psi_{m+1}(s) \quad &= \quad \tfrac{L}{2} \, (t_2 - t_1) \, t^{m+2} + s \, L_0 \, (1 + s + t^2 + \cdots + t^{m+1}) \, (t_2 - t_1) \\ &\quad -(1 - L_0 \, t_1) \, s \\ &= \quad \tfrac{L}{2} \, (t_2 - t_1) \, t^{m+2} + s \, L_0 \, (1 + s + t^2 + \cdots + t^{m+1}) \, (t_2 - t_1) \\ &\quad -(1 - L_0 \, t_1) \, s - \tfrac{L}{2} \, (t_2 - t_1) \, t^m \\ &\quad -s \, L_0 \, (1 + s + t^2 + \cdots + t^m) \, (t_2 - t_1) + (1 - L_0 \, t_1) \, s + \psi_k(s). \end{aligned}$$

Therefore, we deduce that

$$\psi_{m+1}(s) = \psi_m(s) + \frac{1}{2} \, (2 \, L_0 \, t^2 + L \, s - L) \, t^m \, (t_2 - t_1). \tag{12}$$

Estimate Equation (11) is satisfied, if

$$\psi_m(\delta) \leq 0 \quad \text{holds for all} \quad m = 1, 2, \cdots. \tag{13}$$

Using Equation (12) we obtain that

$$\psi_{m+1}(\delta) = \psi_m(\delta) \quad \text{for all} \quad m = 1, 2, \cdots.$$

Let us now define function ψ_∞ on $[0, 1)$ by

$$\psi_\infty(s) = \lim_{m \to \infty} \psi_m(s). \tag{14}$$

Then, we have by Equation (14) and the choice of δ that

$$\psi_\infty(\delta) = \psi_m(\delta) \quad \text{for all} \quad m = 1, 2, \cdots.$$

Hence, Equation (13) is satisfied, if

$$\psi_\infty(\delta) \leq 0. \tag{15}$$

Using Equation (11) we get that

$$\psi_\infty(\delta) = \left(\frac{L_0}{1-\delta}(t_2 - t_1) + L_0 t_1 - 1 \right) \delta. \tag{16}$$

It then, follows from Equations (2.1) and (2.13) that Equation (15) is satisfied. The induction is now completed. Hence, sequence $\{t_n\}$ is increasing, bounded from above by t^{**} given by Equation (6), and as such it converges to its unique least upper bound t^* which satisfies Equation (7). $\square$

Let $S(z, \varrho)$, $\bar{S}(z, \varrho)$ stand, respectively for the open and closed ball in E_1 with center $z \in E_1$ and of radius $\varrho > 0$.

The conditions (A) for the semi-local convergence are:

(A_1) $G : D \subset E_1 \to E_2$ is Fréchet differentiable and there exist $z_0 \in D$, $\eta \geq 0$ such that $G'(z_0)^{-1} \in$ Ł(E_2, E_1) and

$$\|G'(z_0)^{-1} G(z_0)\| \leq \eta.$$

(A_2) There exists $L_0 > 0$ such that for all $x \in D$

$$\|G'(z_0)^{-1}(G'(x) - G'(z_0))\| \leq L_0 \|x - z_0\|.$$

(A_3) $L_0 \eta < 1$ and there exists $L > 0$ such that

$$\|G'(z_0)^{-1}(G'(x) - G'(y))\| \leq L\|x - y\|.$$

for all $x, y \in D_0 := S(z_1, \frac{1}{L_0} - \|G'(z_0)^{-1} G(z_0)\|) \cap D$.

(A_4) There exists $H > 0$ such that

$$\|G'(z_0)^{-1}(G'(z_1) - G'(z_0))\| \leq H\|z_1 - z_0\|,$$

where $z_1 = z_0 - G'(z_0)^{-1} G(z_0)$.

(A_5) There exists $K > 0$ such that for all $\theta \in [0, 1]$

$$\|G'(z_0)^{-1}(G'(z_0 + \theta(z_1 - z_0)) - G'(z_0))\| \leq K\theta \|z_1 - z_0\|.$$

Notice that $(A_2) \implies (A_3) \implies (A_5) \implies (A_4)$. Clearly, we have that

$$H \leq K \leq L_0 \tag{17}$$

and $\frac{L}{L_0}$ can be arbitrarily large [9]. It is worth noticing that (A_3)–(A_5) are not additional to (A_2) hypotheses, since in practice the computation of Lipschitz constant T requires the computation of the other constants as special cases.

Next, first we present a semi-local convergence result relating majorizing sequence $\{t_n\}$ with Newton's method and hypotheses (A).

Theorem 1. *Suppose that hypotheses* (A)*, hypotheses of Lemma 1 and* $\overline{S}(z_0, t^*) \subseteq D$ *hold, where* t^* *is given in Lemma 1. Then, sequence* $\{z_n\}$ *generated by Newton's method is well defined, remains in* $\overline{S}(z_0, t^*)$ *and converges to a solution* $z^* \in \overline{S}(z_0, t^*)$ *of equation* $G(x) = 0$*. Moreover, the following estimates hold*

$$\|z_{n+1} - z_n\| \leq t_{n+1} - t_n \tag{18}$$

and

$$\|z_n - z^*\| \leq t^* - t_n \quad \textit{for all} \quad n = 0, 1, 2, \cdots, \tag{19}$$

where sequence $\{t_n\}$ *is given in Lemma 1. Furthermore, if there exists* $R \geq t^*$ *such that*

$$\overline{S}(z_0, R) \subseteq D \quad \text{and} \quad L_0\,(t^* + R) < 2,$$

then, the solution z^* *of equation* $G(x) = 0$ *is unique in* $\overline{S}(z_0, R)$*.*

Proof. We use mathematical induction to prove that

$$\|z_{k+1} - x_k\| \leq t_{k+1} - t_k \tag{20}$$

and

$$\overline{S}(z_{k+1}, t^* - t_{k+1}) \subseteq \overline{S}(z_k, t^* - t_k) \quad \text{for all} \quad k = 1, 2, \cdots. \tag{21}$$

Let $z \in \overline{S}(z_1, t^* - t_1)$.

Then, we obtain that

$$\|z - z_0\| \leq \|z - z_1\| + \|z_1 - z_0\| \leq t^* - t_1 + t_1 - t_0 = t^* - t_0,$$

which implies $z \in \overline{S}(z_0, t^* - t_0)$. Note also that

$$\|z_1 - z_0\| = \|G'(z_0)^{-1}\,G(z_0)\| \leq \eta = t_1 - t_0.$$

Hence, estimates Equations (20) and (21) hold for $k = 0$. Suppose these estimates hold for $n \leq k$. Then, we have that

$$\|z_{k+1} - z_0\| \leq \sum_{i=1}^{k+1} \|z_i - z_{i-1}\| \leq \sum_{i=1}^{k+1} (t_i - t_{i-1}) = t_{k+1} - t_0 = t_{k+1}$$

and

$$\|z_k + \theta\,(z_{k+1} - z_k) - z_0\| \leq t_k + \theta\,(t_{k+1} - t_k) \leq t^*$$

for all $\theta \in (0, 1)$. Using Lemma 1 and the induction hypotheses, we get in turn that

$$\|G'(z_0)^{-1}(G'(z_{k+1}) - G'(z_0))\| \leq M\|x_{k+1} - z_0\| \leq M(t_{k+1} - t_0) \leq Mt_{k+1} < 1, \tag{22}$$

where

$$M = \begin{cases} H & \text{if} \quad k = 0 \\ L_0 & \text{if} \quad k = 1, 2, \cdots. \end{cases}$$

It follows from Equation (22) and the Banach lemma on invertible operators that $G'(z_{m+1})^{-1}$ exists and

$$\|G'(z_{k+1})^{-1}\,G'(z_0)\| \leq (1 - M\,\|z_{k+1} - z_0\|)^{-1} \leq (1 - M\,t_{k+1})^{-1}. \tag{23}$$

Using iteration of Newton's method, we obtain the approximation

$$\begin{aligned} G(z_{k+1}) &= G(z_{k+1}) - G(z_k) - G'(z_k)\,(z_{k+1} - z_k) \\ &= \int_0^1 (G'(z_k + \theta\,(z_{k+1} - z_k)) - G'(z_m))\,(z_{k+1} - z_k)\,d\theta. \end{aligned} \tag{24}$$

Then, by Equation (24) we get in turn

$$\begin{aligned} &\|G'(z_0)^{-1}\,G(z_{k+1})\| \\ &\le \int_0^1 \|G'(z_0)^{-1}\,(G'(z_k + \theta\,(z_{k+1} - z_k)) - G'(z_k))\|\,\|z_{k+1} - z_k\|\,d\theta \\ &\le M_1 \int_0^1 \|\theta\,(z_{k+1} - z_k)\|\,\|z_{k+1} - z_k\|\,d\theta \le \frac{M_1}{2}\,(t_{k+1} - t_k))^2, \end{aligned} \tag{25}$$

where

$$M_1 = \begin{cases} K & \text{if} \quad k = 0 \\ L & \text{if} \quad k = 1, 2, \cdots. \end{cases}$$

Moreover, by iteration of Newton's method, Equations (23) and (25) and the induction hypotheses we get that

$$\begin{aligned} \|z_{k+2} - z_{k+1}\| &= \|(G'(z_{k+1})^{-1}\,G'(z_0))\,(G'(z_0)^{-1}\,G(z_{k+1}))\| \\ &\le \|G'(z_{k+1})^{-1}\,G'(z_0)\|\,\|G'(z_0)^{-1}\,G(z_{k+1})\| \\ &\le \frac{\frac{M_1}{2}\,(t_{k+1}-t_k)^2}{1-M\,t_{k+1}} = t_{k+2} - t_{k+1}. \end{aligned}$$

That is, we showed Equation (20) holds for all $k \ge 0$. Furthermore, let $z \in \overline{S}(z_{k+2}, t^* - t_{k+2})$. Then, we have that

$$\begin{aligned} \|z - x_{k+1}\| &\le \|z - z_{k+2}\| + \|z_{k+2} - z_{k+1}\| \\ &\le t^* - t_{k+2} + t_{k+2} - t_{k+1} = t^* - t_{k+1}. \end{aligned}$$

That is, $z \in \overline{S}(z_{k+1}, t^* - t_{k+1})$. The induction for Equations (20) and (21) is now completed. Lemma 1 implies that sequence $\{s_n\}$ is a complete sequence. It follows from Equations (20) and (21) that $\{z_n\}$ is also a complete sequence in a Banach space E_1 and as such it converges to some $z^* \in \overline{S}(z_0, t^*)$ (since $\overline{S}(z_0, t^*)$ is a closed set). By letting $k \longrightarrow \infty$ in Equation (25) we get $G(^*) = 0$. Estimate Equation (19) is obtained from Equation (18) (cf. [4,6,12]) by using standard majorization techniques. The proof for the uniqueness part has been given in [9]. □

The sufficient convergence criteria for Newton's method using the conditions (A), constants L, L_0 and η given in affine invariant form are:

- Kantorovich [6]

$$h_K = 2T\eta \le 1. \tag{26}$$

- Argyros [9]

$$h_1 = (L_0 + T)\eta \le 1. \tag{27}$$

- Argyros [3]

$$h_2 = \frac{1}{4}\left(T + 4L_0 + \sqrt{T^2 + 8L_0T}\right)\eta \le 1 \tag{28}$$

- Argyros [11]

$$h_3 = \frac{1}{4}\left(4L_0 + \sqrt{L_0T + 8L_0^2} + \sqrt{L_0T}\right)\eta \le 1 \tag{29}$$

- Argyros [12]

$$\begin{aligned} h_4 &= \tilde{L}_4\eta \leq 1, \\ \tilde{L}_4 &= L_4(T), \quad \delta = \delta(T). \end{aligned} \tag{30}$$

If $H = K = L_0 = L$, then Equations (27)–(30) coincide with Equations (26). If $L_0 < T$, then $L < T$

$$h_K \leq 1 \Rightarrow h_1 \leq 1 \Rightarrow h_2 \leq 1 \Rightarrow h_3 \leq 1 \Rightarrow h_4 \leq 1 \Rightarrow h_5 \leq 1,$$

but not vice versa. We also have that for $\frac{L_0}{T} \to 0:$

$$\begin{aligned} &\frac{h_1}{h_K} \to \frac{1}{2}, \quad \frac{h_2}{h_K} \to \frac{1}{4}, \quad \frac{h_2}{h_1} \to \frac{1}{2} \\ &\frac{h_3}{h_K} \to 0, \quad \frac{h_3}{h_1} \to 0, \quad \frac{h_3}{h_2} \to 0 \end{aligned} \tag{31}$$

Conditions Equations (31) show by how many times (at most) the better condition improves the less better condition.

Remark 1. **(a)** *The majorizing sequence $\{t_n\}$, t^*, t^{**} given in [12] under conditions (A) and Equation* (29) *is defined by*

$$\begin{aligned} t_0 &= 0, \quad t_1 = \eta, \quad t_2 = t_1 + \frac{L_0(t_1 - t_0)^2}{2(1 - L_0 t_1)} \\ t_{n+2} &= t_{n+1} + \frac{T(t_{n+1} - t_n)^2}{2(1 - L_0 t_{n+1})}, \quad n = 1, 2, \ldots \\ t^* &= \lim_{n \to \infty} t_n \leq t^{**} = \eta + \frac{L_0\eta^2}{2(1 - \delta)(1 - L_0\eta)}. \end{aligned} \tag{32}$$

Using a simple inductive argument and Equation (32) *we get for $L_1 < L$ that*

$$t_n < t_{n-1}, n = 3, 4, \ldots, \tag{33}$$

$$t_{n+1} - t_n < t_n - t_{n-1}, n = 2, 3, \ldots, \tag{34}$$

and

$$t^* \leq t^{**} \tag{35}$$

Estimates for Equations (5)–(7) *show the new error bounds are more precise than the old ones and the information on the location of the solution z^* is at least as precise as already claimed in the abstract of this study (see also the numerical examples). Clearly the new majorizing sequence $\{t_n\}$ is more precise than the corresponding ones associated with other conditions.*

(b) *Condition $\bar{S}(z_0, t^*) \subseteq D$ can be replaced by $S(z_0, \frac{1}{L_0})$ (or D_0). In this case condition $(A_2)'$ holds for all $x, y \in S(z_0, \frac{1}{L_0})$ (or D_0).*

(c) *If $L_0\eta \leq 1$, then, we have that $z_0 \in \bar{S}(z_1, \frac{1}{L_0} - \|G'(z_0)^{-1}G(z_0)\|)$, since $\bar{S}(z_1, \frac{1}{L_0} - \|G'(z_0)^{-1}G(z_0)\|) \subseteq S(z_0, \frac{1}{L_0})$.*

3. Numerical Examples

Example 1. *Returning back to the motivational example, we have $L_0 = 3 - p$.*

Conditions Equations (27)–(29) *are satisfied, respectively for*

$$p \in I_1 := [0.494816242, 0.5),$$

$$p \in I_2 := [0.450339002, 0.5)$$

and

$$p \in I_3 := [0.4271907643, 0.5).$$

We are now going to consider such an initial point which previous conditions cannot be satisfied but our new criteria are satisfied. That is, the improvement that we get with our new weaker criteria.

We get that

$$H = \frac{5+p}{3},$$

$$K = 2,$$

$$L = \frac{2}{3(3-p)}(-2p^2 + 5p + 6).$$

Using this values we obtain that condition Equation (4) is satisfied for $p \in [0.0984119, 0.5)$*, However, must also have that*

$$L_0 \eta < 1$$

which is satisfied for $p \in I_4 := (0, 0.5]$*. That is, we must have* $p \in I_4$*, so there exist numerous values of p for which the previous conditions cannot guarantee the convergence but our new ones can. Notice that we have*

$$I_K \subseteq I_1 \subseteq I_2 \subseteq I_3 \subseteq I_4$$

Hence, the interval of convergence cannot be improved further under these conditions. Notice that the convergence criterion is even weaker than the corresponding one for the modified Newton's method given in [11] by $L_0(\eta) < 0.5$.

For example, we choose different values of p and we see in Table 1.

Table 1. Convergence of Newton's method choosing $z_0 = 1$, for different values of p.

p	**0.41**	**0.43**	**0.45**
z_1	0.803333	0.810000	0.816667
z_2	0.747329	0.758463	0.769351
z_3	0.742922	0.754802	0.766321
z_4	0.742896	0.754784	0.766309
z_5	0.742896	0.754784	0.766309

Example 2. *Consider* $E_1 = E_2 = \mathcal{A}[0,1]$*. Let* $D^* = \{x \in \mathcal{A}[0,1]; \|x\| \le R\}$*, such that* $R > 0$ *and G defined on* D^* *as*

$$G(x)(u_1) = x(u_1) - f(u_1) - \lambda \int_0^1 \mu(u_1, u_2) x(u_2)^3 \, du_2, \quad x \in C[0,1], \; u_1 \in [0,1],$$

where $f \in \mathcal{A}[0,1]$ *is a given function,* λ *is a real constant and the kernel* μ *is the Green function. In this case, for all* $x \in D^*$*,* $G'(x)$ *is a linear operator defined on* D^* *by the following expression:*

$$[G'(x)(v)](u_1) = v(u_2) - 3\lambda \int_0^1 \mu(u_1, u_2) x(u_2)^2 v(u_2) \, du_2, \quad v \in C[0,1], \; u_1 \in [0,1].$$

If we choose $z_0(u_1) = f(u_1) = 1$*, it follows*

$$\|I - G'(z_0)\| \le 3|\lambda|/8.$$

Hence, if

$$|\lambda| < 8/3,$$

$G'(z_0)^{-1}$ *is defined and*

$$\|G'(z_0)^{-1}\| \leq \frac{8}{8-3|\lambda|},$$

$$\|G(z_0)\| \leq \frac{|\lambda|}{8},$$

$$\eta = \|G'(z_0)^{-1}G(z_0)\| \leq \frac{|\lambda|}{8-3|\lambda|}.$$

Consider $\lambda = 1.00$, *we get*

$$\eta = 0.2,$$

$$T = 3.8,$$

$$L_0 = 2.6,$$

$$K = 2.28,$$

$$H = 1.28$$

and

$$L = 1.38154\ldots.$$

By these values we conclude that conditions (26)–(29) *are not satisfied, since*

$$h_K = 1.52 > 1,$$

$$h_1 = 1.28 > 1,$$

$$h_2 = 1.19343\ldots > 1,$$

$$h_3 > 1.07704\ldots > 1,$$

but condition (2.27) and condition (4) *are satisfied, since*

$$h_4 = 0.985779\ldots < 1$$

and

$$h_5 = 0.97017\ldots < 1.$$

Hence, Newton's method converges by Theorem 1.

4. Application: Planck's Radiation Law Problem

We consider the following problem [15] :

$$\varphi(\lambda) = \frac{8\pi cP\lambda^{-5}}{e^{\frac{cP}{\lambda BT}}-1} \tag{36}$$

which calculates the energy density within an isothermal blackbody. The maxima for φ occurs when density $\varphi(\lambda)$. From (36), we get

$$\varphi'(\lambda) = \left(\frac{8\pi cP\lambda^{-6}}{e^{\frac{cP}{\lambda BT}}-1}\right)\left(\frac{(\frac{cP}{\lambda BT})e^{\frac{cP}{\lambda BT}-1}}{e^{\frac{cP}{\lambda kT}}-1} - 5\right) = 0, \tag{37}$$

that is when

$$\frac{\left(\frac{cP}{\lambda BT}\right)e^{\frac{cP}{\lambda BT}-1}}{e^{\frac{cP}{\lambda BT}-1}} = 5. \tag{38}$$

After using the change of variable $x = \frac{cP}{\lambda BT}$ and reordering terms, we obtain

$$f(x) = e^{-x} - 1 + \frac{x}{5}. \tag{39}$$

As a consequence, we need to find the roots of Equation (39).

We consider $\Omega = \overline{E(5,1)} \subset \mathbb{R}$ and we obtain

$$\eta = 0.0348643\ldots,$$

$$L_0 = 0.0599067\ldots,$$

$$K = 0.0354792\ldots,$$

$$H = 0.0354792\ldots$$

and

$$L = 0.094771\ldots.$$

So (A) are satisfied. Moreover, as $b = 0.000906015 > 0$, then

$$L_4 = 10.0672\ldots,$$

which satisfies

$$L_4\eta = 0.350988\ldots < 1$$

and that means that conditions of Lemmal 1 are also satisfied. Finally, we obtain that

$$t^* = 0.0348859\ldots.$$

Hence, Newton's method converges to the solution $x^* = 4.965114231744276\ldots$ by Theorem 1.

Author Contributions: All authors have contributed in a similar way.

References

1. Amat, S.; Busquier, S.; Gutiérrez, J.M. Geometric constructions of iterative functions to solve nonlinear equations. *J. Comput. Appl. Math.* **2003**, *157*, 197–205. [CrossRef]
2. Amat, S.; Busquier, S. Third-order iterative methods under Kantorovich conditions. *J. Math. Anal. Appl.* **2007**, *336*, 243–261. [CrossRef]
3. Argyros, I.K. A semi-local convergence analysis for directional Newton methods. *Math. Comput.* **2011**, *80*, 327–343. [CrossRef]
4. Argyros, I.K.; Magreñán, Á.A. *Iterative Methods and Their Dynamics with Applications: A Contemporary Study*; CRC Press: Bocaratón, FL, USA, 2017.
5. Farhane, N.; Boumhidi, I.; Boumhidi, J. Smart Algorithms to Control a Variable Speed Wind Turbine. *Int. J. Interact. Multimed. Artif. Intell.* **2017**, *4*, 88–95. [CrossRef]
6. Kantorovich, L.V.; Akilov, G.P. *Functional Analysis*; Pergamon Press: Oxford, UK, 1982.
7. Kaur, R.; Arora, S. Nature Inspired Range Based Wireless Sensor Node Localization Algorithms. *Int. J. Interact. Multimed. Artif. Intell.* **2017**, *4*, 7–17. [CrossRef]

8. Amat, S.; Busquier, S.; Negra, M. Adaptive approximation of nonlinear operators. *Numer. Funct. Anal. Optim.* *2004*, *25*, 397–405. [CrossRef]
9. Argyros, I.K. On the Newton-Kantorovich hypothesis for solving equations. *J. Comput. Math.* **2004**, *169*, 315–332. [CrossRef]
10. Argyros, I.K.; González, D. Extending the applicability of Newton's method for k-Fréchet differentiable operators in Banach spaces. *Appl. Math. Comput.* **2014**, *234*, 167–178. [CrossRef]
11. Argyros, I.K.; Hilout, S. Weaker conditions for the convergence of Newton's method. *J. Complex.* **2012**, *28*, 364–387. [CrossRef]
12. Argyros, I.K.; Hilout, S. On an improved convergence analysis of Newton's method. *Appl. Math. Comput.* **2013**, *225*, 372–386. [CrossRef]
13. Ezquerro, J.A.; Hernández, M.A. How to improve the domain of parameters for Newton's method. *Appl. Math. Lett.* **2015**, *48*, 91–101. [CrossRef]
14. Gutiérrez, J.M.; Magreñán, Á.A.; Romero, N. On the semi-local convergence of Newton-Kantorovich method under center-Lipschitz conditions. *Appl. Math. Comput.* **2013**, *221*, 79–88.
15. Divya, J. Families of Newton-like methods with fourth-order convergence. *Int. J. Comput. Math.* **2013**, *90*, 1072–1082.

Permissions

All chapters in this book were first published by MDPI; hereby published with permission under the Creative Commons Attribution License or equivalent. Every chapter published in this book has been scrutinized by our experts. Their significance has been extensively debated. The topics covered herein carry significant findings which will fuel the growth of the discipline. They may even be implemented as practical applications or may be referred to as a beginning point for another development.

The contributors of this book come from diverse backgrounds, making this book a truly international effort. This book will bring forth new frontiers with its revolutionizing research information and detailed analysis of the nascent developments around the world.

We would like to thank all the contributing authors for lending their expertise to make the book truly unique. They have played a crucial role in the development of this book. Without their invaluable contributions this book wouldn't have been possible. They have made vital efforts to compile up to date information on the varied aspects of this subject to make this book a valuable addition to the collection of many professionals and students.

This book was conceptualized with the vision of imparting up-to-date information and advanced data in this field. To ensure the same, a matchless editorial board was set up. Every individual on the board went through rigorous rounds of assessment to prove their worth. After which they invested a large part of their time researching and compiling the most relevant data for our readers.

The editorial board has been involved in producing this book since its inception. They have spent rigorous hours researching and exploring the diverse topics which have resulted in the successful publishing of this book. They have passed on their knowledge of decades through this book. To expedite this challenging task, the publisher supported the team at every step. A small team of assistant editors was also appointed to further simplify the editing procedure and attain best results for the readers.

Apart from the editorial board, the designing team has also invested a significant amount of their time in understanding the subject and creating the most relevant covers. They scrutinized every image to scout for the most suitable representation of the subject and create an appropriate cover for the book.

The publishing team has been an ardent support to the editorial, designing and production team. Their endless efforts to recruit the best for this project, has resulted in the accomplishment of this book. They are a veteran in the field of academics and their pool of knowledge is as vast as their experience in printing. Their expertise and guidance has proved useful at every step. Their uncompromising quality standards have made this book an exceptional effort. Their encouragement from time to time has been an inspiration for everyone.

The publisher and the editorial board hope that this book will prove to be a valuable piece of knowledge for researchers, students, practitioners and scholars across the globe.

List of Contributors

Ruben González, Cristina Amorós and Íñigo Sarría
Escuela Superior de Ingeniería y Tecnología, Universidad Internacional de La Rioja, 26006 Logrono, Spain

Á. Alberto Magreñán
Departamento de Matemáticas y Computación, Universidad de La Rioja, 26004 Logrono, Spain
Departamento de Matemáticas y Computación, Universidad de La Rioja, 26006 Logroño, Spain

Lara Orcos
Facultad de Educación, Universidad Internacional de La Rioja, 26006 Logrono, Spain
Departamento de Matemática Aplicada, Universidad Politècnica de València, 46022 València, Spain

Dilan Ahmed
Department of Mathematics, College of Education, University of Sulaimani, Kurdistan Region, Sulaimani 46001, Iraq

Mudhafar Hama and Karwan Hama Faraj Jwamer
Department of Mathematics, College of Science, University of Sulaimani, Kurdistan Region, Sulaimani 46001, Iraq

Stanford Shateyi
Department of Mathematics, University of Venda, Private Bag X5050, Thohoyandou 0950, South Africa

Ramandeep Behl
Department of Mathematics, Faculty of Science, King Abdulaziz University, Jeddah 21589, Saudi Arabia

Eulalia Martínez
Instituto Universitario de Matemática Multidisciplinar, Universitat Politècnica de València, Camino de Vera s/n, 46022 Valencia, Spain

Fabricio Cevallos
Fac. de Ciencias Económicas, Universidad Laica "Eloy Alfaro de Manabí", Manabí 130214, Ecuador

Diego Alarcón
Departamento de Matemática Aplicada, Universitat Politècnica de València, Camino de Vera s/n, 46022 Valencia, Spain

Wachirapong Jirakitpuwapat and Pawicha Phairatchatniyom
KMUTTFixed Point Research Laboratory, Room SCL 802 Fixed Point Laboratory, Science Laboratory Building, Department of Mathematics, Faculty of Science, King Mongkut's University of Technology Thonburi, 126 Pracha Uthit Rd., Bang Mod, Thung Khru, Bangkok 10140, Thailand

Poom Kumam
KMUTTFixed Point Research Laboratory, Room SCL 802 Fixed Point Laboratory, Science Laboratory Building, Department of Mathematics, Faculty of Science, King Mongkut's University of Technology Thonburi, 126 Pracha Uthit Rd., Bang Mod, Thung Khru, Bangkok 10140, Thailand
Department of Medical Research, China Medical University Hospital, China Medical University, Taichung 40402, Taiwan

Yeol Je Cho
Department of Mathematics Education, Gyeongsang National University, Jinju 52828, Korea
School of Mathematical Sciences, University of Electronic Science and Technology of China, Chengdu 611731, China

Kanokwan Sitthithakerngkiet
Intelligent and Nonlinear Dynamic Innovations Research Center, Department of Mathematics, Faculty of Applied Science, King Mongkut's University of Technology North Bangkok (KMUTNB),Wong sawang, Bangsue, Bangkok 10800, Thailand

Shengfeng Li and Xiaofang Zhang
Institute of Applied Mathematics, Bengbu University, Bengbu 233030, China

Xiaobin Liu
Institute of Applied Mathematics, Bengbu University, Bengbu 233030, China
School of Computer Engineering, Bengbu University, Bengbu 233030, China

Alicia Cordero, Cristina Jordán and Juan R. Torregrosa
Multidisciplinary Institute of Mathematics, Universitat Politènica de València, 46022 València, Spain

Esther Sanabria
Department of Applied Mathematics, Universitat Politènica de València, 46022 València, Spain

Yanlin Tao
School of Computer Science and Engineering, Qujing Normal University, Qujing 655011, China

Juan Liang
Data Science and Technology, North University of China, Taiyuan 030051, Shanxi, China
Department of Science, Taiyuan Institute of Technology, Taiyuan 030008, Shanxi, China

Linke Hou
Center for Economic Research, Shandong University, Jinan 250100, Shandong, China

Xiaowu Li, Feng Pan and Lin Wang
College of Data Science and Information Engineering, Guizhou Minzu University, Guiyang 550025, Guizhou, China

Taixia Cheng
Graduate School, Guizhou Minzu University, Guiyang 550025, Guizhou, China

Jian Li and Xiaomeng Wang
Inner Mongolia Vocational College of Chemical Engineering, Hohhot 010070, China

Kalyanasundaram Madhu
Department of Mathematics, Saveetha Engineering College, Chennai 602105, India

Malik Zaka Ullah
Department of Mathematics, King Abdulaziz University, Jeddah 21589, Saudi Arabia

Sergio Amat and Sonia Busquier
Departamentode Matemática Aplicada y Estadística, Universidad Politécnica de Cartagena, 11003 Cádiz, Spain

Ioannis Argyros
Department of Mathematical Sciences, Cameron University, Lawton, OK 73505, USA

Miguel Ángel Hernández-Verón and María Jesús Rubio
Departamento de Matemáticas y Computación, Universidad de La Rioja, Calle Madre de Dios, 53, 26006 Logrono, Spain

Saima Akram, Fiza Zafar and Nusrat Yasmin
Centre for Advanced Studies in Pure and Applied Mathematics, Bahauddin Zakariya University, Multan 60800, Pakistan

Zhinan Wu
School of Mathematics and Computer Science, Yichun University, Yichun 336000, China

Ioannis K. Argyros
Department of Mathematics Sciences Lawton, Cameron University, Lawton, OK 73505, USA

Index

www.ingramcontent.com/pod-product-compliance
Lightning Source LLC
LaVergne TN
LVHW060522210926
848485LV00005B/1130